Emerging Technologies

Edited by Edna F. Einsiedel

Emerging Technologies
From Hindsight to Foresight

UBCPress · Vancouver · Toronto

17 16 15 14 13 12 11 10 09   5 4 3 2 1

Printed in Canada with vegetable-based inks on ancient-forest-free paper (100% post-consumer recycled) that is processed chlorine- and acid-free.

Library and Archives Canada Cataloguing in Publication

Emerging technologies : from hindsight to foresight / edited by Edna F. Einsiedel.

Includes bibliographical references and index.
ISBN 978-0-7748-1548-2 (bound); ISBN 978-0-7748-1549-9 (pbk.)

1. Biotechnology. 2. Genomics. 3. Nanotechnology. 4. Technological innovations—Social aspects. 5. Technology—Social aspects. I. Einsiedel, Edna F.

T14.5.E445 2009   303.48'3   C2008-904298-0

Canadä

UBC Press gratefully acknowledges the financial support for our publishing program of the Government of Canada through the Book Publishing Industry Development Program (BPIDP), and of the Canada Council for the Arts, and the British Columbia Arts Council.

This book has been published with the help of a grant from the Canadian Federation for the Humanities and Social Sciences, through the Aid to Scholarly Publications Programme, using funds provided by the Social Sciences and Humanities Research Council of Canada.

This project was made possible with assistance from the Canadian Biotechnology Secretariat Office and the GE3LS Project (Genomics, Ethics, Environmental, Economic, Legal, and Social Studies) funded by Genome Canada. Support from these organizations is gratefully acknowledged.

UBC Press
The University of British Columbia
2029 West Mall
Vancouver, BC V6T 1Z2
604-822-5959 / Fax: 604-822-6083
www.ubcpress.ca

Contents

Acknowledgments

We thank Genome Canada for its support of this book project. A few of the chapters in this book were developed from papers commissioned by the Canadian federal government's Canadian Biotechnology Secretariat and presented at a 2005 symposium in Ottawa sponsored by the GE3LS Alberta project (Genomics, ethics, economic, environmental, legal and social studies), funded by Genome Canada and the Biotechnology Secretariat.

Abbreviations

BRS	Biotechnology Regulatory Services
CBD	Convention on Biological Diversity
CFIA	Canadian Food Inspection Agency
CIHR	Canadian Institutes of Health Research
CSGA	Canadian Seed Growers' Association
EPA	Environmental Protection Agency
EPC	European Patent Convention
EU	European Union
FDA (US)	Food and Drug Administration
GM	genetically modified
GMO	genetically modified organism
hESC	human embryonic stem cell
IP	intellectual property
IPRs	Intellectual property rights
NGO	nongovernmental organization
OECD	Organisation for Economic Co-operation and Development
PGR	plant genetic resource
PMF	plant molecular farming
PMIP	plant-made industrial product
PMP	plant-made pharmaceutical
PNTs	plants with novel traits
PRBs	plant breeders' rights
REBs	research ethics boards
SCNT	somatic cell nuclear transfer
TCPS	*Tri-council policy statement*
TK	traditional knowledge
TRIPS Agreement	Agreement on Trade-Related Aspects of Intellectual Property Rights
USDA	United States Department of Agriculture
WHO	World Health Organization
WTO	World Trade Organization

Emerging Technologies

Introduction: Making Sense of Emerging Technologies

Edna F. Einsiedel

This book is about emerging transformative technologies. Emerging technologies can be described in several ways. These are technologies in the developmental stage of production, perhaps not yet fully exploited by firms, or in the early stages of commercialization. For a number of these, basic research may still be taking place, and the projections of potential applications remain just that–projected aspirations and hopes. The term *emerging technologies* has been used to describe information and communication technologies (perhaps three decades ago), and more recently, biotechnology, genomics, and nanotechnology.

These emerging technologies are also called "strategic" technologies–so labelled because national investments and aspirations ride on them. They are also strategic in the sense that they involve forward thinking and planning. Numerous countries identify technologies they consider strategic to their national interests and global competitiveness, build national systems of innovation and develop strategic plans around them, develop policy approaches, and finance a variety of activities, from research and development initiatives to human resource activities.

Finally, they are revolutionary or transformative. We do not use this word lightly. A technology is "revolutionary" when it has the capacity to change a wide range of sectors. These transformative technologies involve shifts in traditional relationships, can be socially disruptive as much as they can bring about greater cohesiveness, and can bring about new institutional rules and arrangements. Think about the taken-for-granted things the automobile has spawned, for example, as described in a journalist's feature on automobiles:

> drive-through restaurants, drive-by shootings, drive-up banks, gas stations, suburbs, motels, back-seat boogies, body shops, paved roads, parking lots, traffic cops, truck stops, decent factory wages, smooth-talking car dealers, highway deaths, gridlock, pollution, "are-we-there-yet?" family vacations. (*USA Today* 1996)

Emerging technologies are projected to have broad-ranging impacts on many areas of life. Biotechnology, for instance, has had, and continues to have, impacts on what we eat and how our food is produced, how we view and how we treat disease, how we clean up the environment, even how we carry out justice in our judicial systems with DNA evidence. This wide-ranging set of implications and the nature of impacts contribute to making a technology revolutionary.

When we look back to the histories of various technologies now embedded in society–from vaccines to computers, from electricity to enhanced foods–we see historical trajectories that have led to life and societal changes, from reduced mortality to revolutionized working conditions. These histories also remind us that, once upon a time, these technologies may not always have been greeted with excitement and anticipation.

What is interesting about the introduction of new–and particularly revolutionary–technologies today is the extent of societal attention accorded to them. This is because of the ubiquity of information sources; the desire of governments to make a given technology "happen" (thus creating the conditions for such a happening to occur); the savvy of social groups in society as they try to raise the alarm bells about the potential negative impacts or, alternatively, to push through the development trajectory at a more rapid pace; or the greater attentiveness of the average citizen.

Such attentiveness may develop from experiences with older technologies, when controversies surrounded their introduction, led to their demise, or brought about a redesign more in keeping with public and stakeholder demands or interests. This attention can also be aroused by the media, whose attention span is only as short as the next big controversy but whose multichannel ubiquity can continue to pique public interest.

In this book, we have chosen to investigate technologies that are in precommercialization or early commercialization stages. These include nanotechnologies, pharmacogenomics, molecular farming, stem cell research, and newer biotechnology applications–technologies or their applications that all have connections to the life sciences. Nanotechnology, of course, embraces a range of applications that connect with the life sciences but also goes beyond. We include it here as another example of "the next big thing." In this respect, the choices for our focus are somewhat arbitrary. On the other hand, as I pointed out earlier, these are "revolutionary" and strategic technologies. Nanotechnology, still in its technological infancy, is similarly expected to have impacts on the types of materials we use and how they are applied, how we diagnose and treat disease, how we produce energy, or how we communicate. It is already identified as key to national innovation interests for many industrialized countries.

Perhaps because of the potential impacts of these technologies, they have been taken notice of much earlier in the innovation process. In addition to the scientific community, publics and policy makers are in on the conversations and debates early in their developmental trajectories. "Everyone" includes the scientists working away on various aspects of the technology, the institutions these scientists belong to, the potential and actual venture capitalists ready to jump on the next big (or small) thing, the media who are alerted by early exciting prognostications, other stakeholders who see the potential benefits and the potential risks, and the publics who have become earlier voyeurs, watching the various aspects of the technology as these are being rolled out in fits and starts, or as claims and counterclaims are being made about them in public arenas.

In this book, we focus on a number of actors in the landscape of emerging technologies. The first group includes the various publics who are going to be eventual users, who currently bankroll some of the research through their tax dollars, who sometimes make decisions in the political sphere through the ballot box or through their choice of political decision makers, or who may bear a greater burden of risks or may have a larger stake in the promised benefits than others.

The way publics have been viewed has changed over time. Perhaps the earliest way of envisioning "the public" involved a unidimensional view of a monolithic public, subject to the vagaries of information disseminated from the so-called experts. This simplistic view has changed significantly, with publics (plural emphasis) engaged or inattentive at various times, occupying different roles at different times citizen, consumer, patient, environmentalist–being naive or displaying expertise, becoming active or noncommittal depending on context and circumstance. One important contextual difference has been identified in terms of the confluence of geography and culture, evident in transnational differences on the reception of biotechnology applications (Gaskell et al. 2001; Hallman et al. 2004).

What we have also learned is that other actors' views of publics are also changing. While others have talked about publics as "a second hurdle" (Von Wartburg and Liew 1999, 34) after regulatory development, the increasing prominence given to publics today, if one is to go by public policy pronouncements, is less in terms of hurdle and more in terms of "participant" in the technology development process.

Policy makers constitute another set of important actors in the public arena. These decision makers do not make decisions in a vacuum; rather, they engage with other stakeholder groups, interact with their counterparts in the global arena, and attend to or play active roles in international policy initiatives.

We also focus on processes. Some of these are common across technologies. How do regulations and policies come about? Who gets consulted? What factors were considered for particular policy stances and why? What challenges have been faced and are ongoing in regulatory development?

In examining each group of technologies, we suggest that a technology becomes emergent when it assumes its form in the public sphere–when others not necessarily involved in the technology's direct development are able to examine its gestation, often through the media or through the activities of various actors. What used to be an inside look that few were privy to is now occurring on an open stage. This happens partly because scientific institutions (such as the leading journals or academic institutions) are linked even more directly to popular channels, because scientists have become more strategic in their use of these popular channels, because the media are constantly on the lookout for stories that whet the public imagination, because "life-enhancing" stories are continuing fodder for the public imaginations, and because values of particular groups stoke an oppositional interest.

Given this context, the appearance of new technologies in the public arena is occurring much earlier in the innovation trajectory, many becoming a fixture in the public landscape even as early as the stage of technology design. In some ways, this may be occurring from the benefit of hindsight. That is, when we look back to the experience of older technologies–nuclear power and GM food are particular examples–we see that discussions of these technologies occurred at the commercialization stage when it was "too late." Those engaged in nanotechnology design see this as a key lesson to be learned (see, e.g., Royal Society and Royal Academy of Engineering 2004).

The currency of these public conversations is hope–but hope is meaningful only in the context of fears; risks are meaningful only in the face of uncertainties. And so, the studies and commentaries presented in this book are early explorations of what these emerging technologies look like from the vantage point of representations among publics and stakeholders, discursive activities in legislative rooms, informal and formal deliberations among citizens or policy makers. These are early impressions in some instances, longer term and more developed views in others. We expect these pictures–snapshots at this point in time–to similarly evolve with the technologies' evolution. How these different interactions develop over time remains to be seen.

We use the themes of hindsight and foresight for examining these emerging technologies. There are lessons to be learned from the experience of earlier biotechnologies and, indeed, the experience with GM foods has become the seminal case for emergent technologies.

We also put forward our base assumption–that technology is *both* social and technical–and this assumption is embedded in how we look at the innovation processes behind these emergent technologies. This approach argues that an understanding of innovation and particularly the question of how a technology is accepted or rejected is a social–as well as a technical–one. If a technology flounders, fails, or succeeds, what criteria or factors help to explain this and why, who defines these criteria, who are excluded?

Part 1 provides hindsight perspectives, looking at what we know about publics and agricultural biotechnology and issues around ownership of knowledge. Who owns what under what conditions remains a highly contentious issue. This is particularly so since the US Supreme Court ruled in 1970 that living organisms were patentable. Ever since this ruling, the trajectory of intellectual property processes has followed a contentious path: What weight should be given to social-ethical concerns in relation to determining patentability, and how should these concerns be accommodated, if they should be at all? What limits should be placed on patent holders' rights? For example, should farmers who use patented seeds be allowed to save and reuse those seeds? How should issues around traditional knowledge–which is shared–be addressed, particularly in the context of companies utilizing elements of such knowledge?

By using intellectual property questions and specifically those about patents as a lens, we examine how publics assess one aspect of technology innovation. Chapter 3 summarizes what we have learned from various studies on publics and patents, the metrics used, and the trade-offs made.

From agricultural biotechnology we move in Part 2 to the hybrid biotechnologies–the transgenics that bridge the worlds of agriculture and medicine. These include transgenic plants modified to produce drugs and vaccines, and transgenic animals that are modified for a similar variety of purposes, such as food and pharmaceutical production.

The questions raised about these technologies are a similar mix of public reaction and concern and debates about economic risks and benefits, environmental challenges, and safety issues, not to mention the inchoate concerns raised by some stakeholder groups that may range from revulsion to fear to anxiety. Questions about regulatory efficacy and trust in those at the technology's helm are often bound up with public views. At the same time, the challenges of appropriate and effective policy and regulation have remained with us.

Moving to the world of medicine and health, issues around cells, tissues, and organs are blended in with questions that resonate with those raised around knowledge: whose cells and tissues, derived from what source, for what purpose, and with what impacts? In the case of stem cells, the issue of origin has never been more contentious and has had important impacts on policy positions. By what values are these decisions made? Human dignity has been a frequent watchword in debates involving the body but, in Chapter 9, on stem cells, this taken-for-granted concept is interrogated and challenged.

Pharmacogenomics–the study of the genetic basis of differential drug reactions–offers a window into a world opened up by the mapping of the human genome. The lens gets focused even more tightly on groups and subgroups bound by similarities of "race" or ethnicity. Here, questions are raised about basic biological dogma of genes as major determinants of health. This basic dogma gets murky in the arena of risks as defined by individual and group responses to drugs. The double-edged sword of differential drug responses is nowhere better illustrated than the first pharmacogenomic drug on the market, BiDil, a heart drug said to be more efficacious for African Americans. How are expectations framed around this drug during its early discussions in the regulatory arena? A further procedural and substantive question can be raised around informed consent, with much pharmacogenomic research resting on tissue banks. What challenges are confronted with consent processes based on individual autonomy when genetic information implicates other members of the family?

Finally, in Chapters 13 and 14 we look further into the future with nanotechnology, with its forerunner applications already making their appearance in the market (wrinkle-free pants, anyone? Or deep-cleansing nanomolecules in your face cream?). Like biotechnology, nanotechnology is a bridging technology, an enabler for a variety of older approaches. It offers a bridge between the life and material sciences. The publics of nanotechnology are as much a product of our researcher imaginations as they are products of identifiable individuals or organized groups whose activities are highlighted in the media; the stakeholders for this technological realm are actors that have appeared in plays from other technological pasts (ETC 2004; Einsiedel and McMullen 2004).

Ultimately, the questions we ask are, what kind of world do we prefer to have, and how do we get to this point? Who decides, and how? This is the governance question. In Part 3, we reflect on the complexities of policy development and ownership of knowledge and the question of processes of decision making. Again, these are hindsight and foresight questions.

Many lessons have been learned, quite often by muddling through rather than by systematic and planful approaches. For some in the policy community, the admonition of being safe rather than sorry–by looking before making a leap–(formally known in terms of the precautionary principle) has become increasingly accommodating of risk-analysis metrics. On the other side of the fence, those who have lived by the rule of "sound science" and science-based risk assessment have had telling experiences of bumping up against the realities of economics and social values. They similarly have had to modulate and accommodate.

At the same time, questions about knowledge ownership and control have not been as clear-cut and straightforward as patents systems would have us believe. If the innovator enjoys the protection of and benefits from intellectual property regimes, what responsibilities does he or she carry when technology use goes awry? The controversies over StarLink corn–approved for animal feed but not for human consumption–found in taco shells in the United States, and the ProdiGene case of transgenic corn modified for pharmaceutical production and inadvertently found in the next round of field crops are important reminders of lingering questions on liability and responsibility.

The book concludes by bringing together the key themes of hindsight and foresight from recent and emerging technologies, examining lessons across a variety of technologies. These lessons are as much about the processes of societal and organizational learning as they are about the lessons actually learned.

References

Einsiedel, E.F., and G. McMullen. 2004. Stakeholders and technology: Challenges for nanotechnology. *Health Law Review* 12: 1, 8-12

ETC. 2004. Down on the farm: the impact of nano-scale technologies on food and agriculture. ETC Group. http://www.etcgroup.org/en/materials/publications.html?pub_id=81.

Gaskell, G., N. Allum, W. Wagner, T. Nielsen, E. Jelsøe, M. Kohring, and M. Bauer. 2001. In the public eye: Representations of biotechnology in Europe. In *Biotechnology 1996-2000,* ed. G. Gaskell and M. Bauer, 53-79. London: Science Museum.

Hallman, W., W.C. Hebden, C. Cuite, H. Aquino, and J. Lang. 2004. *Americans and GM food: Knowledge, opinion, and interest in 2004.* New Brunswick, NJ: Food Policy Institute.

Royal Society and Royal Academy of Engineering. 2004. *Nanoscience and nanotechnologies: Opportunities and uncertainties.* London: Royal Society and Royal Academy of Engineering.

USA Today. 1996. Celebrating the automobile. 2 February: 2E.

Von Wartburg, W., and J. Liew. 1999. *Gene technology and social acceptance.* Lanham, MD: University Press of America.

Part 1
Hindsight Learnings

1 GM Foods in Hindsight

William K. Hallman

Worldwide, billions of dollars have been spent on research and development of genetically modified (GM) crops to create new and improved sources of food, feed, fibres, fuels, and pharmaceuticals. Widely touted as a transforming technology, agricultural biotechnology was expected to change agriculture in ways that would rival the gains made during the green revolution. As Shoemaker, Johnson, and Golan (2003, 32) write, "Biotechnology is often associated with promise . . . promise to feed the world, promise to reduce environmental harm, promise to expand agricultural markets and production possibilities, promise to create products that consumers want. Farmers in the United States seem to be sold on these promises."

Indeed, GM crops have been adopted by farmers at an extraordinary rate over the past decade. Since they were introduced in 1996, farmers have consistently increased the acreage planted with biotech crops by double-digit growth rates every year. As a result, the amount of acreage devoted to GM crops has increased more than fifty-fold worldwide in ten years (James 2005).

Significantly, consumers in the countries that produce the majority of GM crops, including the United States, Argentina, Canada, Brazil, and China, seem to have paid little attention to the technology (Gaskell et al. 1999; Hebden and Hallman 2005). At the same time, GM food continues to provoke anxiety among many consumers in other parts of the world. Reflecting perceived consumer sentiment, governments in Europe, Japan, South Korea, and parts of Africa have instituted public policies that effectively restrict the planting or sale of GM food products. The resulting international trade disputes over GM agricultural commodities have highlighted complex issues involving the role of science and consumer sentiment in establishing policies intended to regulate the creation and adoption of new technologies and their products.

Thus, while GM crops have been widely adopted in some parts of the world, they are deeply held in suspicion in others, belying the initial promise of GM foods. As such, reactions to the emergence of GM foods may offer important lessons applicable to the introduction of other new and potentially controversial technologies. Here then, are some of those lessons.

Awareness of and Familiarity with New Technologies Are Often Surprisingly Low

That a new technology becomes controversial does not necessarily mean that it has captured the full attention of most people in society. Stories about biotechnology and GM foods and the controversies they have engendered have been widely reported in the media for more than two decades. So, it is difficult to believe that most people know little about the technology or the issues surrounding it. Yet, a 2005 survey of Canadians and Americans conducted by the Canadian Biotechnology Secretariat confirms that most people in North America don't have a great deal of familiarity with biotechnology as a whole. While the majority now say they have *heard* of biotechnology, fewer than 10 percent of Canadians or Americans say they are "very familiar" with it. In fact, about as many Canadians (13 percent) and Americans (9 percent) say they are "not at all familiar" with biotechnology as say they are very familiar with it (Walker 2005).

Studies conducted in China (Huang et al. 2004) and in South Korea (Hallman et al. 2005) show similar results. Few Chinese (10 percent) say that they know "a great deal" about biotechnology, while most say that they have heard or read "some" (34 percent), "not much" (23 percent), or "nothing at all" (33 percent). In South Korea, where the government has placed emphasis on becoming a leading developer of biotechnology products and where there has been a great deal of media coverage, more people indicate that they are familiar with biotechnology. Still, nearly two-thirds (63 percent) of South Koreans say that they only heard "some" about biotechnology, while 18 percent say that they had heard or read "a great deal." Thirteen percent say that they have heard "not much," and 6 percent say that they had heard nothing at all.

Even in Europe, where the controversy over biotechnology has perhaps been most heated, research suggests that most Europeans are not very familiar with biotechnology either. When asked whether they had heard of each of three applications of biotechnology, on average, the respondents reported having heard of only about half (1.79) of them (Gaskell, Allum, and Stares 2003).

Given that most people say that they have not heard or read very much about GM foods and that they are not very familiar with them, it is not surprising that most Americans and Canadians also say that they have not talked very much about them. More than half (53 percent) of Americans and 41 percent of Canadians say that they have never had a conversation about

GM foods (CBS 2005; Hallman 2005). Similar results have been found in the United Kingdom, Greece, Portugal, Spain, and Belgium, though people in the United States and Canada are considerably less likely to have talked about GM foods than their counterparts in Europe as a whole and substantially less likely than those in Germany and Denmark, where reported discussion is at its highest (Gaskell, Allum, and Stares 2003).

The fact that most people in North America do not seem to be actively discussing GM food suggests that one should not assume that controversy over a technology will necessarily create a place for it on the national agenda, nor will it lead to consumer knowledge, understanding, or attentiveness. As Hallman and Metcalfe concluded in 1994, "While the battle over biotechnology has raged between experts, most of the shots have passed over the heads of the non-combatants" (36).

News Coverage about a Technology Does Not Necessarily Make a Lasting Impression on People

Many of the controversies about biotechnology have been particularly centred on genetically modified food. As a result, hundreds of news stories have appeared in newspapers, on television, and on the internet about agricultural biotechnology since the early 1990s (Nisbet and Lewenstein 2002; Ten Eyck and Williment 2003; Thompson and Dininni 2005). Yet, when asked how familiar they are with GM food, less than a sixth of Canadians and Americans say they are "very familiar." More than half of Americans (54 percent) and Canadians (56 percent) report that they are only "somewhat familiar," and a third of both Canadians and Americans say that they are either "not very familiar" or "not at all familiar" with GM foods (Walker 2005).

In the Canadian Biotechnology Secretariat's study, Canadians and Americans were asked if they had read, seen, or heard "a lot," "a little," or "nothing" about issues surrounding GM foods within the three months prior to being interviewed, in January 2005. Consistent with their answers concerning their familiarity with GM foods, more than half of the Canadians (56 percent) and Americans (52 percent) reported that they had heard only "a little" about GM food issues. Moreover, more than one-third of the Americans (39 percent) and 29 percent of the Canadians responded that they had heard "nothing" about GM food issues (CBS 2005). In fact, when Americans were asked if they could remember events or news stories related to GM foods, fewer than one in five could recall any (Hallman et al. 2004).

Many New Technologies Are Invisible

Many important emerging technologies, especially those targeted at the genetic, cellular, or nano scale, are literally invisible to people. They work at a level that cannot be seen and, because of this, their mechanisms are often misunderstood.

These technologies are often figuratively invisible as well. Unless it offers a specific consumer benefit or some other advantage that can be advertised, it is unlikely that a particular technology will be purposely promoted as part of a product. Thus, consumers may often purchase products brought about through scientific breakthroughs about which they have heard or read little and which they do not fundamentally understand. For example, one of the fundamental problems with public understanding of GM technology is that although the morphological and functional consequences of genetic modification are sometimes apparent, current GM crops are largely indistinguishable from their conventionally bred counterparts. They look, smell, and taste the same as non-GM foods. As such, without special labels, it is generally not possible for people to readily identify even whole grains, fruits, or vegetables that have been genetically modified and virtually impossible for people to recognize processed foods that contain GM ingredients. This helps to explain why, despite the ubiquity of GM ingredients in the food system, national surveys show that most Americans are generally unaware of the presence of GM foods on their own plates (Hallman et al. 2002; Hallman et al. 2003; Hallman et al. 2004). Fewer than half of Americans realize that foods containing GM ingredients are available in supermarkets (Hallman et al. 2004) and fewer than one in four believes they have consumed GM foods (Pew Initiative on Food and Biotechnology 2005). Moreover, many of those who do believe that GM foods are sold in supermarkets are confused about which GM products are actually for sale. While three-quarters of those who believe that GM foods are available in supermarkets say that products containing GM corn are for sale, and 65 percent correctly believe that products containing GM soybeans can be purchased, many more apparently are convinced that they are eating GM tomatoes, GM chicken, and GM rice, none of which is available for purchase in the United States (Hallman et al. 2004). For other "invisible" emerging technologies, this suggests that it is not possible to rely on the ubiquity of the technology or its applications to drive consumer awareness.

Invisible Technologies May Be Perceived as Having Invisible Consequences

That the products of agricultural biotechnology are not implicitly visible means that people are often unaware of their exposure to them, and therefore unable to assess their potential effects. This can have both positive and negative consequences.

Upon learning that they have been eating GM foods for more than a decade, some consumers in North America may decide that because they have been doing so without any apparent adverse consequences, these foods are fundamentally safe and so they will continue to eat them. But it is also possible that some people will feel that they have been tricked into eating GM products without their consent; this may potentially lead to a backlash among angry consumers. Moreover, many consumers understand that scientists cannot categorically assure that there will be no long-term health consequences of consuming GM ingredients. As such, it is possible that some consumers will attribute otherwise medically unexplainable symptoms, illnesses, or other health problems to their unknowing consumption of GM foods. For example, in September 2000, after the discovery of the accidental contamination of taco shells and other corn products by a variety of GM corn not approved for human consumption, at least twenty-eight people reported allergic reactions to the US Centers for Disease Control. However, after testing, none of those reporting unexplained symptoms was found to have demonstrable allergic reactions to the corn (CDC, National Center for Environmental Health 2001).

Because many emerging technologies are both literally and figuratively invisible to consumers, and most visits to physicians involve symptoms without medical explanations (Kroenke and Mangelsdorff 1989), the potential to blame unexplained illnesses or symptom syndromes on the effects of invisible technologies seems particularly great. Without credible alternative explanations, it may be difficult to convince people that these technologies may not be ultimately responsible for their problems.

Making Invisible Technologies Visible May Have Unintended Consequences

In part, the lack of public awareness of GM foods in the United States is likely because they are not required to bear labels that would make their GM content apparent. According to a 1992 US Food and Drug Administration (FDA) policy, special labels for foods derived from new GM plant varieties are required only under several particular circumstances. Specifically, labels are necessary to notify consumers if the food derived from GM plants is no longer generally considered to be equivalent to its non-GM counterpart, in which case the food product also needs to have a new name. A GM food product also must be labelled if it has a new use, if a new nutritional characteristic not customary to the product is introduced, or if a known allergen is introduced that is not inherent to the product (FDA 1992). Significantly, however, although these regulations require that labels be used to alert consumers when the characteristics of a familiar food product have been substantially altered through the use of GM ingredients, the labels do not need to say that the change was produced through the process of genetic modification. As a result, there are no current regulations requiring GM foods be identified as such in the United States. However, voluntary guidelines drafted by the FDA in 2001 permit food manufacturers to voluntarily label their products as containing GM ingredients (FDA 2001). Similarly, food manufacturers can label their products as containing no GM ingredients if they choose to, as long as the label does not state or imply that the product is superior because it does not contain such ingredients.

Canada has adopted a similar voluntary labelling standard, with regulatory oversight focusing on a scientific assessment of the characteristics and associated risks of a GM food product before approval for sale. Like the US policy, it is mandatory in Canada for manufacturers to label *any* novel food product where there is a health or safety concern, such as the introduction of a known allergen, or if the composition or nutritional value of the food has been substantially changed. Moreover, in Canada, as in the United States, it is not mandatory to label any food (including GM food) for the process by which it has been developed. As such, under current regulations, in Canada a producer is not required to label a product produced through a process of biotechnology (CBS 2005; Canadian General Standards Board 2004).

This voluntary labelling approach assumes that the technology is generally safe and that market forces should ultimately determine the fate of GM foods. In doing so, the Canadian and US policies attempt to support consumer sovereignty, allowing people to make choices about the foods they buy (Thompson 1997), without imposing unnecessary regulatory costs. These voluntary labelling policies theoretically give consumers who want to avoid GM foods the ability to do so without forcing additional costs on the majority who are assumed not to have such a preference, and, based solely on scientific assessments of the risks of GM foods, *should not* have such a preference (Hallman and Aquino 2005; Moon and Balasubramanian 2004).

In contrast, the European Union (EU) has taken a different approach, promulgating a law that requires any food product that contains more than 0.9 percent genetically modified material to be specially labelled as such (Alvarez 2003). Moon and Balasubramanian (2004) argue that this EU mandatory labelling policy reflects two regulatory principles. The first permits the separation of scientific assessment of risk from the management of those risks. This allows EU regulatory agencies to consider important economic, political, and societal concerns apart from whether the technology presents particular health or environmental risks. The second is the application of the precautionary principle, which requires ongoing scientific assessments of the technology to resolve any questions about its possible adverse effects on human health or the environment.

Like the US and Canadian policies, the EU labelling policy is based on the idea that the cumulative decisions of individual informed consumers should ultimately determine whether GM foods survive in the marketplace. However, unlike the US and Canadian approaches, the EU policy presumes that although no problems have yet been linked with GM food products, there are uncertainties that need to be recognized. By requiring labels, the EU policy theoretically gives the majority who presumably would prefer to avoid GM foods the ability to do so, while passing the additional costs created by the mandatory labelling system onto those who want to produce or consume GM products.

The two labelling approaches are based on the idea that they will allow informed consumers the ability to make choices in the marketplace that will determine the future of the technology. Indeed, the implementation of the EU labelling policy theoretically ends a de facto regulatory ban on GM food products (Alvarez 2003). Ironically, however, neither policy effectively gives consumers the choices they are intended to provide. Food processors are reluctant to put GM labelled products in the marketplace because they fear that consumers will interpret these labels as warnings that the products are of inferior quality or are unsafe. As a result, manufacturers are concerned that GM food labels will cause consumers to reject any products bearing them (GMA News 2001; Food Safety and Inspection Service 2002). Therefore, while most processed foods in the United States and Canada are likely to contain at least traces of GM ingredients, major North American food manufacturers have decided not to label their products as containing GM ingredients (Hallman and Aquino 2005). In the European Union, most supermarkets have chosen not to stock products containing GM ingredients on the grounds that many clients would decide to shop somewhere else (BBC News 2006).

So, policy makers face a dilemma. Consumer opinion surveys show that, when asked directly, consumers around the world show very high levels of support for the labelling of GM food products (Einsiedel 2002). For example, when Americans are asked if they want GM food labels, more than 90 percent say they do, including 95 percent of those who say that they never pay attention to food labels (Hallman et al. 2003). These results are consistent with the position of advocates of mandatory labelling, who argue that people want to retain the ability to choose between GM and conventional food products.

However, some research also suggests that a declaration of the presence of GM ingredients on a food label is likely to cause the product to be rejected by consumers (Hallman et al. 2003). In a national survey of attitudes toward GM foods, more than half (52 percent) of Americans said that a GM food label would make them less willing to purchase the product. In addition, focus group results reported by Hallman and Aquino (2005) suggest that the consumers most familiar with GM food would use the labels to avoid products bearing them and those least familiar would very likely see GM food labels as warnings that they should be concerned about the contents of the products. These results are consistent with the position of those opposing mandatory labelling, who argue that the likely effect of GM labels would be to lead consumers to reject foods made with GM ingredients, although there is no scientific evidence that GM products are inferior or unsafe. Widespread rejection of GM products would likely cause them to disappear from supermarket shelves, thereby reducing real consumer choice.

The paradox, of course, is that without products on the shelves bearing GM labels, it is unlikely that consumers will become much more aware of the presence of GM foods than they already are. Yet, until consumers become more familiar with the technology, they are likely to reject products associated with it.

People Often Lack the Scientific Background to Understand New Technologies

Given that most people say they are not very familiar with GM foods and have not heard a great deal about them, it is not surprising that most also say they do not know very much about them. When asked how much they know about GM foods, most Americans say they know "very little" (48 percent) or "nothing at all" (16 percent). Far fewer say that they know "some" (30 percent) or "a great deal" (5 percent) (Hallman et al. 2005).

That most Americans know little about biotechnology and GM foods is also borne out by studies of factual knowledge about genetics and biotechnology (Hallman et al. 2003; Cuite, Aquino, and Hallman 2005). To assess respondents' actual knowledge of science and genetic modification, they were quizzed using a set of eleven true-or-false questions based on those originally developed for use in the Eurobarometer surveys of European attitudes toward GM foods (Gaskell, Allum, and Stares 2003). More than half of the respondents (52 percent) received a failing grade of less than 70 percent correct and only 4 percent of the sample answered all the quiz questions correctly. There was only a moderate relationship between what Americans think they know about science and objective measures of their actual knowledge. The correlation between self-assessed knowledge of science and technology and quiz scores was 0.38. The correlation between self-assessed knowledge of genetic modification and the quiz scores was 0.35.

This lack of correlation between what people think they know about the science of genetics and biotechnology and what they actually know is important in understanding how people may approach learning about emerging technologies. Those who overestimate what they know about science and technology are unlikely to recognize the gaps they have in their knowledge about biotechnology and so are unlikely to seek information to fill in those gaps. In contrast, those who underestimate their knowledge are unlikely to recognize that the basics of biotechnology are not as difficult to understand as they may assume. In both cases, because of a lack of understanding of what they already know or do not know about a technology, people may not be particularly motivated to try to learn more about it.

Knowing More about a Technology Is Not Necessarily Related to Public Acceptance

The idea that knowing more about the science behind a specific technology will lead to greater approval of that technology is relatively common throughout the scientific literature. Indeed, some studies have found a positive relationship between increased knowledge and approval of the application of science (Evans and Durant 1995; Hayes and Tariq 2000; Gaskell et al. 2002; Allum, Boy, and Bauer 2003; Sturgis and Allum 2004). However, the findings are not entirely consistent, with other studies finding no relationships between knowledge and acceptance (Pfister, Bohm, and Jungermann 2000) or curvilinear relationships between the two (Jallinoja and Aro 2000; Peters 2000).

Some studies have also shown both positive and negative relationships between knowledge and approval, and that some types of knowledge may be more influential than others. Gaskell et al. (1999) compared American and European data to examine the relationship between knowledge and acceptance of GM foods. They found that Europeans had higher scores than Americans on an objective true/false test of textbook knowledge related to the science behind genetic modification, but they were less approving of GM foods when compared with the American sample. This suggested a negative relationship between scientific knowledge and approval. However, the researchers examined three additional knowledge items focusing on beliefs in fictional threatening images related to GM foods, such as "GM animals are always larger than non-GM animals." The American respondents were more likely than their European counterparts to recognize that these images are false and thus scored higher on this scale of knowledge, suggesting a positive relationship between knowledge and approval of GM foods. The clear implication is that the link between knowledge and approval of new technologies has as much to do with the *kind* of knowledge as the amount of knowledge understood by individuals.

In addition, the strength of the relationship between knowledge and approval of new technologies is often modest. In a study of the relationship between knowledge and approval of GM foods using items identical to those used by Gaskell et al. (1999), Cuite, Aquino, and Hallman (2005) found that although increased knowledge was significantly related to higher levels of approval, the relationship between the two was weak. Together, a model including several measures of knowledge, including their scores on textbook knowledge about the basics of genetics and biotechnology, awareness of GM foods in the marketplace, and self-assessed knowledge, plus demographic variables, explained only 8 percent of the variance in approval ratings. This suggests that while knowledge appears to be related to approval, the practical significance of this relationship is open to question and may not be as important a factor in influencing opinions about GM foods as it is often assumed to be.

People Cannot Simply Be "Educated" into Accepting New Technologies

Unfortunately, many efforts to gain public acceptance of new technologies are based on the faulty premise that if people understood more about the science behind a technology, they would come to understand and share the viewpoints and conclusions about the technology reached by the experts. This knowledge deficit model (Wynne and Irwin 1996; Einsiedel 2000; Hansen et al. 2003) assumes that if people just knew the facts, they would reach the right conclusions or make the right decisions.

It is easy to understand why the knowledge deficit model holds such an appeal for scientific and technical experts–the model takes for granted that people fundamentally approach the world just like experts; it's just that lay people are not as intelligent, educated, or well informed. It also presumes that the conclusions reached by experts are consistently correct and that those of non-experts are fundamentally wrong because lay people do not have the required scientific knowledge, technical experience, or understanding of the issues to make proper decisions. But, the model assumes, given the requisite knowledge, experience, and understanding, lay people can develop the expertise they need to "correctly" evaluate technologies and reach the "right" conclusions; that is, the same conclusions already reached by the experts. Therefore, the solution to public acceptance of new technologies is to "educate" lay people to fill in the gaps in their knowledge and to lead them to adopt the "informed" views of the experts.

Unfortunately, research clearly shows that people are more complex creatures and that merely providing information rarely changes their attitudes or behaviours (Weinstein 1988). This should come as little surprise, as the ineffectiveness of this approach is regularly confirmed in our everyday experience. If "simply providing the facts" were really effective in persuading people to adopt a particular point of view about technologies or to change their behaviours toward them, people would wear seat belts in their cars, install carbon monoxide detectors in their homes, and avoid trans fats in their diets. Moreover, if this strategy of "informing people" were successful, people would all drive the same kinds of cars, use the same kind of mouthwash, and wear the same brand of athletic shoes. But, clearly, this kind of universal consensus does not exist. If it did, there would be no need for advertising, politics, self-help books, or late-night televangelists. There would also be no reason to read this book.

Yet, although people are very rarely swayed by mere presentation of scientific facts, experts continue to labour under the illusion that such efforts *must* be effective. The problem, of course, is that although people *do* use scientific information in their decisions about the risks and benefits of technologies, it is not the *only* information they use when considering a risk. So, although scientific knowledge does seem to have an impact on attitudes about new technologies, the relationship is often modest, and non-linear (Sturgis and Allum 2004).

There are likely several reasons for this. Because most people do not have extensive backgrounds in scientific disciplines, some scholars suggest that the influence of scientific knowledge on attitudes may be directly related to the extent to which scientific information is seen as consistent with personal experience (Jasanoff 2000) and with the specific worldviews, core beliefs, or values held by individuals (Slovic and Peters 1998).

The impacts of scientific information may also be moderated or contextualized by other types of knowledge that may include an understanding of how scientific expertise is developed and how science is organized, financed, and controlled. Each can have an effect on trust in the "truths" developed by science (Wynne 1992).

Finally, people may have concerns about a technology that go beyond those typically addressed by science (Hansen et al. 2003). How scientific information is incorporated into how people think about the risks and benefits of a new technology can be strongly influenced by cultural norms and assumptions about what is good, pure, and useful (Douglas and Wildavsky 1982). As such, as Peters (2000) suggests, there are various reasons people may lack faith in, disagree with, or fail to follow the science-based recommendations made by experts that have little to do with a lack of understanding of the science. Indeed, outside the areas of their own expertise, even experts use sources of information other than science to make decisions.

Still, this does not mean that we should abandon efforts to help people understand the science behind new technologies. At the same time, however, we should discard the idea that it is possible to "educate" people into accepting them.

People Are More Interested in What Technology Can Do Than in How It Works

We do need to explain some of the science behind new technologies, because such an understanding may ultimately be necessary for people to separate fact from fiction as proponents and opponents of the technologies attempt to persuade the public to adopt their positions. The problem is that communications about new technologies are often based on an expert model of what people *need* to know rather than on what people *want* to know. Often, this strategy begins with having a group of experts assess the facts they believe non-experts need to know to understand the science behind the technology. The second step is usually an attempt to translate these scientific facts into simple language and graphics so that non-experts can easily understand them. The last step is usually to create a set of communications materials to be distributed to as many people as possible.

Unfortunately, communication strategies based on an expert model are usually unsuccessful. One reason is that experts tend to overestimate what ordinary people know about science. In the case of communications about GM foods, many educational efforts assume that lay people have a level of understanding of basic biology that is not generally borne out by surveys of the public (Hallman et al. 2002). (For example, in a national survey of 1,203 Americans, half of those interviewed said they had never heard about traditional crossbreeding methods, even when those methods were described to them in simple terms. In addition, despite virtually all the varieties of fruits and vegetables available having resulted from centuries of traditional crossbreeding, 61 percent of respondents said they had never eaten a fruit or vegetable produced using traditional crossbreeding methods; another 11 percent indicated that they were not sure.) Efforts designed to convince people that genetic modification techniques are simply a faster, more versatile, and more precise way of achieving the aims of traditional crossbreeding are likely to have little impact. The lesson is that some efforts to teach people about new technologies are not successful simply because many people are not able to put the new information into any meaningful context. These efforts fail because they try to build on a foundation that just does not exist.

Some efforts to educate people about new technologies fail because the experts who design them discover that many people lack the necessary foundation to understand the complex scientific ideas and information the experts want to get across. The experts therefore attempt to remediate the situation by insisting on efforts to build such a foundation for their audiences, block by boring block. These attempts fail because, while lay people *say* they are interested in science, what they are often more interested in is the excitement of new discoveries and their potential impacts, rather than in learning about the rather tedious *details* of the science (Harp and Mayer 1997, 1998).

For example, in a national survey, Hallman et al. (2004) asked respondents to "imagine that we designed a television show especially for you on the topic of genetically modified foods." They were then asked to rate their interest in each of thirteen shows using a scale of one to ten where one represented "not at all interested" and ten represented "extremely interested." Most respondents expressed interest in all the topics presented, giving each median ratings of eight or better on the ten-point scale. However, they indicated the greatest interest (median ratings of ten) in the two stories related to the potential health consequences of the technology: "the potential dangers of eating GM food on personal and family health" and "whether anyone has ever gotten sick from GM foods." They showed somewhat less interest in stories about whether genetic modification would have an effect on the cost of food for consumers or on farmers' costs of producing food, the companies involved in the production of GM foods, which foods or brands of food specifically *do not* contain GM ingredients, and "the science involved in the genetic modification of food products." Each of these hypothetical shows was given a median rating of eight on the scale. The results suggest that although Americans *are* interested in the science behind GM foods, it is not necessarily their first interest. Yet, many efforts to communicate about GM foods focus predominantly (or exclusively) on how the technology *works,* rather than on what the technology *means* for people.

The Risks and Benefits of New Technologies Matter

Even though many people have difficulty understanding the scientific details of how a technology works, they are often interested in, and better able to grasp, the costs, risks, and benefits of these technologies. As a result, there can be substantial differences in the way people think about new technologies in the abstract and the way they think about specific applications of that technology.

For example, the technology behind genetic modification is an unfamiliar and relatively abstract concept, lacking any real context for most Americans. Because they do not understand the technology and do not feel particularly qualified to evaluate its risks or merits, people's first reactions to the technology itself tend to be rather negative. Yet, people are often readily able to understand, and contextualize, the specific benefits of products created through the use of GM technology and so respond quite positively to the idea of using genetic modification to create products with useful characteristics.

For example, Hallman et al. (2002) found that when asked about GM technology in the abstract, only 58 percent of Americans say they approve of the use of genetic modification to create new kinds of plants. Yet, 88 percent say they would approve of the use of genetic modification to create rice with enhanced vitamin A to prevent blindness, and 85 percent say they would approve of the use of the technology to create more nutritious grain that could feed people in poor countries. Similarly, when asked in the abstract, only 28 percent of Americans say they approve of the use of genetic modification to create hybrid animals. Nonetheless, 84 percent say they would approve of the use of genetic modification to create hormones such as insulin to help diabetics, and more than three-quarters of the public (76 percent) say they would approve of the use of the technology to create sheep whose milk can be used to produce medicines and vaccines.

However, people's approval of GM products is not entirely based on altruism. They also tend to approve of GM products from which they might benefit personally. Indeed, nearly three-quarters of those surveyed (74 percent) said they would also approve of the use of genetic modification to create less expensive or better tasting produce, and more than three-quarters (76 percent) said they would approve of "genetically modified grass that you don't have to mow so often." In fact, many researchers have attributed the lack of acceptance of GM foods around the world to the lack of perceived personal benefits to consumers (Hoban 1998).

The large difference between the way people think about GM technology in the abstract and their perceptions of specific products is illustrated by Hallman et al. (2003). Early in their national survey, they asked respondents whether they approved or disapproved of the use of GM technology to create new plant-based and animal-based food products. Later in the survey, the respondents were asked whether they would *buy* GM food products with particular beneficial characteristics. Matching the answers of individual respondents, it was clear that many who initially said they disapproved of the use of the technology later said they would purchase products of that technology with appealing personal benefits. For example, of those who initially disapproved of plant-based GM food products, 26 percent later said they would purchase such products if they had less fat and 21 percent said they would if it tasted better than ordinary food.

GM foods with environmental benefits were also looked on favourably. About one-third (31 percent) of those who initially disapproved of creating plant-based GM food products said they would be willing to buy a GM product grown in a more environmentally friendly way than ordinary food. Moreover, 44 percent of those who initially disapproved of plant-based GM products said they would be willing to purchase them if they contained less pesticide residues than non-GM food.

This expressed preference for the use of GM technology to reduce pesticide residues in food is particularly interesting since reductions in the use of pesticides is a main benefit currently conferred by existing GM crops (Economic Research Service 2005). However, the reduced use of pesticides on crops is an advantage of GM technology that is not marketed specifically as a direct benefit to consumers. Instead, it is advertised as being better for the environment and for the farmer's bottom line.

Not All Applications of the Technology Are Seen as Equivalent

The large difference between the way people think about GM technology in the abstract and their perceptions of specific products suggests that people may judge the merits of the products of genetic modification on a case-by-case basis, considering the characteristics of specific products of biotechnology rather than categorically deciding that all biotechnology is good or bad. Therefore, just because consumers find one GM product to be acceptable or even desirable does not mean that they necessarily approve of the underlying technology, or of its application to other products. Unlike in the information-technology sector, there is almost certainly no "killer application" in the foreseeable future that will drive consumer acceptance of the basic technology behind GM foods. Therefore, one should be cautious in concluding, for example, that "due to the importance of rice in the developing world and the significant part played by the public sector in providing new rice crop technology, the drive to apply GM technology to rice may well result in faster acceptance of the technology in rice than would be the case for other crops. Rice therefore has the potential to act as a catalyst to the wider adoption and acceptance of GM crop technology" (PG Economics 2002, 1).

Indeed, most studies of public opinion concerning genetic modification confirm that consumers express a hierarchy of approval. That is, they are much more willing to approve of the use of GM technology on plants than on animals, more likely to endorse the use of the technology within rather than across species, and least supportive of the use of GM to alter humans in any way (see Gaskell, Allum, and Stares 2003; Pew Initiative on Food and Biotechnology 2004, 2005). Indeed, many people who say they favour using genetic engineering to create new varieties of plants do not approve of its use to create new breeds of animals. As has been suggested, "people see a real difference between using biotechnology to create better beef and using it to create better beefsteak tomatoes" (Hallman 1995).

Perceived Motivations Matter

Although people may ultimately be interested in the mechanisms that underpin new technologies, many will also admit that because the technical issues involved are so complex, they have a difficult time using this information to make judgments about the risks or the safety of the technology. They must therefore rely on the conclusions reached by those specialists who do have the requisite expertise to evaluate the merits of the technology. Because they lack the ability to interpret the technical information, it is not surprising that many people have less interest in the particular details about the science and technology than in specific details about the scientists and technicians involved. They want to be able to trust that the people behind the technology are using common sense and are making decisions that will appropriately protect the well-being of both people and the planet.

The question of whether those in the scientific community, the food biotechnology industry, and government regulators can be trusted to protect consumers and the environment from unsafe products features prominently in the debate over consumer acceptance or opposition to GM foods (Lang and Hallman 2005). Hallman et al. (2002) found that three-quarters (74 percent) of Americans believe that strict regulation of GM technology and products is needed. Yet, most were skeptical that scientists and the companies involved with food biotechnology have sufficient motivation or competence to protect the public from potentially adverse effects that might arise from the use of genetic modification (Hallman et al. 2002).

Although People Must Rely on Experts to Evaluate the Science, They Still Want to Be Involved in Decision Making

Although most Americans seem to know very little about the science of genetic modification, and many who are well informed will concede that the issues are so complex that they have a difficult time reaching conclusions on their own, they are reluctant to relinquish their involvement in decisions regarding the technology. For example, only one-quarter of Americans agree that "decisions about the issue of genetically modified food are so complicated that it is a waste of time to consult the public on this subject" (Hallman et al. 2002).

Gaskell et al. (2005) reported the results of social surveys in the United States, Canada, and the European Union, examining who the public thinks should make decisions about science policy and on what criteria such decisions should be based. They found that given a choice between having decisions about technology based mainly on the views and advice of experts or on the views of average citizens, nearly three-quarters say they would prefer that the views of experts guide such decisions. Moreover, when asked if these decisions should be made on the basis of scientific evidence or on the basis of moral or ethical considerations, two-thirds preferred that they be guided by scientific principles. The responses to these questions permitted the division of the public into four groups reflecting different preferred principles of governance over technology. The results showed considerable consistency in the attitudes of respondents across the three jurisdictions. About half of the respondents in each country were classified as "scientific elitists" preferring that decisions about technology be guided by expert advice based on scientific evidence. In contrast, about one in five were categorized as "moral elitists" preferring that such decisions be made on the basis of expert advice using moral and ethical criteria. Between 10 percent and 14 percent were identified as "scientific populists" opting for decisions based on the average citizen's views of the scientific evidence, and about 15 percent were tagged as "moral populists" preferring that decisions be based on the average citizen's views of the moral and ethical issues.

Thus, while about half the populations of the United States, Canada, and Europe appear to endorse the status quo regarding the governance of technology (experts making decisions based on scientific evidence), the other half do not seem particularly content with the current state of affairs. Rather than leave such important decisions to be made by unknown experts (who may or may not be trustworthy) on the basis of science that much of the public does not fully understand, they would prefer that decisions about technology be influenced by some combination of moral and ethical principles shared by average citizens.

What is interesting about this is that it suggests that people make some implicit assumptions about the ability or inability of average citizens to understand science and the ability or inability of scientific experts to understand ethics and morality. Yet, while they may question the ability of average people to understand science and the ability of scientists to understand ethical and moral issues, the assumption seems to be that scientific experts are uniquely qualified to judge science, but that average people are perhaps better qualified to judge ethics and morality. So, in deciding who should govern science, the choice comes down to "Although I assume that they are competent to understand the science, do I trust scientific experts to reflect the values of society?" versus "Although I trust average citizens to reflect the values of society, do I believe that they are competent to understand the science?"

Given that the scientific populists make up the smallest group identified in the United States, Canada, and the European Union, the answer to the second question appears to be no. Average citizens do not appear to judge each other as understanding enough about science and technology to have sufficiently informed views.

In the case of GM foods, this may be because each respondent perceives himself or herself to be an average citizen, and because they judge themselves as not well informed enough about GM technology to make decisions about it, this must therefore be true of all average citizens. In contrast, some people may consider themselves to be particularly knowledgeable about the technology, with expertise on the subject well beyond that of the average citizen. Realizing the complexities involved and the knowledge and effort necessary to understand them, these people may realistically decide that average citizens do not have the time, scientific background, or proclivity to undertake what is required. Either way, people do not seem to be particularly optimistic about the abilities of their fellow citizens to participate in complex decisions about technology.

The remaining question then is, how are people to decide whether scientific experts share the same values, morals, and ethics as average citizens, and how can they be confident that the decisions made by these strangers will be consonant with societal wishes? The GM debate suggests that those in the scientific community have a responsibility to deliberately counter the view that science and ethics are incompatible.

Consensus about Which Problems Are Worth Solving Is Important

Most people are decidedly *not* anti-technology; the public wants science and technology to solve problems. The issue is that they want some say in deciding which problems are worth solving and whether the solutions proposed are acceptable. Indeed, the final and perhaps most important lesson to be drawn from the GM food experience is that, for new technologies, success is not about attracting public trust and consent for an agenda already established by science, scientists, industry, or government officials but, rather, achieving some shared societal vision of what needs to be done and how. Rather than simply being seen as obstructionists who need to be convinced of the acceptability of new technologies, perhaps members of society should be seen as *investors* who want to have some influence on the direction of development.

References

Allum, N., D. Boy, and M. Bauer. 2003. European regions and the knowledge deficit model. In *Biotechnology: The making of a global controversy,* ed. M. Bauer and G. Gaskell, 224-43. Cambridge: Cambridge University Press.

Alvarez, L. 2003. Europe acts to require labelling of genetically altered food. *New York Times*, 2 July: A3.

BBC News. 2006. Q&A: Trade battle over GM food. February 8. http://news.bbc.co.uk/1/hi/world/europe/4690010.stm.

Canadian General Standards Board. 2004. Voluntary labelling and advertising of foods that are and are not products of genetic engineering. Gatineau, QC: Canadian General Standards Board.

CBAB (Canadian Biotechnology Advisory Board). 2005. The status of labelling of genetically engineered/genetically modified (GE/GM) foods in Canada. http://www.cbac-cccb.ca/epic/internet/incbac-cccb.nsf/en/ah00539e.html.

CBS (Canadian Biotechnology Secretariat). 2005. International public opinion research on emerging technologies: Canada-US survey results. Government of Canada BioPortal. http://www.bioportal.gc.ca/CMFiles/E-POR-ET_200549QZS-5202005-3081.pdf.

CDC (Centers for Disease Control and Prevention), National Center for Environmental Health. 2001. *Investigation of human health effects associated with potential exposure to genetically modified corn: A report to the US Food and Drug Administration from the Centers for Disease Control and Prevention.* Atlanta, GA: Centers for Disease Control and Prevention.

Cuite, C.L., H.L. Aquino, and W.K. Hallman. 2005. An empirical investigation of the role of knowledge in public opinion about GM food. *International Journal of Biotechnology* 7 (1, 2, and 3): 178-94.

Douglas, M., and A. Wildavsky. 1982. *Risk and culture: An essay on the selection of technical and environmental dangers*. Berkeley, CA: University of California Press.

Economic Research Service. 2005. Adoption of genetically engineered crops in the US. http://www.ers.usda.gov/Data/BiotechCrops/adoption.htm.

Einsiedel, E.F. 2000. Understanding "publics" in the public understanding of science. In *Between understanding and trust: The public, science, and technology,* ed. M. Dierkes and C. von Grote, 205-16. Amsterdam: Harwood Academic Publishers.

–. 2002. GM food labelling: The interplay of information, social values, and institutional trust. *Science Communication* 24 (2): 209-21.
Evans, G., and J. Durant. 1995. The relationship between knowledge and attitudes in the public understanding of science in Britain. *Public Understanding of Science* 4 (1): 57-74.
FDA (US Food and Drug Administration). 1992. Statement of policy: Foods derived from new plant varieties. *Federal Register* 57 (104): 22984. http://www.cfsan.fda.gov/~acrobat/fr920529.pdf.
–. 2001. Guidance for industry: Voluntary labelling indicating whether foods have or have not been developed using bioengineering: Draft guidance. Washington, DC: US Food and Drug Administration, Center for Food Safety and Applied Nutrition. Food Safety and Inspection Service. http://www.fda.gov/cvm/Guidance/001598gd.pdf.
Food Safety and Inspection Service (FSIS), USDA. 2002. CODEX Committee on Food Labeling thirtieth session, USA comments (May). http://www.fsis.usda/oa/codex/biotech02.htm.
Gaskell, G., N. Allum, M. Bauer, J. Durant, A. Allansdottir, H. Bonfadelli, D. Boy, et al. 2002. Biotechnology and the European public. *Nature Biotechnology* 18 (9): 935-38.
Gaskell, G., N. Allum, and S. Stares. 2003. Europeans and biotechnology in 2002: Eurobarometer 58.0. Brussels: European Commission.
Gaskell, G., M.W. Bauer, J. Durant, and N.C. Allum. 1999. Worlds apart? The reception of genetically modified foods in Europe and the US. *Science* 285 (5426): 384-87.
Gaskell, G., E. Einsiedel, W. Hallman, S.H. Priest, J. Jackson, and J. Olsthoorn. 2005. Social values and the governance of science. *Science* 310 (5756): 1908-9.
GMA News. 2001. GMA says Massachusetts mandatory labelling bill "unnecessary and redundant." Press release, May. http://www.gmabrands.com/news/docs/newreleaase.cfm.
Hallman, W.K. 1995. Public perceptions of agri-biotechnology. *Genetic Engineering News* 15 (13): 4-5.
–. 2005. Predicting approval and discussion of genetically modified foods in Canada and the United States. In *First impressions: Understanding public views on emerging technologies*, ed. E. Einsiedel, 20-42. Calgary: University of Calgary Press.
Hallman, W.K., A.O. Adelaja, B.J. Schilling, and J. Lang. 2002. Public perceptions of genetically modified foods: Americans know not what they eat. Food Policy Institute report no. RR-0302-001. New Brunswick, NJ: Food Policy Institute, Rutgers University.
Hallman, W.K., and H.L. Aquino. 2005. Consumers desire for GM labels: Is the devil in the details? *Choices* 20 (4): 217-22.
Hallman, W.K., H.M. Jang, C.W. Hebden, and H.K. Shin. 2005. *Consumer acceptance of GM food: A cross-cultural comparison of Korea and the United States*. New Brunswick, NJ: Food Policy Institute, Rutgers University.
Hallman, W.K., and J. Metcalfe. 1994. *Public perceptions of agricultural biotechnology: A survey of New Jersey residents*. New Brunswick, NJ: Food Policy Institute, Rutgers University.
Hallman, W.K., W.C. Hebden, H.L. Aquino, C.L. Cuite, and J.T. Lang. 2003. Public perceptions of genetically modified foods: A national study of American knowledge and opinion. Food Policy Institute report no. RR-1003-004. New Brunswick, NJ: Food Policy Institute, Rutgers University.
Hallman, W.K., W.C. Hebden, C.L. Cuite, H.L. Aquino, and J.T. Lang. 2004. *Americans and GM food: Knowledge, opinion and interest in 2004*. Food Policy Institute report no. RR-1104-007. New Brunswick, NJ: Food Policy Institute, Rutgers University.
Hansen, J., L. Holm, L. Frewer, P. Robinson, and P. Sandoe. 2003. Beyond the knowledge deficit: Recent research into lay and expert attitudes to food risks. *Appetite* 41 (2): 111-21.
Harp, S.F., and R.E. Mayer. 1997. The role of interest in learning from scientific text and illustrations: On the distinction between emotional and cognitive interest. *Journal of Educational Psychology* 89 (1): 92-102.
–. 1998. How seductive details do their damage: A theory of cognitive interest in science learning. *Journal of Educational Psychology* 90 (3): 414-34.
Hayes, B.C., and V.N. Tariq. 2000. Gender differences in scientific knowledge and attitudes toward science: A comparative study of four Anglo-American nations. *Public Understanding of Science* 9 (4): 433-47.

Hebden, W.C., and W.K. Hallman. 2005. Consumer responses to GM foods: Why are Americans so different? *Choices* 20 (4): 243-46.
Hoban, T.J. 1998. Trends in consumer attitudes about agricultural biotechnology. *AgBio-Forum* 1 (1): 3-7.
Huang, J., J. Bai, C. Pray, and F. Tuan. 2004. *Public awareness, acceptance of and willingness to buy genetically modified foods in China.* Rutgers University, unpublished report.
Jallinoja, P., and A.R. Aro. 2000. Does knowledge make a difference? The association between knowledge about genes and attitudes toward gene tests. *Journal of Health Communication* 5(1): 29-39.
James, C. 2005. Executive summary of global status of commercialized biotech/GM crops: 2005. ISAAA Briefs no. 34. Ithaca, NY: International Service for the Acquisition of Agri-biotech Applications.
Jasanoff, S. 2000. The "science wars" and American politics. In *Between understanding and trust: The public, science, and technology,* ed. M. Dierkes and C. von Grote, 39-60. Amsterdam: Harwood Academic Publishers.
Kroenke K., and A.D. Mangelsdorff. 1989. Common symptoms in ambulatory care: Incidence, evaluation, therapy, and outcome. *American Journal of Medicine* 86 (3): 262-66.
Lang, J.T., and W.K. Hallman. 2005. Who does the public trust? The case of genetically modified food in the United States. *Risk Analysis* 25 (5): 1241-52.
Moon, W., and S.K. Balasubramanian. 2004. Public attitudes toward agrobiotechnology: The mediating role of risk perceptions on the impact of trust, awareness, and outrage. *Review of Agricultural Economics* 26 (2): 186-208.
Nisbet, M.C., and B.V. Lewenstein. 2002. Biotechnology and the American media: The policy process and the elite press, 1970 to 1999. *Science Communication* 23 (4): 359-91.
Peters, H.P. 2000. From information to attitudes? Thoughts on the relationship between knowledge about science and technology and attitudes toward technologies. In *Between understanding and trust: The public, science, and technology,* ed. M. Dierkes and C. von Grote, 265-86. Amsterdam: Harwood Academic Publishers.
Pew Initiative on Food and Biotechnology. 2004. Overview of findings: 2004 focus groups and poll. http://pewagbiotech.org/research/2004update/overview.pdf.
–. 2005. Public sentiments about genetically modified food (2005). http://www.pewtrusts.org/news_room_detail.aspx?id=32804.
Pfister, H.R., G. Bohm, and H. Jungermann. 2000. The cognitive representation of genetic engineering: Knowledge and evaluations. *New Genetics and Society* 19 (3): 295-316.
PG Economics. 2002. *GM rice: Will this lead the way for global acceptance of GM crop technology?* Dorset, UK: PG Economics.
Shoemaker, R., D.D. Johnson, and E. Golan. 2003. Consumers and the future of biotech foods in the United States. *Amber Waves* 1 (5): 30-36.
Slovic, P., and E. Peters. 1998. The importance of worldviews in risk perception. *Journal of Risk Decision and Policy* 3 (2): 165-70.
Sturgis, P., and N. Allum. 2004. Science in society: Re-evaluating the deficit model of public attitudes. *Public Understanding of Science* 13 (1): 55-74.
Ten Eyck, T.A., and M. Williment. 2003. The national media and things genetic: Coverage in the *New York Times* (1971-2001) and the *Washington Post* (1977-2001). *Science Communication* 25 (2): 129-52.
Thomson, J., and L. Dininni. 2005. What the print media tell us about agricultural biotechnology: Will we remember? *Choices* 20 (4): 247-52.
Thompson, P.B. 1997. Food biotechnology's challenge to cultural integrity and individual consent. *The Hastings Center Report* 27 (4): 34-39.
Walker, J. 2005. Report on a study of emerging technologies in Canada and the US: Prevailing views, awareness and familiarity. In *First impressions: Understanding public views on emerging technologies,* ed. E. Einsiedel, 20-42. Calgary: University of Calgary Press.
Weinstein, N.D. 1988. The precaution adoption process. *Health Psychology* 7 (4): 355-86.
Wynne, B. 1992. Public understanding of science research: New horizons or hall of mirrors? *Public Understanding of Science* 1 (1): 37-43.
Wynne, B., and A. Irwin. 1996. *Misunderstanding science? The public reconstruction of science and technology.* Cambridge: Cambridge University Press.

2
Patentable Subject Matter: Who Owns What Knowledge?

Chika B. Onwuekwe

Inequality in the world–among nation states, individuals, corporations, and organizations–is neither a secret nor a recent development (Wallerstein 1979). The astronomical rate of development in the North, especially in the area of technology and Western science, perpetually confines the South (developing countries) to playing catch-up. This is exacerbated because Western epistemology is mostly regarded, erroneously, as the preferred and the only way of knowing. Consequently, knowledge founded on other epistemologies is treated as non-scientific, indigenous, traditional, non-methodological, and archaic. With increased commercialization of knowledge, developing countries are eager to gain both recognition and some form of economic reward for their traditional knowledge, most of which was hitherto freely shared. Incidentally, the criteria established by the North for commercialization of knowledge, and enforced through the World Trade Organization (WTO), exclude prior art or knowledge in the public domain. Wallerstein (1999) attributes this artificial segregation to the nature of capitalism, which inherently thrives on inequality.

The Agreement on Trade-Related Aspects of Intellectual Property Rights (TRIPS Agreement) consolidates the North's position on the kind of knowledge that will qualify for proprietary protection through intellectual property rights. The WTO became the international institution responsible for enforcing compliance with the TRIPS Agreement's minimum international standard of intellectual property rights. Because of its history, the WTO appears to be an exclusive multilateral club for championing the agenda of the technology-advanced countries. Its capacity to play this role buttresses Amin's contention (1997) that states with five identifiable monopolies are stronger than others; also, such states are well positioned to influence the outcome of world order. These monopolies are technological monopoly, financial control of worldwide financial markets, monopolistic access to the planet's natural resources, media and communication monopolies, and monopolies over weapons of mass destruction. Although debatable as to which state in the North currently holds all these monopolies, it is a fact that none of the developing countries in Africa, Asia (excluding Japan and, to some extent, China), and Latin America possesses any of these monopolies.

It is against this backdrop that I challenge in this chapter the erroneous notion that plant genetic resources (PGRs) and the associated traditional knowledge (TK) are part of the common heritage of humankind and therefore not patentable subject matters. I also submit that patentable subject matter, although locally defined, has an extra-territorial effect. This is because each country monitors the development of intellectual property rights in other countries. With increased competition among developed countries on technology advancement, each technology-rich country perceives intellectual property as an instrument for encouraging and rewarding innovation. Thus, no amount of ethical or moral objection from the South will persuade the North to abandon gene and cell patenting, especially as it applies to lower life forms. Like the controversy over genetically modified organisms (GMOs) in agricultural biotechnology, the North is not ready to accommodate the concerns of the South in its bid to further widen the gap in the structural power of knowledge through advanced technology while rejecting other systems of knowledge. This explains why I canvass in the chapter for a redefinition of the criteria for patents, a type of intellectual property available for inventions, to accommodate PGRs and the associated TK. Such changes, it is envisaged, will be in line with the spirit of paragraph 19 of the 2001 WTO Ministerial Declaration at Doha, Qatar, currently under negotiation. To achieve this objective, developing countries will have to persuade technology-rich countries to abandon their sole Eurocentric basis for patents. Besides being a daunting task for the South, any changes along that line will undermine the interests the current TRIPS Agreement projects. As the analysis below demonstrates, there is no likelihood that the envisaged changes will occur, because of developing countries' lack of essential structural powers.

The Structural Power of Knowledge

Susan Strange (1994) identified four distinct but related structural powers–the security structure, the production structure, the finance structure, and the knowledge structure–the possession of which gives an entity immense advantage over others. Strange submits that structural power "is the power to determine and shape the structure of the global political economy within which other states, their political institutions, their economic enterprises and (not least) their scientists and other professional people have to operate" (Strange 1994, 24-25). Structural power is not restricted to the power to set the agenda for discussion but "confers the power to decide how things shall be done, the power to shape frameworks within which states relate to each other, relate to people, or relate to corporate enterprises" (Strange 1994, 25). Strange identified the United States as the current hegemon in the world system controlling the four structural powers. In another forum, she (1987) argues that rather than diminishing, the United States' structural power has increased despite that it now shares these powers with its home multinational corporations.

Accordingly, Strange (1994) acknowledged that the era of "*total* control and monopoly by the state (outside the Soviet Union and China) has seemingly gone for good" (135).[1] The ability of multinational corporations to influence how either a domestic or a global agenda develops is attributed to their technical know-how. By way of example, Strange refers to the surprising inclusion and successful negotiation of intellectual property rights as a trade-related issue at the Uruguay Round. The manner of this item's inclusion and the manipulation of other WTO member states to accept the TRIPS Agreement, particularly those in developing countries that had little intellectual property to protect, support Strange's analysis of the immense influence multinational corporations currently have on the global trade issues. Kuanpoth's examination of the political economy of the TRIPS Agreement clarifies Strange's assertion: "The introduction of TRIPS into the Uruguay Round of global trade talks, combined with the use of GSP [generalized system of preferences] benefits and Section 301 of the Trade Act, indicates the determination of world economic superpowers to defend their national interests regardless of the international rules and the resulting damage to the economies of other countries" (2003, 55).

What then is the nature of power exercised through the knowledge structure? Strange (1994) points out that unlike the other structural powers, "the power in the knowledge structure often lies as much in the negative capacity to deny knowledge, to exclude others, rather than in the power to convey knowledge" (19).[2] Historically, knowledge is a public good and "its possession by any one person does not diminish the supply to any other" (Strange 1994, 31). However, Strange agrees that the historical position of knowledge has changed with the proliferation of modern technology. Knowledge has over the years acquired proprietary interest, with states and their home multinational corporations investing huge funds to acquire or improve any knowledge associated with technology (such as biotechnology, genetic engineering, or nanotechnology). This approach confirms Strange's argument (1994, 121) that "a knowledge structure determines what knowledge is discovered, how it is stored, and who communicates it by what means to whom and on what terms."

Specifically, the current criteria for patents under the TRIPS Agreement make it difficult to protect other knowledge systems that also contribute to biodiversity and modern biotechnology. The existing dominant intellectual property regime is a mere proprietary construct that enables the current structural power holders–the North–to perpetuate their hold on the structural power of knowledge through commercialization of Eurocentric knowledge or Western technology. At a time when developing countries are no longer willing to freely allow access to their PGRs and the associated TK, an impasse on what patentable subject matter covers appears to be looming. These countries are now demanding compensation for and recognition of the uses of their PGRs and TK, at least within the regime of the Convention on Biological Diversity's (CBD's) benefit-sharing initiative. It is no longer a secret that their international competitiveness will be enhanced if they are able to control access to or the utilization and manipulation of the genetic information embodied in PGRs found within their borders (Onwuekwe 2007). As a result, "ownership and control of PGRs is becoming a critical element in the trade policies of developing countries" (Low 2001, 326).

The 2001 WTO Doha Ministerial Declaration

With the failure of the 1999 World Trade Organization's ministerial conference in Seattle, there was a conscious effort by biotechnology-rich states, particularly the United States and its key Organisation for Economic Co-operation and Development (OECD) allies,[3] to ensure the success of any subsequent WTO ministerial conference (Subedi 2003). These countries were united in their efforts to assure pessimists and critics of the WTO that the institution is still intact and relevant, and capable of resolving issues including those important to developing countries (Jordan 2001; Ostry 1997). Furthermore, in view of the constant and coordinated pressure from biotechnology nongovernmental organizations (NGOs), there was a need to reassure this constituency that the WTO is an inclusive rather than exclusive institution for both the developed and developing countries (Ricupero 2001). It was therefore not surprising that the subsequent 2001 WTO ministerial conference held in Doha, Qatar, resembled the development-driven conference that was not achieved at the 1999 WTO ministerial conference in Seattle, Washington.

In a declaration issued at the end of the Doha ministerial conference, core issues for negotiation by WTO member states were outlined. The one on intellectual property rights was contained in paragraph 19. The thrust of paragraph 19 of the Doha Declaration is that most member states of the WTO from developing countries are dissatisfied with the intellectual property regime of the TRIPS Agreement. Although the TRIPS Agreement provides a minimum international standard from which nation states are to draw their national intellectual property laws, its article 27.3 contains provisions for patents that are of utmost concern to these countries. For instance, patents on plants and life forms, including lower life forms, remain objectionable to developing countries on various grounds. Some of the objectionable grounds include economic, religious (the playing God syndrome), moral, ethical, and social biases (Menikoff 2001; Canadian Biotechnology Advisory Committee 2002).[4] The outcome of the Doha ministerial conference, which overcame initial suspicion amid last-minute concessions, was dubbed the Doha Development Agenda. The agenda set the tone for the ongoing round of multilateral trade negotiations. The relevance of paragraph 19 of the Doha Declaration, and its impact on the proliferation of biotechnology (gene) patents, especially life patenting, are also of paramount importance.

Suffice it to say that there is no doubt that the Doha ministerial conference was successful because the United States and its OECD allies ensured that the Seattle failure was not repeated. Otherwise, the rancour that marred the fifth WTO ministerial conference in Cancun, Mexico,[5] suggests that the North will not easily yield to the demands of developing countries for a radical overhaul of the TRIPS Agreement, especially on the scope of patentable subject matter.[6]

Patentable Subject Matter

Articles 27(1) and 27(2) of the TRIPS Agreement are the guiding provisions on what patentable subject matter means. Article 27.1 directs WTO members to provide patents for "any inventions whether products or processes, in all fields of technology, provided that they are new, involve an inventive step and are capable of industrial application." Article 27.2 permits the denial of patents to certain innovations judged to be morally unacceptable by countries. The article provides that "members may exclude from patentability inventions, the prevention within their territory of the commercial exploitation of which is necessary to protect *ordre public* or morality, including to protect human, animal or plant life or health or to avoid serious prejudice to the environment."

This provision empowers each country to determine what is patentable within its jurisdiction. On this note, patent laws of most countries, including the European Patent Convention, exclude morally unacceptable inventions from patentability. Unfortunately, neither the United States patent regime, which supports the patenting of anything under the sun and made by man, nor its Canadian counterpart incorporates moral limitations on what is patentable. Where it exists, as in the European Patent Convention, the rejection of patents on moral grounds gives the patent commissioner a basis to make a value judgment on each patent application. In the absence of such provision, the Supreme Court of Canada stated in *Harvard College v. Canada (Commissioner of Patents)* that the "Commissioner of Patents was given no discretion to refuse a patent on the grounds of morality, public interest, public order, or any other ground if the statutory criteria are met."[7]

In December 2002, the Canadian Supreme Court blazed a different trail when it held (by 5-4 majority) that the meaning of *manufacture* and *composition of matter* in the country's Patent Act does not cover higher life forms. According to the court, "the best reading of the words of the Act supports the conclusion that higher life forms are not patentable . . . I do not believe that a higher life form such as the OncoMouse is easily understood as either a 'manufacture' or a 'composition of matter' . . . I am not satisfied that the definition of 'invention' in the Patent Act is sufficiently broad to include higher life form."[8]

By this judgment, the Canadian Supreme Court distanced itself from the expansive interpretation in the United States and the European Union on what constitutes patentable subject matter. In these other jurisdictions–the United States and the European Union–the Harvard OncoMouse patent had been upheld. As such, a different judgment by the Canadian Supreme Court on the same Harvard OncoMouse patent shocked the biotechnology world. Nevertheless, the controversy generated by the judgment was yet to settle when the court had another bite at the cherry two years later, in *Monsanto Canada Inc. v. Schmeiser.*[9]

Monsanto is the owner and licensor of the patent that disclosed the invention of chimeric genes. These genes are tolerant to glyphosate herbicides such as Roundup and cells containing those genes. Monsanto markets a special kind of canola containing the patented genes and cells under the trade name Roundup Ready canola. Percy Schmeiser, on the other hand, grows canola commercially in Saskatchewan. Mr. Schmeiser neither purchased Roundup Ready canola nor obtained a licence to plant it. Following a suspicion of infringement of its kind of canola, Monsanto tested and found that 95 to 98 percent of Mr. Schmeiser's canola was Roundup Ready canola. Monsanto sued Schmeiser for patent infringement. At trial, the judge found the patent to be valid and allowed the action, concluding that the appellants knew or ought to have known that they saved and planted seed containing the patented gene and cell and that they sold the resulting crop also containing the patented gene and cell. The Federal Court of Appeal affirmed the decision but made no finding on patent validity. Upon appeal to the Canadian Supreme Court, by a split of 5-4 decisions, the court held that Monsanto's patent (on a life form) was valid. According to the court: "The respondents did not claim protection for the genetically modified plant itself, *but rather for the genes and the modified cells that make up the plant*. A purposive construction of the patent claims recognizes that *the invention will be practised in plants regenerated from the patented cells, whether the plants are located inside or outside a laboratory.* Whether or not patent protection for the gene and the cell extends to activities involving the plant is not relevant to the patent's validity."[10]

This decision clearly validated gene and cell patenting of either higher or lower life forms.[11] After all, as the court (per Binnie J.) earlier noted in the *Harvard* OncoMouse case, the Canadian Patent Act "does not distinguish . . . between subject matter that is less complex ('lower life forms') and subject matter that is more complex ('higher life forms')."[12] Since cells are patentable subject matter, it follows that direct products of such inventions would automatically be infringed upon if used without the consent of the patent owner. It does not matter that the end product, such as a plant, was not originally part of the patent claim. Once again, Justice Binnie's minority opinion in the *Harvard* OncoMouse case (at paragraph 68) is instructive in understanding the court's subsequent reasoning in the *Monsanto* case: "What it did was to modify the genome of the OncoMouse so that every cell in its body contained a modified gene. It is not like adding a new and useful propeller to a ship. The oncogene is everywhere in the genetically modified OncoMouse, and it is this important modification that is said to give the OncoMouse its commercial value, which is what interests the Patent Act."

Consequently, what the Canadian Supreme Court shied away from doing in the *Harvard* OncoMouse case it did in the subsequent *Monsanto* case. Unlike in the *Monsanto* case, the Supreme Court was critical in the earlier *Harvard* OncoMouse case of the expansive approach adopted by the United States regarding the meaning of patentable subject matter. The court clearly disapproved of such expansive interpretation as was also applied in the earlier United States' case of *Diamond v. Chakrabarty*.[13] Criticizing the United States' approach in *Diamond,* the court stated that "the majority attributed the widest meaning possible to the phrases *composition of matter* and *manufacture* for the reason that inventions are, necessarily, unanticipated and unforeseeable."[14] With the rejection of a broad interpretation of invention under the Canadian Patent Act in a manner adopted in *Chakrabarty* and subsequently in the Harvard OncoMouse patents by other respective competing jurisdictions–the European Union and the United States–the court held that animals (including a mouse that was genetically modified to make it susceptible to cancer), seeds, and plants were unpatentable higher life forms. This explains why the *Monsanto* case is perceived, and rightly so, as a summersault by the court. It is therefore safe to conclude that the Canadian Supreme Court overruled its earlier decision in the *Harvard* OncoMouse case when it held by a 5-4 majority that Monsanto's patent was valid.

Since what constitutes patentable subject matter may be expanded at will by each country, similar to what happened in the *Harvard* OncoMouse and *Monsanto* cases, why has PGRs and the associated TK of indigenous and local communities in the South remained unpatentable? As argued elsewhere (Onwuekwe 2004, 2007), there is no legal or juridical basis for describing either state-occurring PGRs or their associated TK as part of the commons. Non-recognition of the proprietary character of these resources robs them of their intrinsic economic, medicinal, and sometimes spiritual values. Further, it makes them freely available to entities outside the source community or state through free riding.

With the proliferation of innovation in biotechnology, the grant of patent protection may no longer depend solely on proof of the three essential criteria for patents, especially the requirement of novelty and inventive step. For instance, it appears that the distinction in patent law between either "invention" or "inventive step" on the one hand and "discovery" on the other has completely disappeared in most jurisdictions in the North. This is despite provisions in the European Patent Convention and the Canadian Patent Act that discourage the patenting of discoveries.[15] The importance of this development is underscored by the increase in the patenting of genes, mainly discovered or isolated from their natural surroundings.[16] Fowler (1994, 226-27) queries the basis for this development when he states: "Has the plant breeder who finds a mutation in the field (perhaps a flower with a new colour) discovered a new plant or invented one? Is the new plant patentable or is it simply a product of nature? What, then, is the nature of the inventive process associated with living things?"

While supporting each country's capacity to determine what is patentable in its jurisdiction within the framework of article 27.3 of the TRIPS Agreement, such a capacity will be of no value if powerful countries can influence what qualifies as patentable subject matter. For instance, the persistent opinion by commentators and policy makers from most industrialized countries that state-occurring PGRs and the associated TK are part of the common heritages of humankind is deliberate. Since there is no binding legal instrument to support this wide but erroneous claim that these resources are common heritages, it can safely be asserted that if PGRs and their associated TK are not useful to the powerful biotechnology countries in the North, this persistent one-sided categorization will not arise. Consequently, there should be a better understanding of the different cultures–Eurocentric and indigenous epistemological knowledge respectively–without attempt by one to eclipse the other. Biotechnology-rich countries should embrace diversity because it is more inclusive than imposing Eurocentric positions and cultures on developing countries through an international institution such as the WTO. PGRs occurring within the territorial threshold of nation states and the related TK do not fit into the commons paradigm.

On a similar note, the potential environmental hazard posed by gene patenting is another reason why developing countries are concerned about the proliferation of biotechnology without adequate national and international regulations. If most countries were to apply the outcome of the Canadian decision in *Monsanto v. Schmeiser* in their jurisdictions to remain competitive in biotechnology advancement, they may undermine the thrust of the CBD, especially its articles 1, 2, and 3. Essentially, article 3 rubberstamps the charter of the United Nations and the principles of international law on sovereign rights of states to control and exploit their resources, such as PGRs, in accordance with each country's environmental policies. Furthermore, the injunction in article 15 of the CBD suggests that states are free to enact local laws and regulations that support genetic resources' use without undermining their environment and biodiversity conservation.

In view of these provisions, developing countries are more interested in ensuring that the pursuit of gains through commercialization of modern biotechnology is not carried out without a full consideration of the environmental and ethical concerns about IPRs on life forms (Blakeney 2001a, 2001b). The successful negotiation of the Biosafety Protocol in 2000 at the New Delhi meeting was a first step toward this goal. The protocol established minimum international standards for the environmental safety of releases of GMOs. With the recent coming into force of this protocol, one hopes that its protective reach will not be undermined through a subsequent international instrument orchestrated by the outcome of cases such as *Monsanto v. Schmeiser.*[17]

The Nature of Plant Genetic Resources and Traditional Knowledge

The misleading idea from the North that PGRs and the traditional knowledge about their uses are common heritages of humankind has exacerbated the controversy between the North and the South on the proprietary status of these resources and the associated TK. It also affects any efforts by the South to declassify PGRs and TK from the commons and accord them the required proprietary protection. The commons concept reduces PGRs and TK to *res nullius*.[18] This makes them freely accessible and incapable of private ownership, contrary to the accepted position that "apart from an unclaimed portion of Antarctica, there is no area of the planet earth which can today be characterized as *res nullius*" (Kindred et al., 397; Battiste and Henderson 2000). The implication of this persistent notion of the commons about PGRs and TK is that no entity can successfully claim exclusive control or ownership over them (Bragdon 2000). Furthermore, the commons concept gives no consideration to past and present labour of indigenous or local communities in conserving biodiversity, which is essential for the maintenance of their PGRs (Shiva 1997).

Because of the foregoing, PGR and TK source countries and indigenous communities reject the commons concept attached to their resources and related TK. Rather, they contend that source communities have proprietary interests in their PGRs and the TK on their uses within the communal proprietary structure practised in centres of origin (Cohen and Noll 2001).

The structure gives rise to a communal property right, which may be different in form but has the same bundle of rights implication as private proprietary interest recognized and revered in the North. In the former, the unit of ownership could be the individual, the family, or the community, while for most developed countries it is usually individuals or corporations.

Considering that the diverse nature of the world necessitates the existence of different proprietary structures, I support a formal binding institutionalization of the proprietary rights of source communities or states over their PGRs and the associated TK. The institutionalization should contain an enforcement mechanism similar to that under the TRIPS Agreement in order to cure the present limitations of the CBD in this regard. It is indeed ironic and inequitable that "elite" crops developed from PGRs and TK could be subject to private ownership when neither PGRs nor TK enjoy a similar proprietary protection. Besides epistemological differences, which the CBD attempts to address, there is no justification for this dichotomy.[19] This is why I view the controversy as a clash between the liberal concept of property promoted by the North and the communal property concept, widely accepted in most developing countries. I concur with Dutfield that if developing countries could accept the Western-oriented system of patents, which is now institutionalized in the TRIPS Agreement of the WTO, there is no reason why their own concept of communal property rights to their TK and PGRs should not be accorded equal treatment (Dutfield 1999, 2001).

Moreover, propertizing PGRs and TK will meet one of Hardin's suggestions (1993) on how to avoid the tragedy of the commons. Although Hardin favoured the use of a "private property" arrangement to fend off overuse, he nevertheless recognized that other systems of property ownership might also be used to achieve the same thing. This can be gleaned from his concluding statement, when, in reference to private property as a tool for policing the recklessness associated with commons property, he says, "or something formally like it" (133). Accordingly, formally institutionalizing the proprietary interest of source communities over their PGRs and TK will only give legal backing to what already exists. Its potency lies in the fact that source communities–either on their own or through their governments–can regulate uses through formal arrangements similar to the technical use agreement in the seed industry. For practical purposes, each source country will sort out how to realize the full benefits of any changes to the persistent concept that state-occurring PGRs and the related TK are part of the commons. This is because any institutionalization through a social arrangement that recognizes source communities or states as rightful owners will formally acknowledge that PGRs and TK are not common heritage resources but communally owned resources.

Furthermore, the insistence by the industrialized capitalist countries of the North that PGRs and TK are common heritages of humankind undermines the international law concept of sovereignty of nations and inviolability of territorial integrity. Indeed, if states still have sovereign control over natural resources found within their territory, there is no basis for treating PGRs occurring within states differently. Moreover, since property rights are social constructs, the non-recognition of the proprietary interest of source communities over their PGRs and the TK about their uses in the North appears discriminatory and somehow portrays an imperialist attitude. This is particularly the case in view of the specific provisions of the CBD and the 2001 International Treaty on Plant Genetic Resources for Food and Agriculture (popularly known as the Global Seed Treaty), which recognize the labour of source communities in the conservation of biodiversity through traditional knowledge and practices.

It should be pointed out that the non-recognition of the proprietary interests in PGRs for patent purposes presents different problems to the various stakeholders in biotechnology development.[20] For instance, one of the key debates on this issue revolves around the purported difficulty in attaching economic value to PGRs (Wood 1988).[21] This debate is without merit, considering the potential economic and medicinal values of PGRs in the hands of multinational corporations. The politicization of ownership of plant germplasm has shown that rather than allow market forces to determine the worth of these resources, the powerful industrialized states have shielded their multinational seed and pharmaceutical corporations from adverse market impacts that may arise from the propertization of PGRs. This supports Kloppenburg's assertion that "the utility of plant genetic resources for the maintenance and improvement of the elite commercial cultivars of the industrial North is not mere theoretical proposition, it is historical fact" (Kloppenburg 1988, 167; Pinel and Evans 1994).

Consequently, the effort by some indigenous communities to establish their own system for protecting TK is commendable.[22] However, until institutionalized by a binding international treaty with a built-in enforcement mechanism, these various declarations lack extra-territorial effect. Although the social (including the aesthetic and spiritual) utilities of TK and PGRs are important to these communities, they are nevertheless entitled to some form of compensation for the utilization of their resources in biotechnology research and development. A compensatory regime in economic terms will largely be in compliance with the yet to be fully explored CBD's benefit-sharing regime. Such a development will not diminish the aesthetic and spiritual connections of source communities to some of these resources. In any event, unless an acceptable model of compensatory regime is agreed upon, source communities should not feel obligated to accept the unaccounted and uncompensated access to, and sharing of, their knowledge on the uses of PGRs. In fact, an acceptable regime of compensation (either in economic terms or otherwise) will provide traditional communities with alternatives when compared with the current position under which neither their communal labour in conserving and improving PGRs nor their usable knowledge of these resources are rewarded.

In addition, the development will minimize the incidence of biopiracy as evidenced by patents granted through the United States Patent and Trademark Office to W.R. Grace on the *Indian neem tree;*[23] to the University of Mississippi on *Use of Turmeric in Wound Healing;* and to Larry Proctor of POD-NERS on *Enola (Mexican) Bean*. These patents are classical examples of the appropriation of TK and PGRs without recompense, acknowledgment, or recognition by profit-seeking institutions and biotechnology multinational corporations. With the denial of any enforceable proprietary protection to PGRs and the associated TK, it appears as if "patents allow global corporate actors to appropriate medicinal treatments used widely in developing countries" (Matthews 2002, 116).

At the moment, developing countries will continue to feel marginalized within the multilateral trading system of the WTO until issues surrounding the extension of patent protection to plant and animal life forms, genetic information, and other items of discovery are resolved. Of utmost importance in this debate is the unequivocal request from the South for a redefinition of the international patent regime in order to recognize non-Western innovative practices. As highlighted earlier, developing countries are concerned about the ownership or proprietary control of their plant genetic resources, especially with the continued notion, reference, or treatment of these resources as common heritages of humankind by industrialized countries. This is in addition to whether or not traditional knowledge of source communities and indigenous peoples concerning the uses of germplasm should be excluded from patent protection for lack of Western-style documentation or publication, despite the incontrovertible evidence that such knowledge belongs to known and identifiable communities in the South. Similarly, since the CBD recognizes the proprietary interests of source communities over PGRs occurring in their states, and their associated TK and practices, is there any basis for persistently describing them as part of the commons?

The extension of patents to life forms in the North has both negative and positive implications for developing countries. On the negative side, there are claims that it offends the morality of most developing countries to see the increased commodification of life. As such, there is an innate fear that it will not be long before patents over humans or their progenies will be granted in these countries, and enforced through the WTO mechanism. In other words, the boundary between what is human and what is not for purposes of patents is becoming thinner. For instance, while the Canadian Biotechnology Advisory Committee reported back to government with proposals to clear the ambiguity on the patentability of human life, its European counterpart took a different approach. In 1998, the European Parliament issued an EU Biotechnology Patents Directive advising that life patents are proper, except in relation to patents pertaining to the human body.[24] Similarly, there is no indication that the principle that the polluter pays will be extended to patented GMOs in the event of contamination of the environment or non-GMO farms. The *Monsanto v. Schmeiser* case is an indication that, in patent infringement cases, nuisance or contamination is not a strong defence.

On the positive side, life patenting of any nature makes the debate over the non-propertization of PGRs and the TK on their uses mute. The controversy over the proprietary status of PGRs and TK about their uses resembles a class struggle in the Gramscian sense (Hoffman 1988). If sustained, the controversy may lead to the WTO losing its leading role as the international institution driving free trade. Developing countries, like the oppressed class in any society, organization, or arrangement, appear to understand that their interests may not after all be in the WTO. Their walk-out from the 2003 WTO ministerial conference in Cancun, Mexico, when they did not receive the anticipated concession on various key demands, is a confirmation of this awareness. However, the Cancun approach suggests a lack of the structural power, influence, and bargaining strength required to obtain favourable conditions at the WTO negotiating table. The controversy and the subsequent events that marred the Cancun conference may be described in Antonio Gramsci's words:

> Events with the inexorable logic of their development give the worker and peasant masses, which are conscious of their destiny, these lessons. The class struggle at a certain moment reaches a stage in which the proletariat no longer finds in bourgeois legality, i.e. in the bourgeois State apparatus (armed forces, courts, administration), the elementary guarantee and defence of its elementary right to life, to freedom, to personal safety, to daily bread. It is then forced to create its own legibility, to create its own apparatus of resistance and defence. At certain moments in the life of the people, this is an absolute historical necessity. (in Hoare 1978, 15)

There is no doubt that the common position that developing countries maintained at Cancun is a step toward creating their own legitimacy at the WTO, similar to how Gramsci's oppressed classes unite to pursue their common goals. Nevertheless, the unity achieved by developing countries at Cancun will be tested in the course of the ongoing multilateral negotiation of paragraph 19 of the Doha Declaration. These countries have been known to abandon their common resolve in pursuit of peculiar interests, which the North dangles in the form of economic or military aids. Under such circumstances, it may not be out of place to ask whether the controversy is actually about patenting of life forms (higher or lower) or merely about who benefits from such patents? Despite the mantra of technology transfer and benefit sharing, the needs of developing countries go beyond playing catch-up each time a new technology enters the lexicon of innovation and development (Onwuekwe and Phillips 2007; Phillips and Onwuekwe 2007). A more inclusive approach before commercialization of these technologies is required to ensure that the fears and concerns of developing countries are properly documented and reviewed in order to find appropriate workable solutions. With great optimism, this may require an institution with a more human face than the current WTO.

Conclusion

Human beings are at the threshold of innovation and development in genomics and other deep sciences, through biotechnological advancements. Working together on the merits of each kind of innovation and technology, the international community is better placed to determine which innovations should be encouraged through the instrument of patents or other *sui generis* (unique) forms of proprietary protection. But it is important that each country continues to define within its borders what patentable subject matter means, to avoid undermining its sovereignty. While a universal guideline is appropriate, there should not be any binding regime on patentable subject matter. To do otherwise will be to impose other countries' morality and economic nuances on weaker nations.

Originally, knowledge was mostly a public good, but this changed with the patenting of biotechnology inventions and gene sequencing. Unfortunately, the Eurocentric-driven patent regime recognizes only certain kinds of knowledge as capable of private ownership through patents, while those deemed incapable of patents are said to belong to the commons. The current criteria for patents provided Western scientific knowledge with proprietary character, while other cultural ways of knowing, such as traditional knowledge, were confined to the public domain. These different treatments of epistemology are difficult to explain under the existing proprietary jurisprudence. This is especially the case when TK has become an indisputable source of most modern biotechnological advancements.

The notion that TK is part of the common heritage of humankind is not only a misnomer but legally, socially, and economically incorrect. TK in its present form has proprietary characteristics, and there is no juridical foundation for denying it some form of enforceable proprietary protection. Hopefully, developing countries will maximize the opportunity presented by the ongoing Doha multilateral trade negotiations orchestrated by paragraph 19 of the Doha Declaration to remedy this anomaly. On their part, the biotechnology-rich countries of the North should not only support this initiative but also reorganize the WTO by making it an inclusive institution for all state members. It is by such actions that the WTO will cease to be an exclusive institution for the technology-rich countries and become a true multilateral trade institution for all member countries.

The above, including the thrust of the two recent Canadian cases on life patenting, recognizes the truism that the enjoyment of patent monopolies, like any other monopoly in the modern world system, exists at the pleasure of the state. Being a human construct, an acceptable model of TK rights, similar to the bundle of rights applicable to its Eurocentric counterpart, should be crafted to accommodate the interests embodied in PGRs and the TK of source communities. A study of the models of traditional resources rights and the community intellectual property rights articulated by Posey and Dutfield (1996) could serve as a first step toward understanding how this can be achieved.

Notes

Many thanks to my wife (Chinwe) and children (Chiamaka, Chiedozie, and Chidubem) for their understanding and generosity with family time during the process of writing this chapter.

1 Emphasis in original.
2 The other three structural powers, namely security, production, and finance (credit), are positively driven.
3 These are mainly Canada, Europe, and Japan. These countries have substantial investment and support for medical and agricultural biotechnologies. The United States also maintains good trade relationships with strong agricultural biotechnology countries in the South, namely Argentina, China, and South Africa.
4 However, the political reasons for this objection are often missed. There is a masked fear on developing countries' part of bio-colonialism, which may not be unfounded or far-fetched.
5 This ministerial conference was held between September 10 and 14, 2003.
6 It was reported that the African Group and other developing countries walked out of the meeting on the last day because the OECD countries were unwilling to accede to most of their requests, particularly on the Singapore issues. See the note of the last day of the conference by the WTO entitled "Conference ends without consensus" at http://www.wto.org/english/thewto_e/minist_e/min03_e/min03_14sept_e.htm. It appears, therefore, that the success recorded at Doha was a face-saving event and image-management opportunity rather than a true success founded on consensus building for which the WTO wishes to be noted. Thus, the failure of the Cancun ministerial conference was bound to happen and so did not come as a surprise to those who have followed the evolution of the WTO as a multilateral trade institution, which has become a tool for the furtherance and protection of interests of powerful states.
7 [2002] 4 S.C.R. 45, at para. 11, Binnie J.
8 Ibid. at para. 155, Bastarache J.
9 [2004] 1 S.C.R. 902, 2004 SCC 34.
10 Ibid. at 903 (emphasis added).
11 Prior to this case, patenting of "lower life forms" was not an issue, as the patent commissioner would issue such patent application if it met the criteria for patents.

12 Minority in *Harvard College v. Canada (Commissioner of Patents)*, [2002] 4 S.C.R. 45 at para. 47.
13 447 U.S. 303 (1980).
14 *Harvard College v. Canada (Commissioner of Patents)*, [2002] 4 S.C.R. 45 at para. 157.
15 See the European Patent Convention (July 2002) at http://www.european-patent-office.org/epc/pdf_e.htm.
The earlier (Canadian) Patent Act contained the word *discovery,* but this was removed by subsequent amendments to the act. See Act Respecting Patents of Invention, S.C. 1869, c. 11, s. 6. Compare with section 2 of the Patent Act, R.S.C. 1985, c. P-4, which defines invention without "discovery" as "invention means any new and useful art, process, machine, manufacture or composition of matter, or any new and useful improvement in any art, process, machine, manufacture or composition of matter."
16 See, for instance, the US cases of *Diamond v. Chakrabarty,* 447 U.S. 303 (1980) and *Moore v. Regents of the University of California* 793. P.2d. 479 (1990). See also the Canadian cases of *Pioneer Hi-Bred Ltd. v. Canada (Commissioner of Patents)* (1982) C.P.R. LEXIS 587, [1987] 3 F.C. 8, [1989] 1 S.C.R. 1623; *Monsanto Co. v. Commissioner of Patents,* [1979] 2 S.C.R. 1108 and *Re Application of Abitibi Co.,* [1982] C.P.R. LEXIS 587. In *Re Abitibi,* the Patent Appeal Board held that "patentable subject-matter includes microorganisms, yeast, moulds, fungi, bacteria, actinomycetes, unicellular algae, cell lines and viruses or protozoa."
17 The Biosafety Protocol to the CBD came into force on September 11, 2003.
18 Black's Law Dictionary (7th edition) defines this Latin phrase to mean "thing of no one." In other words, *res nullius* refers to something that belongs to no one; "an ownerless chattel."
19 At the moment, the World Intellectual Property Organization (WIPO) is working on developing model contract clauses for benefit sharing. WIPO was asked by its members in June 2001 to "develop the model clauses in order to address a perceived resource imbalance between private companies seeking to secure the rights to use genetic resources for commercial purposes, and those governments or government agencies which claim ownership over the resources. The model clauses are intended to serve as 'best practices' guidelines and would not be binding on members." The non-bindingness of WIPO's guidelines or adopted recommendations makes efforts through international institutions to harmonize an intellectual property regime, especially in relation to traditional knowledge, a Herculean task. States' discretion to choose what binds them in international law has both a positive and a negative side. On paper, this principle provides a level of comfort to the not-too-powerful nations. That is, it assures them that they are still independent and have the freedom of decision making at international forums. In reality, this is not the position. The powerful nations always have their way.
20 These include multinational seed companies, farmers, indigenous and local communities, governments (developed and industrial countries), and environmental nongovernmental organizations.
21 Wood (1988) argues that lack of unanimity on the compensation to pay source countries for the use of their germplasm is because there is no "workable mechanism for valuing germplasm." He contends rightly that "failure to define such a mechanism could lead to long drawn-out bargaining over access to germplasm" (285). Wood fails to discuss how such a mechanism can be fashioned in view of the existing mutual suspicion between the North and South on this issue. Thus, while the Food and Agriculture Organization's Global Seed Treaty focuses on PGRs used in food and agriculture, there is no mention of what happens to medicinal plants used by pharmaceuticals. Moreover, the North has consistently refused to yield IPRs on its elite seeds.
22 See, for instance, the Mataatua Declaration on Cultural and Intellectual Property Rights of Indigenous Peoples, 1993; Charter of the Indigenous-Tribal Peoples of the Tropical Forests, signed at Penang, Malaysia, on February 15, 1992, http://www.mtnforum.org/resources/library/citpt92a.htm; Kari-Oca Declaration and the Indigenous Peoples' Earth Charter, 1992; UN Draft Declaration on the Rights of Indigenous Peoples, E/CN.4/Sub.2/1994/2/Add.1 (1994); Declaration of Principles of the World Council of Indigenous

Peoples ratified by the IV General Assembly of the Council at Panama, September 23-30, 1984.

23 The European Patent Office granted W.R. Grace & Co. patent for the fungicidal effects of the neem oil.

24 See Directive 98/44/EC of the European Parliament and of the Council of July 6, 1998, on the legal protection of biotechnological inventions, art. 5(1), (2), 6.

References

Amin, S. 1997. *Capitalism in the age of globalization: The management of contemporary society.* London: Zed Books.

Battiste, M., and J.Y. Henderson. 2000. *Protecting indigenous knowledge and heritage: A global challenge.* Saskatoon: Purich Publishing.

Blakeney, M. 2001a. Intellectual property aspects of traditional agricultural knowledge. In *IP Biodiversity and agriculture: Regulating the biosphere,* ed. P. Drahos and M. Blakeney, 29-32. London: Sweet and Maxwell.

–. 2001b. Intellectual property in the Millennium Round: The TRIPs Agreement after Seattle. In *Foundations and perspectives of international trade law,* ed. I. Fletcher, L. Mistelis, and M. Cremona, 537-40. London: Sweet and Maxwell.

Bragdon, S.H. 2000. Recent intellectual property rights controversies and issues at the CGIAR. In *Agriculture and intellectual property rights: Economic, institutional and implementation issues in biotechnology,* ed. V. Santaniello, R.E. Evenson, D. Zilberman, and G.A. Carlson, 77-90. New York: CABI.

Canadian Biotechnology Advisory Committee. 2002. Patenting of higher life forms and related issues (June 2002)—A report to the Government of Canada Biotechnology Ministerial Coordinating Committee. http://cbac-cccb.ca/epic/site/cbac-cccb.nsf/en/ah00188e.html.

Cohen, L.R., and R.G. Noll. 2001. Intellectual property, antitrust, and the new economy. *University of Pittsburgh Law Review* 62 (3): 453-63.

Dutfield, G. 1999. Protecting and revitalising traditional ecological knowledge: Intellectual property rights and community knowledge databases in India. In *Intellectual property aspects of ethnobiology,* ed. M. Blakeney, 101-22. London: Sweet and Maxwell.

–. 2001. Indigenous peoples, bioprospecting, and the TRIPS Agreement: Threats and opportunities. In *IP in biodiversity and agriculture: Regulating the biosphere,* ed. P. Drahos and M. Blakeney, 135-49. London: Sweet and Maxwell.

Fowler, C. 1994. *Unnatural selection: Technology, politics, and plant evolution.* Yverdon, Switzerland: Gordon and Breach.

Hardin, G. 1993. The tragedy of the commons. In *Valuing the earth: Economics, ecology, and ethics,* ed. H.E. Daly and K.N. Townsend, 127-44. Cambridge, MA: MIT Press.

Hoare, Q. ed. and trans. 1978. *Antonio Gramsci: Selections from political writings (1921-1926).* London: Lawrence and Wish.

Hoffman, J. 1988. *The Gramscian challenge: Coercion and consent in Marxist political theory.* Oxford: Basil Blackwell.

Jordan, B. 2001. Building a WTO that can contribute effectively to economic and social development worldwide. In *The role of the World Trade Organization in global governance,* ed. G.P. Sampson, 243-58. Tokyo: United Nations University Press.

Kindred, H.M., K. Mickelson, R. Provost, L.C. Reif, T.L. McDorman, A.L.C. deMestral, and S.A. Williams. 2000. *International law: Chiefly as interpreted and applied in Canada.* 6th ed. Toronto: Emond Montgomery.

Kloppenburg, J.R. 1988. *First the seed: The political economy of plant biotechnology, 1492-2000.* Cambridge: Cambridge University Press.

Kuanpoth, J. 2003. The political economy of the TRIPS Agreement: Lessons from Asian countries. In *Trading in knowledge: Development perspectives on TRIPS, trade, and sustainability,* ed. C. Bellmann, G. Dutfield, and R. Meléndez-Ortiz, 45-56. London: Earthscan.

Low, A. 2001. The third revolution: Plant genetic resources in developing countries and China; Global village or global pillage? In *International Trade and Business Law*

Annual, ed. J. Kinsler, R. Jones, and G.A. Moens, 323-60. Newport, New South Wales: Cavendish.

Matthews, D. 2002. *Globalising intellectual property rights: The TRIPs Agreement.* London: Routledge.

Menikoff, J. 2001. *Law and bioethics: An introduction.* Washington, DC: Georgetown University Press.

Onwuekwe, C.B. 2004. The commons concept and intellectual property rights regime: Whither plant genetic resources and traditional knowledge? *Pierce Law Review* 2 (1): 65-90.

–. 2007. Ideology of the commons and property rights: Who owns plant genetics resources and the associated traditional knowledge? In *Accessing and sharing the benefits of the genomics revolution,* ed. P.W.B. Phillips and C.B. Onwuekwe, 21-48. Dordrecht: Springer/Kluwer.

Onwuekwe, C.B., and P.W.B. Phillips. 2007. Conclusion: New paths to access and benefit sharing. In *Accessing and sharing the benefits of the genomics revolution,* ed. P.W.B. Phillips and C.B. Onwuekwe, 199-208. Dordrecht: Springer/Kluwer.

Ostry, S. 1997. *The post-Cold War trading system: Who's on first?* Chicago: University of Chicago Press.

Phillips, P.W.B., and C.B. Onwuekwe. 2007. Introduction to the challenge of access and benefit sharing. In *Accessing and sharing the benefits of the genomics revolution,* ed. P.W.B. Phillips and C.B. Onwuekwe, 3-17. Dordrecht: Springer/Kluwer.

Pinel, S.L., and M.J. Evans. 1994. Tribal sovereignty and the control of knowledge. In *Intellectual property rights for indigenous peoples: A sourcebook,* ed. T. Greaves, 43-55. Oklahoma City: Society for Applied Anthropology.

Posey, D.A., and G. Dutfield. 1996. *Beyond intellectual property: Towards traditional resource rights for indigenous peoples and local communities.* Ottawa: IDRC.

Ricupero, R. 2001. Rebuilding confidence in the multilateral trading systems: Closing the legitimacy gap. In *The role of the World Trade Organization in global governance,* ed. G.P. Sampson, 37-58. Tokyo: United Nations University Press.

Shiva, V. 1997. *Biopiracy: The plunder of nature and knowledge.* Toronto: Between the Lines.

Strange, S. 1987. The persistent myth of lost hegemony. *Int'l Organization* 41 (2): 551-74.

–. 1994. *States and markets.* London: Pinter.

Subedi, S.P. 2003. The road from Doha: The issues for the Development Round of the WTO and the future of international trade. *International and Comparative Law Quarterly* 52 (2): 424-46.

Wallerstein, I. 1979. *The capitalist world-economy.* Cambridge: Cambridge University Press.

–. 1999. *The end of the world as we know it: Social science for the twenty-first century.* Minneapolis: University of Minnesota Press.

Wood, D. 1988. Crop germplasm: Common heritage or farmers' heritage? In *Seeds and sovereignty: The use and control of genetic resources,* ed. J.R. Kloppenburg Jr., 274-91. Durham, NC: Duke University Press.

3
Patents in the Public Sphere: Public Perceptions and Biotechnology Patents

Edna F. Einsiedel

One of the foundation principles of liberal democracy is ownership and protection of property. A subsidiary principle is the right to develop and profit from one's ideas, a right enshrined in intellectual property legislation. Patents are one of several tools designed to protect this right of inventors, tools that have been viewed as an important mechanism to spur innovation.

With the rapid pace of discoveries in the life sciences and the expansion of biotechnology as a commercial venture, there is increasing interest in and competition to patent inventions, which include human and non-human genetic sequences, genes, and even entire microorganisms, plants, and animals. These life science discoveries and their associated intellectual property (IP) protections used to be the arcane domain of specialist lawyers and biotechnology industry experts in the 1970s and 1980s. The cloaks that shrouded IP debates began to lift in the mid-1980s and increased in the 1990s with public interest stoked by advocacy groups concerned about the fences erected around knowledge, making the 1990s "the decade of lost innocence about intellectual property" (Hilgartner 2002). The move from patenting non-living inventions to seeking to control innovations involving organisms or their parts began to elicit public interest and concern, and IP questions increasingly became regarded as political questions rather than simply legal issues to be decided by patent bodies and the courts.

Several cases illustrate this broader spectrum of interests and concerns surrounding patents that have shone a more public spotlight on this intellectual property tool:

- A multinational corporation, Monsanto, alleged that a Saskatchewan farmer, Percy Schmeiser, infringed on the company's patent rights to a genetically engineered seed variety by planting Roundup Ready canola without licence or permission. The Supreme Court of Canada held that the patent was valid and that the defendant, Schmeiser, had infringed

on the patent. The case became a *cause célèbre* among advocacy organizations and was also of interest to the wider public.
- Gene-use restriction technologies, more commonly known as "terminator technology," were also another case of patenting controversy. The technology involved genetically induced sterility so the seed could not be replanted–a genetically induced form of patent protection. The controversy generated turned on the violation of poor farmers' traditional practice of saving and sharing seeds. International organizations such as the Consultative Group on International Agricultural Research and the Rockefeller Foundation were moved to publicly declare that the technology should not be adopted.
- People with the Canavan disease gene donated blood and tissue samples to researchers who secretly patented and restricted the availability of testing after discovering the Canavan gene. A lower court in the United States ruled that the patent challenge on the basis of "unjust enrichment" could move forward.
- The breast cancer genes, BRCA1 and BRCA2, were patented by Myriad Genetics, and the company broadened its interest in protection of these patents through its diagnostic tests. Provincial governments in Canada (Ontario and British Columbia) refused to honour these patents, preferring to have the testing done in local laboratories that charged less than the Myriad labs. At the same time, the company was also involved in direct marketing to US consumers of its breast cancer tests.

These cases illustrate that patent controversies have expanded well beyond the traditional expert communities to become social issues of interest to the general public.

In this chapter, I examine public perceptions of patenting biotechnology as one window into how publics have made judgments about biotechnology applications generally and, more specifically, how publics have viewed questions of ownership of knowledge and living matter. I base this description on various research reports of public opinion surveys and focus groups carried out among Canadian publics, at the behest of provincial and federal agencies.

Public Awareness about Patent Protection

Most Canadians possess a general knowledge of human genomics research, usually expressed in terms of specific applications they have heard or read about. Particularly among involved Canadians (those identified as politically active in terms of media exposure and contact and civic engagement), higher levels of awareness of genomics and genetic research and sophistication on issues that surround these topics are more evident than they have been in the past. Canadians associate human genomic discoveries with substantial benefits, and majorities expect these benefits to result in better health treatment and health care in Canada. However, concerns have also been expressed, typically clustering around issues of risk, regulation, and ethics (Earnscliffe Research and Communications 2005).

When the specific topic of patenting is broached, awareness levels about the purpose of patenting and some of its most fundamental elements tend to be low (Earnscliffe Research and Communications 2002). However, focus groups have revealed that most people have a notional sense that patenting provides some form of rights to inventors (Earnscliffe Research and Communications 2003a). More specifically, while participants in focus groups have some awareness that applications of genetic technologies such as diagnostic tests and gene therapies could be patented, most were unaware that genes or gene sequences could also be patented (Earnscliffe Research and Communications 2005).

Interestingly, once people in focus group discussions were informed about what patenting is and the advantages and disadvantages of such a system, positions on patenting were often made situationally, a finding that reflects general trends in public opinion studies (Gaskell et al. 2001a; Gaskell et al. 2001b). In terms of gene patenting, there was about a 65-35 split between support and opposition after participants were informed about patenting issues and processes (Earnscliffe Research and Communications 2002). However, in the case of patenting higher life forms, provision of detailed information demonstrated even increased negativity because of the basic moral resistance people have to the idea of commodifying living beings (Earnscliffe Research and Communications 2003a).

Patenting and Public Opinion: General Acceptance

Both Canadian and US consumers have generally expressed support for the principle of patenting (Earnscliffe Research and Communications 2000a), and many believe that the idea of patent protection is necessary in the field of biotechnology to encourage inventions for all the benefits they can bring (about four in ten and about half in Canada and the United States, respectively). However, a similar or greater proportion in both countries say that they are uncomfortable with the idea of providing patent protection because there is something wrong with the idea of patenting parts of a life form such as an animal or plant or because the benefits of new inventions might be available only to those who can afford to pay more (Earnscliffe Research and Communications 2003b).

A survey of Ontarians on patents in connection with genetic tests showed the majority expressing favourable attitudes toward allowing companies to patent these tests–over six in ten indicated they had a favourable attitude. However, when asked whether such patent rights should be extended to genes or genetic material that are used to develop the tests, the number of those who express favourable attitudes drops to about half (Ipsos Reid 2001).

Hierarchies of Acceptance

When judging the acceptability of patenting biotechnology applications, it is clear from various studies that publics perceive a hierarchy of acceptability that follows the same general pattern as approval for the use of the applications themselves (Earnscliffe Research and Communications 2000a).

First is the hierarchy of purpose. The intended uses for which the patent is granted play a significant factor in determining acceptability. Many publics, Canadians included, are usually more accepting of granting patents for applications that solve medical, environmental, or crop challenges and less accepting of those for industrial or aesthetic purposes. For example, respondents in a 2000 survey indicated that patenting is more acceptable in the context of human health and environmental applications, with two-thirds supporting the patenting of genetically modified bacteria that clean up toxics and six in ten supporting the patenting of disease-resistant rodents bred to find cures for humans. Agricultural and commercial applications were met with less enthusiasm, with 55 percent supporting altered trees that grow faster and 38 percent agreeing with patenting cows that are modified for greater milk production (Earnscliffe Research and Communications 2000a).

Views are also modulated by the perception of the patent contributing to or detracting from societal or public interest. When the purpose achieves broader societal benefits (e.g., encouraging research for a more competitive agricultural sector), greater support for the application is elicited, in contrast to cases where the benefits are restricted to a few (Environics Research Group 1998).

The second observable hierarchy concerns the object of the patent. Acceptability declines as the object of the patent involves increasingly higher orders of life, crosses species boundaries (e.g., mixing plant and animal genes), or alters the organism itself (Earnscliffe Research and Communications 2000a). This public sensitivity to the patenting of higher life forms and, specifically, the patenting of human genetic material is reflected in some expert communities, as seen in the Nuffield Bioethics Council's position that patents asserting rights over human DNA sequences should not be allowed on the basis of the special status of human DNA material (Nuffield Council 2002).

In general, it is more acceptable to Canadian publics to patent the process or technology that creates a novel organism or genetic sequence rather than the novel creation itself. In discussions about the Harvard OncoMouse, people were much more comfortable with the idea of patenting the process that developed the OncoMouse than patenting the mouse itself, because of their resistance to patenting living beings (Earnscliffe Research and Communications 2003a).

On the issue of patenting genetic tests and the genetic material used to develop such tests, a survey of Ontarians showed that close to two-thirds (64 percent) supported the patenting of the tests while only half (51 percent) had a favourable opinion of patenting the genetic material itself (Ipsos Reid 2001). Interest in the provision of genetic tests to the public and in accessing such tests was also high, with seven in ten Ontarians agreeing that genetic testing should be available to the public (Hay Health Care Consulting Group 2002). This interest springs from the perception of direct health benefits that can be derived from genetic testing. At the same time, concerns about delivery of these services revolve around issues of privacy, ethics, and lack of control over who is using what test. The question of gene patents is directly related in the public mind to these areas of concern (Ipsos Reid 2001; Hay Health Care Consulting Group 2002).

Human Genes and Higher Life Forms

Human Genes

The success of the mapping of the human genome has led numerous organizations to apply for patents on genes with particular traits within the newly discovered human DNA (Earnscliffe Research and Communications 2002). The public has strongly supported the mapping of the human genome and, with the success of this project, has shown increased support for the idea of patenting genes for the purposes of developing genetic therapies or drugs (Einsiedel and Sheremeta 2005). However, relative to DNA mapping, patenting human genes with particular traits was met with resistance, with roughly half of the sample expressing discomfort (in a forced choice on comfort or discomfort about patenting genes). Affordability concerns took precedence over ethical concerns, and, overall, Canadians expressed more discomfort than Americans (Earnscliffe Research and Communications 2003a).

Genes, as the raw material in our bodies, are not patentable. Only when human genes or sequences have been described, isolated, and purified can they be patentable, provided they meet the criteria of being new, non-obvious, and useful (Gold 2000). It is acknowledged in most jurisdictions that human (and non-human) genes and genetic sequences are acceptable objects of patents. However, debate continues, especially among publics, about the ethical appropriateness of patentability, with specific concerns about the following issues:

- Is a gene or gene sequence truly an invention? Some members of the public have expressed skepticism about the patentability of genes, which are considered an element of nature or the body (Earnscliffe Research and Communications 2005).
- Will allowing patents over human genetic material create a demand for such biological materials and increase the likelihood of individuals being exploited (Einsiedel and Sheremeta 2005)?
- Will allowing patents of genes or sequences actually inhibit research? In group discussions, doubts have been expressed about the rationale for patent protection as a means of furthering research and innovation, with some suggesting that because the patent holder has the right to control access to the patented material, this could mean that other researchers might be unable to carry out research to develop new products (Earnscliffe Research and Communications 2005).

Higher Life Forms and Their Parts

While patenting is not a new concept, its use to claim ownership over higher life forms or their parts raises new public concerns over the commodification of life and the ethics of patenting.

The discomfort associated with the view that there is something wrong with the idea of patenting parts of a life form weighs in the minds of Canadians: a greater proportion of Canadians–at least half–are uncomfortable with patenting on this basis, compared with those four in ten who say the idea of patent protection is necessary to encourage biotechnology inventions for the benefits they can bring. Similar conclusions are reached in considering entire organisms rather than their parts (Earnscliffe Research and Communications 2000a). Americans appear to be more evenly split on this issue (Earnscliffe Research and Communications 2003b).

Patenting Plants versus Animals

Assessments of public views on genetic modification of organisms have demonstrated hierarchies of preferences. This is particularly evident in judgments about plants versus animals, "lower" life forms such as microorganisms versus other non-human animal organisms, and the latter as compared with humans (Einsiedel 2005). In general, there is less support for patenting genetically modified animals than plants (CBAC 2001), as the public tends to see animal patenting as breaching an ethical standard that plant patenting does not (Earnscliffe Research and Communications 2003a).

Substantial majorities have indicated agreement with granting patents on new plants as necessary to encourage globally competitive agricultural research. However, this support is tentative and conditional (Environics Research Group 1998). Specific concerns have been raised in the context of agricultural practices. Some people fear that patents will provide an incentive for the creation of monocrops to maximize profits and that granting patents on products of biotechnology would send a signal to the public that the product is socially acceptable. Others oppose patenting on grounds of the unknown risks involved in biotechnology or of the potential abuses of power that may result (CBAC 2001).

In the case of animals, the patenting of an entire animal is offensive for many at an emotional level (Earnscliffe Research and Communications 2000b). In discussions, the issue of patenting whole animals (and substantial human body parts such as organs) prompted substantial opposition (Earnscliffe Research and Communications 2003a). Objections to patenting new animal species are strong, and the type of animal does not seem to matter in the minds of Canadians. That is, there does not seem to be a differentiation between animals that are closely related to man and animals commonly associated with medical experimentation. When asked whether it is "permissible to patent a new species of chimpanzee or guinea pig that included human genes," seven in ten were opposed to patenting either new species, with close to half strongly opposed to this idea (Environics Research Group 1998, 50).

Close to two-thirds say that granting a patent on an animal modified through the use of biotechnology is different from granting a patent on a consumer product (Environics Research Group 1998). Some argued that humans and primates should be excluded from patentability. Those who were prepared to consider the patenting of animals felt that this right should be limited. Consideration was given to the suffering imposed on animals, and balanced with the potential use of the modified animal, with medical uses having the highest priority (CBAC 2001).

What Are the Issues of Concern?

The issues raised among publics include the commodification of life and the social consequences of patenting on equity and accessibility of products and services.

1A Commodification

The patenting of genetic information, including genes and gene sequences, blurs the distinction that we have traditionally maintained between life forms and material objects. Particularly in the context of higher life forms, many fear that providing patent protection to inventions involving living material or beings may result in the commodification of life (CBAC 2001). When Canada's Federal Court of Appeal agreed with the patentability of the Harvard OncoMouse (a decision since overturned by the Supreme Court of Canada), about half of Canadians surveyed said they were not comfortable with the Appeals Court decision (Earnscliffe Research and Communications 2000b, 2003a).

In any case, an overwhelming majority of Canadians agree that ethical considerations should be taken into account when determining whether or not to grant a patent, with over nine in ten rating ethical considerations as "important," if not "very important" (Environics Research Group 1998).

Social Concerns

While moral and ethical considerations weigh in the minds of Canadians, the majority of those who are troubled by patenting issues raise objections on the grounds of access and affordability. Although various surveys have shown publics understanding and accepting the argument that patenting creates incentives and rewards innovation, and considering these outcomes as important, there were also concerns expressed that patenting drives up pricing and reduces accessibility (Earnscliffe Research and Communications 2000b). At least two-thirds maintained that patents allow multinational companies to charge higher than necessary prices for new products (Environics Research Group 1998).

Genomics research has generally been associated by publics with medical products and, in this instance, equality of access was underscored as a primary guiding principle in commercialization efforts, including the patenting of products (Einsiedel and Sheremeta 2005). Most express strong views that the cost of pharmaceuticals cannot dictate who receives them, and that those of average means should not suffer financial hardship to obtain them. A majority believes the patenting of genetic information will lead to both of these problems (Earnscliffe Research and Communications 2000b).

Despite knowing that some sort of incentive is required for those who innovate, a substantial minority "reject the economic paradigm and maintain that those who require therapies not be disadvantaged because of patent protections" (Earnscliffe Research and Communications 2000b, 48). Some members of the public argue that researchers are motivated by finding cures rather than by money and that the government should ensure that important research is undertaken if not done in the private sector (Earnscliffe Research and Communications 2000b). Mistrust of large corporations and worry over restricted access to important products are explanations given by some who in addition argue that patents may even "hold back innovation because no one else can develop a similar product" (50). Often-cited examples include patented drugs and seeds with price tags too high for millions to afford, such as the cost of AIDS drugs in Africa and the patenting of seeds by agricultural corporations to force annual repurchase (Earnscliffe Research and Communications 2003a).

Discussions on patents have shown the tension between the individual right to protect property and profit from its use and the wider society's rights to ensure the protection and well-being of its citizenry (Environics

Research Group 1998). In group discussions, such patents have been seen to restrict broader societal benefits and more than two-thirds say that some technologies should not be patented, as they are not in the public interest. This demonstrates a strong interest in using a public interest test as part of the patent process (Environics Research Group 1998).

Some discussants have also expressed concern that the costs of obtaining or protecting patents give an advantage to large industries, and small companies may be provided with an incentive to relocate to countries where it is easier to obtain patents. In addition, some are worried that patent interests promoted by corporate financing may sideline other important research, encouraging only that which is economically viable, rather than what might be viewed as best for society (CBAC 2001).

Another key social issue considers how Canada should identify who has a legitimate claim to the benefits, economic or otherwise, of innovation. Many participants in surveys and focus groups expressed unease over patents disadvantaging indigenous peoples and other cultures in less developed countries. If traditional knowledge is used to develop a patented product, are the people or country of origin entitled to a portion of the benefits yielded by the patented product? If a legal battle is required to gain access to benefits, indigenous people may be disadvantaged because of the high costs involved. Lastly, it has been suggested that communities may suffer from a lack of traditionally used food or resources because of the increased value of such materials as a result of patenting (CBAC 2001).

In summary, questions persist about the impacts of patenting genetic material on social welfare, prompting many to question whether the balance between encouraging innovation and meeting societal interests has been struck fairly.

Policy Implications

Patenting in the area of biotechnology is an issue that affects all Canadians. As such, discussion group participants felt that informed public debate was required to solve the key questions associated with patenting and that the issues should not be left to the experts to decide (CBAC 2001), despite the current limitations in public awareness. Further, such debates should consider the following:

- What should be the proper limits of patentable subject matter?
- What should belong to all of society and therefore should not be patentable?
- Should the sequencing process and involvement of a gene sequence in a specific biological process be considered an invention?
- Who should be entitled to the benefits of patents?

- How should society balance the encouragement of innovation with the promotion of a social good?
- Do the benefits that patents provide outweigh the anti-competitive behaviour that might also result from patenting? What is the actual relationship between patenting and innovation?
- Does anyone have the right to prevent or stifle innovation that leads to societal benefits? Who decides what is right? on what basis?
- Does the public good override the rights of individuals or vulnerable groups?

Proposed Patent Reforms

How do Canadian publics view options that might be utilized to address patent issue challenges? Recent focus group discussions around the country considered a variety of options.

First, there was strong public interest in the notion of price regulation for patented genetic health technologies along the lines of the Patented Medicine Prices Review Board (PMPRB) model. This was viewed as one important means to address concerns that were voiced about potential price increases and problematic access to products if they were under patent protection (Earnscliffe Research and Communications 2005). This price-control regime (possibly through extending the PMPRB remit to include genetic applications and products) had considerable support in these public discussions.

Second, discussion group participants were in favour of excluding certain products or processes from patentability. This exclusion gave discussants a sense that authorities would have another tool for setting limits on development and patenting of controversial genetic technologies (Earnscliffe Research and Communications 2005). Other surveys and focus group discussions have demonstrated public concern over patenting of higher life forms, and it is possible this is the type of exclusion that some people would consider acceptable. It is also quite possible that this support for limits, although not articulated in specific terms, is an expression of concern that different elements (concern for the collective interest, ethical considerations) are being taken into account and appropriately balanced.

Third, there was support for the policy option of allowing government to infringe on a patent, but only in extreme or emergency situations, for example, an epidemic, or where patent rights might be abused in a way that could endanger lives (Earnscliffe Research and Communications 2005).

Fourth, the idea of compulsory licensing received public support. While there was widespread preference for not patenting genes or gene sequences, recognizing this reality made the idea of compulsory licences an acceptable alternative and a means to ensure that innovation would not be hampered by patents (Earnscliffe Research and Communications 2005).

As the Patent Act is a statute dealing with property rights, some view it as an inappropriate mechanism to address the social and ethical issues associated with patenting, believing that the very wide range of factors involved in biotechnology patenting may be too broad for management through one process. As such, some have suggested the creation of a separate regulatory review mechanism to address these issues (CBAC 2001). Focus group participants around the country were supportive of the view that the Patent Act or the patent office were not necessarily the best arenas to deal with moral or ethical issues around patenting; many were comfortable with leaving this responsibility to adjudicate these issues with Parliament or the courts (Earnscliffe Research and Communications 2005), or vesting other bodies (bodies overseeing price controls) with appropriate regulatory powers.

In response to the concerns over gene patents and the patenting of higher life forms, academics have called for substantial reforms to patent laws, including:

- creating a statutory definition of "patentable subject matter" that includes or excludes certain biotechnological inventions
- adding an *ordre public* (public order) or morality clause to the Patent Act
- adding a statutory opposition procedure similar to that which exists in Europe
- creating a narrow compulsory licensing regime that would facilitate access by others to key patented technologies
- creating a specialized court to ensure that only judges with expertise in technology and patent law can hear intellectual property cases. (Einsiedel and Sheremeta 2005)

The "public order" or morality clause is the most contentious of the recommendations, as it would enable patent examiners or another ruling body to determine patentability on the basis of morality. Some recommend the creation of an independent, transparent, and responsible tribunal of specialists in ethics, research, and economics, with power to suspend or withhold patents in limited circumstances. Emphasis would be given to avoiding delays in the patent-granting process. The advantage of such an approach is that it leaves ethical decisions to specialists, preventing frivolous complaints against patentees (Einsiedel and Sheremeta 2005).

Alternatively, the Canadian Biotechnology Advisory Committee asserts that the existing range of mechanisms available to restrict or prevent activities determined to be socially or morally undesirable is quite extensive, and the social and ethical considerations raised by biotechnology should continue to be addressed primarily outside of the Patent Act. If new limits are required, it is more efficient to modify or expand current regulations than to introduce a completely new mechanism into the act (Einsiedel and Sheremeta 2005).

Benefit Sharing

To address the unequal distribution of economic and other benefits, the concept of benefit sharing has emerged in international law. It is argued that the human genome is a unique natural resource, possessing qualities that may render it a common heritage resource. Characterized as such, there is a moral and ethical obligation for researchers and exploiters to promote the equitable sharing of this resource and any information gleaned from its use. There is, however, no clear or crystallized legal imperative to insist on or enforce benefit sharing (Einsiedel and Sheremeta 2005).

Benefit sharing may be used as a means to correct inequity or promote more equal distribution of the benefits created by a patent system. As a supplement to the current patent system, the two should provide an incentive to innovators and a mechanism to achieve equitable and sustainable development of human biological resources (Einsiedel and Sheremeta 2005).

Conclusion

Investigations on the opinions, values, and priorities of publics have shown that personal and public interests are taken into account as part of the calculus for judging the importance of patents. However, these utilitarian interests are further leavened by concerns about equity and accessibility, oversight over the patent system, and the balancing of innovation priorities and inventor rewards with public interests (Einsiedel and Sheremeta 2005; Willison and MacLeod 2002). As Canadians have expressed an interest in participating in informed public discussion, the public must be provided with information about the patenting system–the process, implications, expected benefits for society (including benefit-sharing programs), issues associated with access and affordability of health products and technologies, and descriptions of how the risks of the technology in question are managed. Planners would be well advised to adequately address concerns. Such attention can be used to foster and maintain public trust in the commercial process and in the patent system in particular.

Acknowledgments

Parts of this chapter have been drawn from Edna F. Einsiedel and Julie A. Smith, "Canadian views on patenting biotechnology," a report prepared for the Canadian Biotechnology Advisory Committee, June 2005, http://www.cbac-cccb.gc.ca/epic/site/cbac-cccb.nsf/vwapj/FINAL_inseidel_e.pdf/$FILE/FINAL_inseidel_e.pdf.

References

CBAC (Canadian Biotechnology Advisory Committee). 2001. *Summary of consultations on biotechnological intellectual property and the patenting of higher life forms: Integrated summary report.* http://cbacccb.ca/epic/internet/incbacccb.nsf/vwapj/IPPHL_integrated_summary_e.pdf/$FILE/IPPHL_integrated_summary_e.pdf.

Earnscliffe Research and Communications. 2000a. *Public opinion research into biotechnology issues: First wave executive summary.* Report prepared for the Biotechnology Assistant Deputy Minister Coordinating Committee (BACC), Government of Canada, by Pollara Research and Earnscliffe Research and Communications.

–. 2000b. *Public opinion research into biotechnology issues: Third wave report.* Report prepared for the Biotechnology Assistant Deputy Minister Coordinating Committee (BACC), Government of Canada, by Pollara Research and Earnscliffe Research and Communications.

–. 2002. *Public opinion research into biotechnology issues: Seventh wave report.* Report prepared for the Biotechnology Assistant Deputy Minister Coordinating Committee (BACC), Government of Canada, by Pollara Research and Earnscliffe Research and Communications.

–. 2003a. *Patenting of higher life forms: Research findings.* Report prepared for Industry Canada, Ottawa.

–. 2003b. *Public opinion research into biotechnology issues in the United States and Canada: Eighth wave summary report.* Report prepared for the Biotechnology Assistant Deputy Minister Coordinating Committee, Government of Canada, by Pollara Research and Earnscliffe Research and Communications.

–. 2005. *Patenting in the area of health genetics.* Ottawa: Health Canada.

Einsiedel, E.F. 2005. Public perceptions of transgenic animals. *Revue scientifique et technique OIE (Office International des Epizooties)* 24 (1): 149-57.

Einsiedel, E., and L. Sheremeta. 2005. Biobanks and the challenges of commercialization. In *Handbook of genome research, genomics, proteomics, metabolomics, bioinformatics, ethical and legal issues*, ed. C.W. Sensen, 537-59. Weinheim, Germany: Wiley-VCH GmbH and Co. KGaA.

Environics Research Group. 1998. *Renewal of the Canadian biotechnology public opinion research.* Report prepared for the Canadian Biotechnology Strategy Task Force, Government of Canada.

Gaskell, G., N. Allum, W. Wagner, T. Nielsen, E. Jelsoe, M. Kohring, and M. Bauer. 2001a. In the public eye: Representations of biotechnology in Europe. In *Biotechnology 1996-2000: The years of controversy,* ed. G. Gaskell and M. Bauer, 53-79. London: Science Museum.

Gaskell, G., E. Einsiedel, S. Priest, T. Ten Eyck, N. Allum, and H. Torgersen. 2001b. Troubled waters: The Atlantic divide on biotechnology policy. In *Biotechnology 1996-2000: The years of controversy,* ed. G. Gaskell and M. Bauer, 96-115. London: Science Museum.

Gold, E.R. 2000. *Patents in Genes.* Report prepared for the Canadian Biotechnology Advisory Committee Project Steering Committee on Intellectual Property and the Patenting of Higher Life Forms. http://www.cbac-cccb.ca.

Hay Health Care Consulting Group. 2002. *Policy issues related to genetics.* Report prepared for Industry Canada, Toronto.

Hilgartner, S. 2002. Acceptable intellectual property. *Journal of Molecular Biology* 319 (4): 943-46.

Ipsos Reid. 2001. *Attitudes toward human genetic testing.* Report prepared for the Ministry of Health and Long-Term Care.

Nuffield Council on Bioethics. 2002. The ethics of patenting DNA: A discussion paper. Nuffield Council of Bioethics, London.

Willison, D.J., and S.M. MacLeod. 2002. Patenting of genetic material: Are the benefits to society being realized? *Canadian Medical Association Journal* 167 (3): 259-62.

Part 2
Foresight Applications

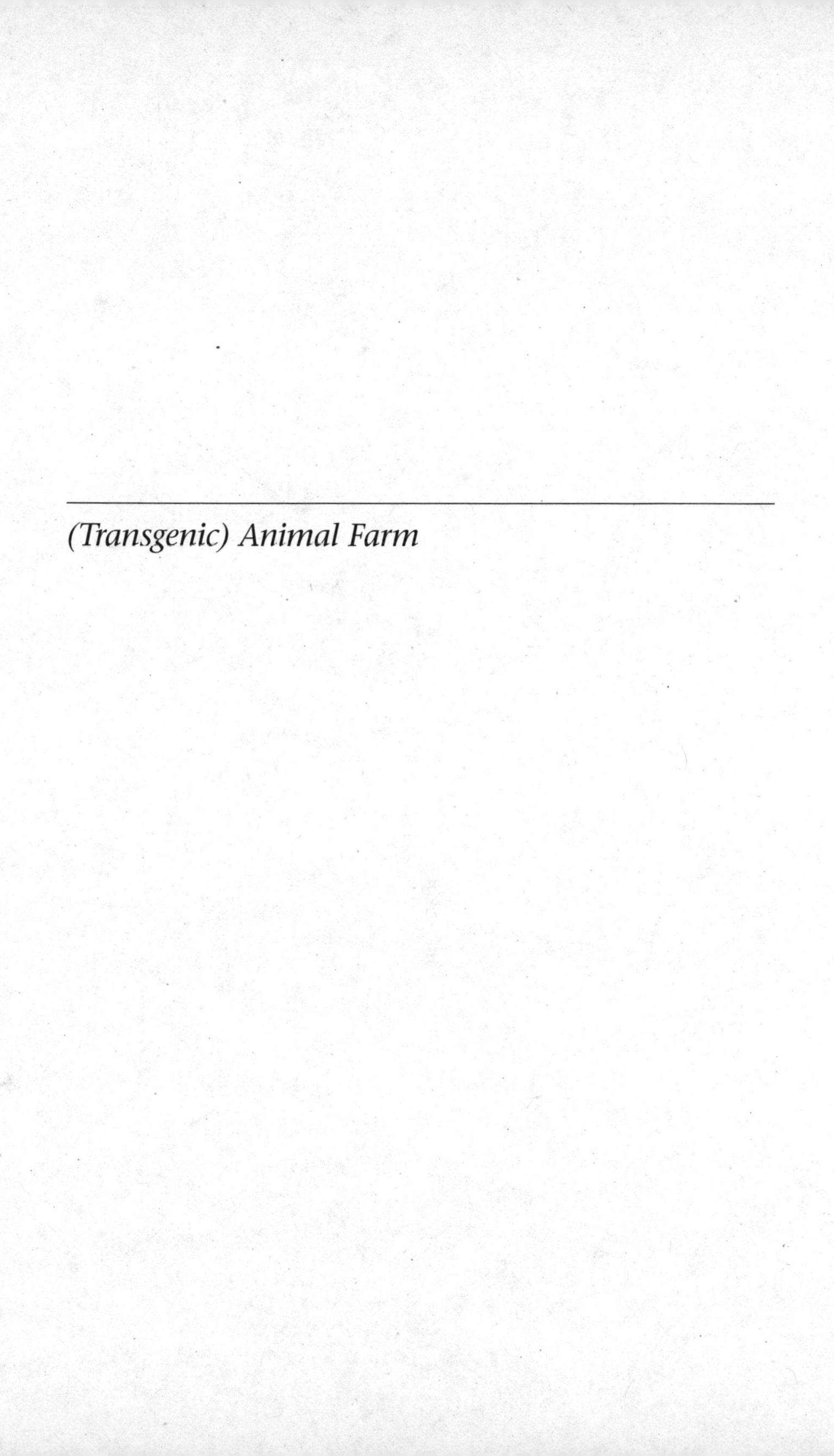

(Transgenic) Animal Farm

4
Of Biotechnology and Blind Chickens

Paul B. Thompson

In 2001, I summarized some ideas from a Danish study on the ethics of animal breeding (Sandoe et al. 1999) for a National Public Radio broadcast of the *Morning Edition* show. The radio program was a report on a conference to discuss issues in animal biotechnology and was widely broadcast throughout the United States and Canada. Here is what I said:

> There's a strain of chickens that are blind, and this was not produced through biotechnology. It was actually an accident that got developed into a particular strain of chickens. Now blind chickens, it turns out, don't mind being crowded together so much as normal chickens do. And so one suggestion is that, "Well, we ought to shift over to all blind chickens as a solution to our animal welfare problems that are associated with crowding in the poultry industry." Is this permissible on animal welfare grounds?
>
> Here, we have what I think is a real philosophical conundrum. If you think that it's the welfare of the individual animal that really matters here, how the animals are doing, then it would be more humane to have these blind chickens. On the other hand, almost everybody that you ask thinks that this is an absolutely horrendous thing to do. (Kastenbaum 2001)

The reaction to this broadcast was overwhelming. I heard from acquaintances I had not been in touch with for twenty years. I received angry telephone calls from representatives of the US poultry industry. An animal protection group created a website claiming that I had advocated blinding chickens and urged people to complain to their local public radio stations. Websites on which I was described as a "philosopher"–note the quotation marks–intimated that I was pimping for industrial animal producers, for the biotechnology industry, or worse.

Perhaps my critics thought that I was promoting the use of blind chickens, though the original David Kastenbaum story makes it clear that I was calling attention to the ethical reaction that many people experience when they hear the blind chicken experiment described. Perhaps the critics were objecting to my description of blind chickens as a conundrum: How could he think there is *anything* to be said for this? But seeing that blind chickens do present a conundrum is critical to any appreciation of the ethics of animal biotechnology. Blind chickens were not products of cloning, genomics, or gene transfer, but we can expect to see an ever lengthening list of animal biotechnologies that mimic the ethical tensions of the blind chicken problem. What should we think about them?

Beyond Blind Hens

Blind chickens signify a potentially large class of science-based strategies that use biotechnology in response to so-called production disease. Production diseases are animal pathologies that occur as a result of or in association with livestock production practices. Hens confined at high density in battery cages are prone to aggressive behaviours, such as feather pecking and cannibalism, that may have a territorial function in wild relatives. In egg or broiler production, such behaviour harms other hens and causes injuries that reduce production yields and require veterinary care. Beak-trimming is one response to this production disease, but trimming each individual hen's beak to limit pecking is itself harmful and costly. Large-breasted broiler chickens are also susceptible to leg and muscle problems because of the difficulty of physically supporting their bulk. High-producing dairy cows experience mastitis. Animals in confined spaces can also exhibit obsessive, repetitive movements called stereotypies. Animal researchers engage in a constant search for responses to these problems.

Not all of their responses involve biotechnology. Beak-trimming is a surgical response, and blind hens themselves were developed through classical breeding. But researchers are increasingly turning to the techniques of genetic engineering, cloning, and cellular manipulation–modern biotechnology–in search of ways to reduce both the suffering and economic cost associated with production disease (see USDA 2006). Some interventions in the animals' genetics would involve little or no change in an animal's cognitive or physiological capacities and may pose few ethical problems. For example, genomics approaches are being used to identify the sequences of base pairs that are associated with genetic diseases. Classical breeding or screening might then be used to reduce the number of animals having susceptibility to genetic diseases in production herds. However, other approaches involve the creation of organisms that produce familiar animal food products but lack the capacity to experience suffering, perhaps because, like blind hens, they are less likely to engage in hurtful behaviour that is typical of conspecifics. In more ambitious strategies, the aim is to undertake a series of scientific investigations that eventually conclude with an animal that does not experience anything at all.

Although there are numerous biological means that might be deployed in pursuit of this end, the ethics of such research can be illuminated by describing two broad strategies that would involve biotechnology. One might be called the dumb-down approach. Here researchers identify the genetic basis for certain characteristics or abilities (such as sight) and produce animals that lack these characteristics and abilities by removing or otherwise disabling the corresponding genetic code in the animal's DNA. Such organisms might, for example, lack an ability to feel pain or any consciousness at all of their external environment. The end result of this process might be the headless commodity-producing organism described as "football birds" by philosopher Fred Gifford (2002). An alternative might be called the build-up approach. Here, researchers work with cells in vitro, designing scaffolding and other mechanisms that might be produced according to instructions encoded in DNA, to wind up with an organism that yields the animal products (meat, milk, and eggs) currently produced using pigs, cows, and chickens (Edelman et al. 2005). Dumb-down and build-up approaches would both involve *research* with animals, but they would logically converge on an organism that we might hesitate to call a pig, cow, or chicken in virtue of its dissimilarity in all familiar respects.

So the ethics question is, should we pursue either dumb-down or build-up lines of research with the intention of eventually using the results of this research in animal production? There is a philosophical conundrum here because our leading theories of animal ethics tell us that this would be a good thing to do, but our moral intuitions tell us that it is an absolutely horrendous thing to do. Philosophers use the word *intuition* for seemingly immediate and involuntary cognitive experiences. *Perceptual intuitions* are raw sensations, such as the sounds I hear outside my hotel window. I recognize them as trucks making their way down Houston Street in San Antonio, but this recognition involves additional processing that I can defer at will, attuning myself to only the growling rumble itself. Although I can "not hear" them in one sense by ignoring my environment and focusing on my work, it is hard to imagine how I could fail to be presented with these auditory intuitions, so long as I can hear at all. *Linguistic intuitions* are the sense that we make of words and sentences when they are spoken or visually presented to us. Here, too, there is a compulsive character that cannot be resisted. If someone says "Move over, loser," I can *pretend* that I have not heard, but I cannot actually choose whether or not I want to understand (though I must, of course, understand English idioms to have this linguistic intuition). *Moral intuitions* are similar in that they are immediate, seemingly involuntary, and do not involve any conscious or thoughtful judgment. When confronted with a given situation (either in practice or, as in the case of blind chickens, by description), we just react to it as "wrong." It is quite possible that, as in the case of language, we are culturally educated into our moral intuitions, but this does not alter the fact that we seem unable to choose whether or not we will have them.

The normative case is that intuitions are supported by ethical principles that could be applied in an explicit act of deliberation and judgment. So acting on the basis of intuitions is usually consistent with acting ethically, a feature that leads people to speculate that moral intuitions are an evolutionary adaptation. But despite their immediacy and compulsive character, intuitions of all sorts can be unreliable. Sometimes we realize that what we thought we saw or heard was not in fact what was there, or that what was actually said was not at all what we thought. The same is true for moral intuitions. In many cases, where our first reaction is to think that something is morally wrong, we may be brought around to the idea that it is not wrong after all, by reasoning carefully about the situation and considering all of the relevant details. In these situations, judgment prevails over intuition. Yet, some moral intuitions are exceedingly robust, and our feeling that they are telling us something ethically important remains even after thinking more carefully about the situation fails to support our initial intuitive reaction. Sometimes intuitions stand at odds with moral principles. Such intuitions produce conundrums.

The thought of blind chickens producing our table eggs is repulsive to many people; it just strikes them as wrong. By extension, build-up and dumb-down approaches strike people as morally problematic. But leading theories of animal ethics do not support this judgment. Peter Singer's approach (1975, 2002) to animal welfare, for example, tells us that we should give equal consideration to interests, without regard to the animal that has these interests. We should take the suffering of animals into account in making our decisions and should not favour choices that produce trivial human benefits simply because the harm or suffering these choices cause happen to occur in non-human animals. Relevant in the present case are interests in avoiding the suffering that is associated with production disease. Conventional animals have these interests and experience the suffering. Modified animals lack the interests and do not experience the suffering. If our goal is to minimize the unnecessary suffering in the world, as utilitarian philosophers have advocated for over two hundred years, the choice seems clear. Organisms that lack the capacity to suffer cannot be harmed, so taking steps to create such organisms seems to be what a utilitarian would have us do.

Perhaps, one might think, a stronger animal rights view would not support this. The position advocated by Tom Regan (1983, 2003), for example, stresses that it is the individual animals themselves that matter, not their experiences. Each animal is, as Regan would have it, a subject-of-a-life, and it is wrong to treat any subject-of-a-life solely for our own purposes. Rather, we must respect them as subjects that lead lives of their own. For the purpose of argument, let us set aside the blind chickens described by Peter Sandoe et al. (1999), which do indeed continue to be capable of having an internal conscious life. More radical approaches like build up or dumb down might eliminate this capability altogether. By Regan's own reasoning, animals (such as insects or protozoa) that lack any conscious capability altogether are *not* subjects-of-a-life. If we can develop an animal that produces meat, milk, or eggs and is not a subject-of-a-life, there is nothing or no one to be harmed by placing these animals in production environments. Further, if doing that is a step toward removing ordinary pigs, cattle, and chickens from the production circumstances in which their rights are, in Regan's view, being violated, it would seem that his ethic of "empty cages" weighs in on the side of developing such literally mindless animals.

To summarize, the question is, should we pursue either build-up or dumb-down research? My claim is that both animal welfare and animal rights positions provide reasons to think that we should, though our intuitions tell us that we should not. There are really two conundrums here. One is the philosophical conundrum that arises when we try to account for our moral intuitions in theoretical terms, to reconcile what we feel to be right or wrong with what reason tells us to be right or wrong. The second conundrum is practical. Both animal welfare and animal rights versions of animal ethics provide rationales for a broad program of research in animal biotechnology, but a quick reality check suggests that this research program will spark serious resistance from consumers. Should we respond by terminating or sharply curtailing the research program, or by trying to convince the consumers that their gut reactions are off base?

Animal Biotechnology and Animal Ethics

The basic conceptual elements of the dumb-down approach were described twenty years ago by Bernard Rollin (1986), who gave an extended discussion of them in his 1995 book *The Frankenstein Syndrome*. Rollin was thinking primarily of laboratory animals that would be genetically engineered to exhibit particularly devastating forms of human disease for the purpose of biomedical research, a possibility then that is now the reality (Crenson 2006). In Rollin's view, the fact that human beings suffer, sometimes horribly, from these diseases provides a compelling ethical reason to conduct research on them, and animal models of genetically-based diseases are useful in developing therapies. However, the suffering that such animals endure is tremendous. Because the entire point of creating these animals is to exhibit the disease, there is no escape. How can this kind of genetic engineering for medical research be ethically justified, Rollin asked?

In partial answer to his own question, Rollin speculated that it might be possible to perform additional genetic engineering as a palliative to the suffering that animals created to model disease might endure. It might, for example, be possible to genetically modify the animal's pain receptors, so that the animal would not experience the torturous and continual pain associated with the disease process. He wrote that it might even be possible to create totally decerebrate animals, animals that experience no conscious life at all. Such animals would have brain functions necessary to maintain breathing, blood circulation, and other automatic life support processes but would lack any capacity for conscious sensory stimulation or awareness. Such animals would be biologically incapable of suffering. Rollin raised this possibility in the context of querying the circumstances under which it would be permissible to change an animal's *telos,* a term he coined to indicate the genetically-based needs, drives, and behaviours characteristic of species, subspecies, and breeds.

Rollin's general answer to this question was the Principle of Welfare Conservation: Genetically modified animals should not have worse welfare (susceptibility to disease and experience of pain or frustration) than unmodified animals of the same species or breed. Applying the Principle of Welfare Conservation, Rollin argued that there is nothing intrinsically wrong with changing the genetic makeup of animals, so long as this change did not create animals that were more likely to experience pain, suffering, or other deprivations of welfare as a result. Following the same animal ethics logic sketched above, Rollin presumed that organisms biologically incapable of conscious experience or awareness of pain cannot have compromised welfare. This logic was implicit in the first edition of Peter Singer's *Animal Liberation* (1975) and was made explicit by many animal advocates as they explained why their concern for animals did not also extend to plants. Plants do not have a welfare in the relevant sense, though clearly they can be made better or worse off (Varner 1990).

In short, Rollin concluded that (1) genetic engineering is acceptable as long as the transgenic animal is not made worse off than comparable non-transgenic animals. But the compelling human needs addressed by transgenic animals developed to study disease presented him with a challenge. How can attempts to alleviate the horrible suffering of human victims of genetic disease be denied? So he further concluded that (2) it is not only acceptable but desirable to render animals that would suffer under these conditions genetically incapable of experiencing suffering, that is, to undertake genetic engineering that places them into something very much like a persistent vegetative state. If vegetables do not have a welfare that can be harmed, vegetative animals do not either. To put this as succinctly as possible, if we are to choose between two possibilities, one of which involves experimental animals enduring constant pain and suffering and the latter which involves creating animals incapable of suffering (hence enduring none), the latter is, on Rollin's logic, obviously the preferable course of action.

Rollin's writings have spawned numerous critics, but they have mostly focused on element (1), his claim that genetic engineering is acceptable in cases where the welfare of animals is not compromised. What seems to trouble the critics is the possibility of genetic engineering at all, which is said to violate species integrity or the dignity of the creature (Colwell 1989; Sapontzis 1991; Balzer, Rippe, and Schaber 2000; Warkentin 2006). Some critics have followed Rollin's use of the term *telos* to describe the genetic basis of species-characteristic proclivities and drives but have argued that it would be wrong to alter this basis (Mauron 1989; Fox 1990). It is possible that these objections indeed capture the essence of a widely felt aversion to biotechnology (see McNaughton 2004). At the least, the language of *telos* and species integrity gives people something to latch on to. The terminology of integrity permits a public airing of the issues, and Bernice Bovenkirk, Frans Brom, and Babs van den Bergh (2002) have argued that this is itself an argument for using expressions such as "Biotechnology violates the integrity of animals (or species)" to articulate the intuitions we experience in response to dumb-down strategies. But these responses fail to address several key points that remain in Rollin's favour.

One is that biotechnology can be used to help people *and* animals to better their lives. Appeals to integrity and dignity can become pompous when thrown in the face of creatures (of whatever species) that are actively enduring suffering right now. Few people would object to gene therapies or even germline genetic modifications of humans if they promise to eliminate debilitating genetic disease and the risks are acceptable. If it is okay to do that to humans, why would it be forbidden in veterinary medicine? Rollin explicitly limited the scope of his argument to just such compelling cases, a point that is seldom mentioned by his critics. Second, the claim that it is wrong to violate species integrity, the dignity of the creature, or *telos* seems to overstate the case. At a minimum, critics would need to explain whether such arguments would also forbid routine forms of animal breeding. Bovenkirk, Brom, and van den Bergh (2002, 20) admit this point, describing the concept of integrity as "flawed but workable."

Finally, the rhetoric of the critics often sounds as if they think actual animals are harmed by manipulations of DNA in a petri dish. But as Rollin (1998) himself has argued vehemently, species integrity, dignity, and *telos* are abstractions graspable (if at all) only by human beings well versed in biology and philosophy. If offence is taken when assaults on dignity, integrity, or *telos* are proposed, it is offence to human beings who understand these ideas. The organisms that are the end result of cell culture and genetic manipulations described in the dumb-down and build-up approaches are like Rollin's decerebrate mice: they lack an experiential welfare altogether, much less a consciousness or sense of self. And it is not as if animals that *once had* the ability to experience their surroundings are having that ability taken away. The organisms being developed through biotechnology *will never have had* that ability. Even if it makes sense in some way to say that there is something to debate about species integrity or dignity, it is a confusion to presume that this has anything at all to do with the integrity or dignity of individual animals.

The Ethics of Livestock Biotechnology

Although the debate over medical uses of animal biotechnology is important and fascinating in its own right, in the present context it is necessary to ask what this debate teaches us about animal biotechnology in the context of traditional livestock production. How does this debate suggest that we should understand our conundrum? If we take the cue from Rollin's critics, the intuition that dumb down or build up might be ethically wrong focuses on the way that biotechnology violates the animal's dignity or integrity. But it is important to note that Rollin's arguments on transgenic animal disease models differ from the example of blind chickens in important respects. First, there are no human patients anxiously awaiting a cure that provide the ethically compelling force that motivates his entire argument. Second, Rollin's mouse models for human disease were being made worse off through genetic engineering, and decerebration was offered as a compensating response. In the case of production disease, there are alternative ways of improving animal welfare, namely, improve the environment. Thus, these cases are not strictly comparable. On the one hand, blind chickens or football birds seem to be more problematic than decerebrate mice because there is no reason to think that creating them is the only way to address a compelling need.

On the other hand, it is doubtful that either the dumb-down or build-up strategies violate Rollin's criterion of welfare conservation in the first place. These animals have *improved* welfare relative to their counterparts in the comparison class, albeit because they are incapable of having their welfare compromised in the usual way. The welfare of a football bird may be zero, but zero is better than less than zero. This point is particularly relevant to those who, like Bovenkirk, Brom, and van den Bergh (2002), suggest that integrity is a construct intended to call attention to an animal's entire life, or to the way that it fits within its farm environment. They are certainly right to insist on calling our attention to how farm animals actually fare, but does integrity provide a useful way to conduct a public debate about build up or dumb down? Not only are these organisms (I hesitate to call them animals) faring *better* for not faring at all, but their creation also allows us to remove animals that would (and currently do) suffer in production environments. Here it may also be worth emphasizing again that blind chickens themselves are not products of biotechnology. Their blindness is a natural phenomenon, just as there are people who are born blind. Yet, the response to my radio interview suggests that people think that it would be wrong to exploit this natural phenomenon as a response to production disease.

All this suggests that the emphasis on integrity (and by extension, the focus on genetic engineering) is actually leading public debate astray. Perhaps Bovenkirk, Brom, and van den Bergh wish to use the language of integrity to bring ethical claims about the farm environment, rather than animals themselves, to public attention. In fact, the intuition about these strategies for addressing production disease may strike us vividly because we think there is a better option, one better than either biotechnology *or* blind chickens. Here, it is not cloning, genomics, or biotechnology that is at the heart of the problem but a human tendency to solve problems by altering the environment instead of altering our own conduct. Here, the ethical focus is on the need for human beings to adopt virtues and principles that are expressive of a mindful, attentive, and morally responsive outlook on the world around us. To "solve" a problem by making it go away deprives us of an opportunity to address it morally by adjusting our own conduct. If we think that there are other options to production disease that would involve adjusting our own consumptive habits and our conduct with regard to animals, perhaps the moral choice is to pursue those options, rather than taking the easy path of the technological fix.

It is worth noting, however, that views on what the options are vary considerably. For some, the alternative is to restore traditional farming practices, for others it is to eliminate farming altogether. The former reaction seems to be what Bovenkirk, Brom, and van den Bergh have in mind. But Karen Davis, the founder of United Poultry Concerns, makes it clear that, in her view, genetic engineering is wrong not because it is new but because it continues a centuries-long human tendency to use animals in cruel and disrespectful ways: "Genetic engineering carries these attitudes and practices technically further, but does not break moral continuity with a past in which nonhuman animals have repeatedly been denied possession of a soul, reason, or some other vaunted human quality, and used without apology . . . I believe that chickens and other domestic fowl do not have a future worth living in their encounter with the human species and that genetic engineering furthers a drive in our species to eliminate not only diversity and autonomy, but joy and happiness in other creatures" (Davis 1996). Davis decries the suggestion that more traditional poultry practices represent morally acceptable alternatives.

This suggests that although there may be widely shared intuitions about blind chickens and dumbing-down our farm animals, the experiential or cultural basis for situating these intuitions within an ethics for agriculture and food may not be either shared widely or experientially deep. The ethical justification of either dumb down or build up may depend a great deal on what one thinks the most likely alternatives really are. If one thinks that simply banning a production system such as the battery cage will solve the problem of aggression and cannibalism among laying hens, an expensive long-run research program to develop biotechnologically altered animals in response to production disease seems ill-advised, and my blind chicken conundrum seems like just another philosopher's thought experiment. In reply, I assert that such simplistic responses do not represent a politically *or* biologically viable response to welfare problems in contemporary agriculture. As Davis (1996) notes, it is not obvious that traditional production systems are more humane, but it is even more important to recognize that the path to change is complex. The concluding chapter of Peter Singer's *Animal Liberation* (2002) offers a realistic assessment of the obstacles that advocates for animals face in reforming abusive practices, much less in eliminating the use of animals for food. Singer acknowledges that cultural preferences for eating animal products are entrenched and that policy imperatives for cheap food are powerful.

Although I would not go as far as Singer (or Davis) in my willingness to revolutionize agriculture, I do believe that the cultural and policy obstacles to any systematic change in our production systems are significant, and I believe that incremental, even piecemeal, changes in the way we raise and treat farm animals will have the largest impact on the actual suffering of actual animals for a long time to come. Many of these changes will be technologically based. What is more, the degree to which university programs in animal science as well as private industry are currently mobilizing research in genomics to pursue biotechnologically-based responses to production disease is striking. At the same time, research that would address the alternative of changing the environment has been very limited, especially concerning the economic and policy basis for such an alternative, and especially in Canada and the United States. For this reason, the practical conundrum of biotechnology and blind chickens is real. We need to bring scientists who pursue biotechnology and genomics responses to production disease together with members of the public who experience the ethical intuition that pursuing this response is an ethically horrendous thing to do.

The philosophical conundrum is real, as well. The analysis above suggests that we should not look for a justification of the intuition that build-up, dumb-down, or even less radical strategies are ethically wrong in ideas about animals or their nature. It is not what is being done *to* animals that makes this kind of research seem wrong, nor is it solely the fact that cloning or genetic engineering is being used, though it is possible that concern about biotechnology causes an auxiliary reaction that makes build up or dumb down seem especially bad. The key intuition is that human beings should adapt *to* the needs of animals to a much greater extent than build-up and dumb-down thinking would suggest. But identifying this intuition only underlines the tension between some widely shared moral attitudes and practices that have been going on in livestock production for a very long time. Much more (and more difficult) work will be needed to chart a path toward ameliorating production disease and the welfare problems associated with modern farming methods.

References

Balzer, Phillipp, Klaus Peter Rippe, and Peter Schaber. 2000. Two concepts of dignity for humans and non-human organisms in the context of genetic engineering. *Journal of Agricultural and Environmental Ethics* 13 (1): 7-27.

Bovenkirk, Bernice, Frans W.A. Brom, and Babs J. van den Bergh. 2002. Brave new birds: The use of integrity in animal ethics. *Hastings Center Report* 32 (1): 16-22.

Colwell, Robert K. 1989. Natural and unnatural history: Biological diversity and genetic engineering. In *Scientists and their responsibilities,* ed. W.R. Shea and B. Sitter, 1-40. Canton, OH: Watson Publishing International.

Crenson, Matt. 2006. "Designer" mice find a monster niche. *USA Today,* March 6.

Davis, Karen. 1996. The ethics of genetic engineering and the futuristic fate of domestic fowl. United Poultry Concerns. http://www.upc-online.org/genetic.html.

Edelman, P.D., D.C. McFarland, V.A. Mironov, and J.G. Matheny. 2005. *In vitro*-cultured meat production. *Tissue Engineering* 11 (5 and 6): 659-62.

Fox, Michael W. 1990. Transgenic animals: Ethical and animal welfare concerns. In *The bio-revolution: Cornucopia or Pandora's box,* ed. P. Wheale and R. McNally, 31-54. London: Pluto Press.

Gifford, Fred. 2002. Biotechnology. In *Life science ethics,* ed. Gary Comstock, 191-224. Ames: Iowa State Press.

Kastenbaum, David. 2001. Analysis: Debate over genetically altered fish and meat. *Morning Edition,* December 4. http://www.npr.org.

Mauron, Alex. 1989. Ethics and the ordinary molecular biologist. In *Scientists and their responsibilities,* ed. W.R. Shea and B. Sitter, 249-65. Canton, OH: Watson Publishing International.

McNaughton, Phil. 2004. Animals in their nature: A case study on public attitudes to animals genetic modification and "nature." *Sociology* 38 (3): 533-51.

Regan, Tom. 1983. *The case for animal rights*. Berkeley: University of California Press.

–. 2003. *Animal rights, human wrongs: An introduction to moral philosophy.* Lanham, MA: Rowman and Littlefield.

Rollin, Bernard. 1986. The Frankenstein thing. In *Genetic engineering of animals: An agricultural perspective,* ed. J.W. Evans and A. Hollaender, 285-98. New York: Plenum Press.

–. 1995. *The Frankenstein syndrome: Ethical and social issues in the genetic engineering of animals*. New York: Cambridge University Press.

–. 1998. On *telos* and genetic engineering. In *Animal biotechnology and ethics,* ed. Allan Holland and Andre Johnson, 156-87. London: Chapman and Hall.

Sandoe, P., B.L. Nielsen, L.G. Christensen, and P. Sörensen. 1999. Staying good while playing God: The ethics of breeding farm animals. *Animal Welfare* 8 (4): 313-28.

Sapontzis, Steve F. 1991. We should not manipulate the genome of domestic hogs. *Journal of Agricultural and Environmental Ethics* 4 (2): 177-85.

Singer, Peter. 1975. *Animal liberation*. New York: Avon Books.

–. 2002. *Animal liberation*. Rev. ed. New York: HarperCollins.

USDA (United States Department of Agriculture). 2006. ARS Project: Identification and manipulation of genetic factors to enhance disease resistance in cattle. http://www.ars.usda.gov/research/projects/projects.htm?ACCN_NO=405817&showpars=true&fy=2003.

Varner, Gary. 1990. Biological functions and biological interests. *Southern Journal of Philosophy* 27: 251-70.

Warkentin, Traci. 2006. Dis/integrating animals: Ethical dimensions of the genetic engineering of animals for human consumption. *AI and Society* 20 (1): 82-102.

5
Transgenic Salmon: Regulatory Oversight of an Anticipated Technology

Emily Marden, Holly Longstaff, and Ed Levy

Transformative–or revolutionary–technologies may emerge in very different ways. They may emerge rapidly, with no anticipation by government or regulators. Or they may be highly anticipated by the public and regulatory authorities, as is the case with animal cloning. An interesting question is whether the social implications of anticipated technologies of this kind are addressed in a more systematic and constructive manner by virtue of their being subject to a regulatory regime.

We do not attempt in this chapter to compare various transformative technologies or to definitively answer the question of whether anticipated technologies are addressed differently. Instead, we focus on a case study of how the US and Canadian regulatory systems have struggled to handle one example of an anticipated–and controversial–technology: transgenic salmon.[1] The possibility of transgenic animals has long been contemplated, and their actual development has been proceeding long enough for a fifth-generation animal to be submitted for initial regulatory review. Given the controversies about genetically modified organisms, it is perhaps not surprising that a transgenic animal food product–namely one with potentially broad implications for aquaculture–has been widely anticipated and discussed. Although there is debate over *how* regulatory systems should handle transgenic salmon, there has been widespread agreement that *some* regulatory position should be taken.

Regulation relating to transgenic salmon has been carved out of an existing framework that predates the potential commercialization of transgenic animals. The result is that there are overlaps and gaps within regulatory jurisdictions, with certain issues being redundantly addressed, while others are not addressed at all. In addition, regulatory agencies appear to be constrained in their reviews by their mandates not to examine broader social concerns flowing from technologies they oversee. Ultimately, this prompts the question of where and how such issues ought to be raised, and how decisions having momentous social implications should be made in democratic societies.

In the following section, we briefly address the scientific innovation underlying transgenic salmon. In the subsequent section, we present the respective regulatory contexts in the United States and Canada. In so doing, we are acutely conscious that the governmental structures of the two countries differ in many ways. Nevertheless, the two jurisdictions have comparable federal regulatory bodies that are among the key institutions developing, and certainly implementing, policy on the transgenic salmon application.[2] In the final section, we provide a series of conjectures and hypotheses that lead to suggestions for further study and analysis.

The Underlying Innovation

The innovation[3] on which this paper is based is best exemplified by a fifth-generation transgenic animal developed by Aqua Bounty Technologies known as AquAdvantage salmon. We will confine our attention to what are commonly called genetically modified (GM) or transgenic salmon. We use both terms to denote the transplantation of a foreign gene (the transgene) into the germ line of an organism.[4]

This salmon, currently being considered for approval by the US Food and Drug Administration (FDA) and Canadian regulators, is a candidate for becoming the first transgenic animal available to Americans and Canadians for human consumption. AquAdvantage salmon are produced by a twofold modification of the Atlantic salmon genome: a growth hormone gene from a Pacific chinook salmon and a promoter sequence derived from the ocean pout (both cold-water fish) are inserted. This is referred to as an "all fish transgene" process because all the material is derived from the DNA of fish, which are, in this case, already available to consumers. The pout promoter gene is used because it acts to stimulate the production of the inserted growth hormones year-round. The new promoter thus disrupts the salmon's normal growth cycle, which produces growth hormones only during the warmer summer months.

As a whole, the modifications work by making the salmon growth cycle continuous rather than seasonal, as is the case in unaltered varieties. The result is that the fish grow to a marketable size within eighteen months, which is about half the time required by unaltered farmed salmon. The process does not actually produce a bigger fish; the AquAdvantage salmon end up being about the same size as their non-transgenic counterparts by the time they are ready for market. However, the 50 percent reduction in time to market represents a significant market advantage. The introduction of this novel fish is intended to increase the potential profitability of salmon farming and could eventually make on-land systems an economic possibility as overall production costs are reduced.[5] In addition, Aqua

Bounty has designed the system so that these transgenic fish are sterile (all female, triploid), and thus may be less harmful to wild stocks if they escape from ocean net pens.

While the science behind the Aqua Bounty salmon may be relatively straightforward, the regulatory status of the fish in both the United States and Canada has been more problematic. How has the potential introduction of the transgenic salmon thus far been handled by regulatory structures in both countries?

Regulation

General

Both the United States and Canada organize regulatory systems according to area of expertise, with different bodies responsible for identified areas such as human, environmental, agricultural, food, and animal health. A single technology or product may impact human health, environmental health, and animal health. Each of these aspects will–by dint of this organizing principle–be subject to the regulatory jurisdiction of different agencies. The net result, often, is that policy making occurs in a piecemeal fashion, with the issues broken down and assigned to bodies with very different mandates. In some cases, the sum of the regulatory parts may not cover the whole range of issues being raised.[6] In other situations, as noted below with respect to Canada, the net result may be that analogous issues are addressed in different fashion by multiple regulatory bodies. Ultimately, while the compartmentalized approach to regulation certainly has benefits in directing expertise, it can result in a situation where issues that are of concern are not satisfactorily addressed because of the very structure of the governing regulatory systems. These phenomena can be discerned in the ways the US and Canadian regulatory systems handle the matter of transgenic salmon.

Salmon genomics raises issues that cross local, regional, and even national regulatory regimes. Moreover, ocean net pen-farmed transgenic salmon potentially have environmental, economic, and social impacts in addition to those traditionally attributable to farmed fish.[7] In North America, there is confusion as to which national and subnational bodies are actually regulating cultured fish and what the relationships among these bodies are or ought to be. In addition, the impacts of transgenic salmon cross national boundaries; yet, there is currently no joint international US-Canadian institution to regulate these transboundary consequences (McDaniels, Dowlatabadi, and Stevens 2005). Although the international regulatory dimensions of transgenic salmon are critical and should be addressed, in this chapter we focus on national regulatory regimes.

The regulatory systems in both the United States and Canada are struggling to keep pace with the fierce rate of change maintained within the agricultural and animal biotechnology industry. Although the case that motivates this chapter is the AquAdvantage salmon, North America has already had a transgenic fish slip through the regulation system, with little reaction. The GloFish is a pet, not intended for human consumption. It glows in the dark, thanks to an artificially introduced sea coral gene. In 2003, the FDA reported that these GM zebra danio fish do not pose "any more threat to the environment than their unmodified counterparts, which have long been widely sold in the US" (FDA 2003, 1), and they were released into the consumer environment. They were also imported into Canada and sold to consumers despite that Environment Canada had not approved this endeavour. Once Environment Canada was alerted to the fact that transgenic fluorescent fish were being sold in Canada, it asked all importers or producers of these animals to freeze dead specimens and retain fish that had not been sold (Environment Canada 2006). The GloFish example illustrates the ease with which transgenic animals can enter the environment and the necessity of focusing on regulatory provisions as well as public awareness of the developments.

United States

In the United States, transgenic salmon are subject to statutes and regulations that predate the emergence of biotechnology. This approach stems from the decision made by the US government in the 1980s to use existing law to address commercial development. This decision allowed the industry to grow but at the same time instituted a system that is imperfect and often unclear in its application.

Currently, there are no US laws or regulations specifically applicable to the transgenic salmon–or even transgenic animals. Some have even suggested that a strict reading of the existing law would not subject transgenic salmon to any pre-market review at all. Despite this, the FDA has "read" the law to require that pre-market review applies.

Regulatory Background

The regulatory regime applicable to transgenic salmon stems from the 1980s, when GM agricultural technologies, such as soy, cotton, corn, and canola, began to reach the commercialization stage. Early genetic technologies had already sparked controversy in the United States. For example, during the 1970s, the development of recombinant DNA (rDNA) techniques sparked public concerns that mutant organisms might be released into the environment, causing serious damage. To counter the threat of further local and national government regulation of this technology, scientists opted to introduce responsible self-regulation (Nelkin 1978; Krimsky 1991).

Against the backdrop of self-regulation by scientists of rDNA techniques

(Rogers 1977; Wade 1977) and an emerging US lead in GM techniques, the Reagan and Bush Sr. administrations took steps toward outlining a federal regulatory policy that would ensure safety and preserve the competitive edge. As with rDNA, the theory was that effective industry and scientific self-regulation could preclude burdensome or inhibitory legislation (Krimsky 1991). Thus, through a series of working groups and policy statements begun in the mid-1980s, the Reagan and Bush administrations developed three tenets of US policy designed to ensure the development of the industry: (1) US policy would focus on the product of GM techniques, not the process itself; (2) only regulation grounded in verifiable scientific risks would be implemented; and (3) GM products were deemed to be on a continuum with existing products and, therefore, existing statutes are sufficient to review the products (US Congress 1983, 1984, 1985).

In 1986, the White House put forth its Coordinated Framework for Regulation of Biotechnology, which proposed regulating genetically engineered products according to measurable risks only.[8] In the legal context, the Coordinated Framework proposed that new biotechnology products be regulated under the existing web of federal statutory authority and regulation: *"Existing statutes seem adequate to deal with the emerging processes and products of [genetic engineering]"* (US *Federal Register* 1986, 23306). Ultimately, the Coordinated Framework sketched broad outlines of the jurisdiction of existing regulatory agencies over GM products.

The agency assignments outlined were consistent with existing federal exercise of jurisdiction. Thus, FDA was to have responsibility for regulating food and feeds modified via genetic modification. The United States Department of Agriculture (USDA) would regulate importation, interstate movement, and environmental release of transgenic plants or animals with an aim to protect existing crops or livestock from hazards. Finally, the Environmental Protection Agency (EPA) would register certain pesticidal aspects (components) of products in transgenic organisms before their distribution and sale and would establish pesticide tolerances for residues in foods (McGarity 1987). The federal government outlined the division of responsibilities as shown in Table 5.1.

Agricultural biotechnology was one of the major GM technologies to emerge in the context of the Coordinated Framework, and there has been abundant discussion of whether the system managed to adequately monitor the risks involved (Marden 2003). Animal biotechnology is the latest challenge and has only recently begun being considered by the US federal government. Consistent with the Coordinated Framework, no laws have been passed to directly regulate the use or release of GM fish or other transgenic animals (CEQ and OSTP 2001). In fact, the bodies addressing the area directly have narrowed as both the EPA and USDA determined that they lack regulatory authority under their authorizing statutes.[9] The

Table 5.1

Division of US regulatory responsibilities

Regulatory body	Products regulated	Reviews for safety
Food and Drug Administration (FDA)	Food, feed, food additives, veterinary drugs	Safe to eat
United States Department of Agriculture (USDA)	Plant pests, plants, veterinary biologic*	Safe to grow as agriculture or livestock
Environmental Protection Agency (EPA)	Microbial/plant pesticides, new uses of existing pesticides, novel microorganisms	Safe for the environment; safety of a new use of a companion herbicide

* A biologic is any virus, therapeutic serum, toxin, antitoxin, vaccine, blood, blood component or derivative, allergenic product, or analogous product applicable to the prevention, treatment, or cure of diseases or injuries to humans (21 U.S. Code Sections 151 and following, "Animal Virus, Serum and Toxin Act of 1913").

FDA is the only regulatory agency to have asserted authority, which it has done within its oversight over the Federal Food, Drug, and Cosmetic Act (FFDCA).

Several other federal agencies and state environmental controls also apply to farmed fish, regardless of whether those fish are transgenic. Coastal zone management authorities in the states, the Army Corps of Engineers, the Fish and Wildlife Service, and the National Marine Fisheries Service are all involved with site selection and permitting of net pens and hatcheries. The EPA and individual states enforce the Clean Water Act, regulating the potential harm that may be caused by fish wastes and disposal of the new animal drugs used on fish (CEQ and OSTP 2001). Some states have been more active, with a number passing legislation barring transgenic fish from being grown, or requiring labelling. As of 2003, fifteen US states had adopted regulations concerning uncontained uses of transgenic fish and other transgenic marine organisms (Pew Initiative on Food and Biotechnology 2003). Although the FDA has stated that it is working together with these agencies, there is little evidence of a cooperative arrangement.

Transgenic Salmon

Within the US Coordinated Framework, even the FDA's asserted regulatory authority had to depend on a somewhat novel application of the FFDCA. The FDA's authority extends to human and animal foods, human and animal drugs, medical devices, biologics, and cosmetics. Thus, to be subject to FDA regulation, a transgenic animal would need to fit within one of these categories, and category determination is dependent on the intended use of the product.

On its face, it seems clear that, if anything, GM salmon and other GM animals intended for human consumption should be regarded as foods and considered under food laws and regulations. Many in and outside the FDA, however, were concerned that the food category would not provide adequate opportunity for regulatory review.[10] Indeed, in US law, under 21 US Code, section 342(a)(1), foods are not subject to pre-market review. Instead, a food is subject to regulatory action only if it "contains any poisonous or deleterious substance which may render it injurious to health." Moreover, new food ingredients are not subject to pre-market government review or approval unless they are characterized as a food additive that is not *"generally recognized as safe"* ("GRAS"). Importantly, the manufacturer of the food makes the first determination of whether the food would be subject to FDA food additive review (Marden 2003). In the absence of ingredients subject to food additive review, no FDA pre-market review is required.

In light of this reality, the FDA wanted to find a broader authority for review. The agency thus took the position that a transgenic fish–or any transgenic animal–would instead be regulated as an animal drug (Miller 1996). To reach this conclusion, the FDA identified the transgenic fish as an "article(s) (other than food) intended to affect the structure or function of the body of man or other animals" (21 US Code, section 321[g][1][C]). Subject to regulation under the New Animal Drug Application regulation, the FDA then stated it was regulating the substance produced by a genetic modification, "not the altered fish itself because the genetic modification changed the function of the salmon's genome" (CEQ and OSTP 2001). This regulatory contortion was taken–in the FDA's words–in order to subject the novel fish to the most stringent regulatory review available and to ensure that the genetic modification was safe for the fish and for humans.[11] The FDA has also taken the position that certain environmental impacts come within the FDA's jurisdiction over animal drugs (CEQ and OSTP 2001).

There is one major limitation of being considered as a new animal drug, with respect to the public: there is no transparency. In fact, the FDA is not authorized to disclose that a New Animal Drug Application has been filed unless this fact is publicly acknowledged first by the sponsor of the product. Moreover, the FDA's authority to consider environmental issues is not well established and has not typically been a significant aspect of New Animal Drug Application approvals either of human or of animal drugs (Pew Initiative on Food and Biotechnology 2002).

The FDA made its most revealing statements on how it planned to carry out its regulatory authority in a 2001 report issued for the Clinton White

House's Council on Environmental Quality (CEQ), an executive branch advisory body. In the CEQ case study on GM salmon, the FDA stated that it was aware of the lack of transparency and the concerns about environmental issues. Consequently, the FDA promised that it would ensure that the review of GM salmon was as broad as possible, would undertake public consultations, and would work with the sponsor of the product to disclose as much information as possible (CEQ and OSTP 2001).

Nongovernmental organizations (NGOs) have pressed the FDA on these promises in a citizen petition filed with the agency by the Center for Food Safety in Washington, DC, in cooperation with a large number of other consumer, environmental, and fisheries NGOs (Center for Food Safety 2001). The petition asks that the FDA adequately conduct a thorough environmental review and that public concerns about impacts beyond health and safety be addressed. Under FDA regulations, the agency has an obligation to respond to citizen petitions within 180 days of their filing. However, no response has been forthcoming,[12] despite that there were news reports suggesting that the FDA would act on the salmon New Animal Drug Application as early as 2005 (Volz 2005).

In fact, many within and outside of government have voiced significant concerns about whether the FDA has the capacity to consider environmental, social, and ethical issues that are not technically in its mandate, and what the impact will be. For example, the National Academy of Sciences (NAS), an independent government-funded science advisory body, was asked by the FDA to produce a science-based risk assessment of animal biotechnology (National Academy of Sciences 2002). The NAS issued a book-length report on its findings, which included a thorough review of the science, with the conclusion that there needs to be further scientific study on certain aspects of the technology. On its own initiative, the NAS panel expressed concern that there were significant ethical, social, and religious issues tied up with transgenic animals, and that these issues did not appear to have a place in the consideration of various regulatory bodies. The FDA has acknowledged receipt of the report but has not responded to it. Similarly, the Pew Initiative on Food and Biotechnology held a conference in 2002 to foster discussion of a variety of viewpoints on transgenic animals. Panellists included government regulators, lawyers, public interest groups, and scientists. Publications of the meetings show that significant concerns were raised about the gaps in the regulatory structure and the lack of consideration being given to ethical issues (Pew Initiative on Food and Biotechnology 2002).

Canada

Canada, too, has a multi-faceted policy landscape concerning transgenic salmon. Canadian and US regulatory authorities wrestle with many similar issues, such as notions of substantial equivalence and the regulation of product, rather than of process. The key distinction is that many Canadian agencies are actively involved in the regulation of transgenic salmon, whereas in the United States, the FDA is the lead agency.

Regulatory Background

Parliament established the National Biotechnology Strategy in 1983, two years after biotechnology was first identified as a priority technology for Canada's economic and industrial development (Natural Resources Canada 2005). The strategy focused primarily on research and human resources development (Government of Canada Biostrategy 2005), received nearly Cdn$10 million each year in support, and underwent multiple arm's-length reviews.

A decade later, in 1993, the government introduced the Federal Regulatory Framework for Biotechnology (the Canadian Framework) for the regulation of biotechnology products, seven years after the United States developed its Coordinated Framework for Regulation of Biotechnology. Just as did its counterpart in the United States, the Canadian Framework encouraged regulatory bodies to use "existing laws and regulatory departments to avoid duplication" in order to protect the health and safety of Canadians and the Canadian environment (Health Canada 2006, 1). This framework resulted from an agreement among federal regulatory departments on six principles for an efficient and effective approach to regulating Canadian biotechnology products (Fisheries and Oceans Canada 2005). The principles of the Canadian Framework were intended to "ensure the practical benefits" and competitiveness of biotechnology products while also protecting the environment and fostering the health and safety of Canadians (CFIA 2003, 1). They also recognized that consultations with Canadians should be central to the development of biotechnology regulations in Canada (Fisheries and Oceans Canada 2005). The six principles of the Canadian Framework are:

- Maintain Canada's high standards for the protection of the health of workers, the general public and the environment.
- Use existing legislation and regulatory institutions to clarify responsibilities and avoid duplication.
- Continue to develop clear guidelines for evaluating products of biotechnology which are in harmony with national priorities and international standards.
- Provide for a sound scientific database on which to assess risk and evaluate products.
- Ensure both the development and enforcement of Canadian biotechnology regulations are open and include consultation.

- Contribute to the prosperity and well-being of Canadians by fostering a favourable climate for investment, development and adoption of a sustainable Canadian biotechnology products and processes. (CFIA 2003, 1)

The Canadian Biotechnology Strategy Task Force coordinated a series of stakeholder consultations that contributed to the formation of the Canadian Biotechnology Strategy (CBS) of 1998, which expanded beyond the National Biotechnology Strategy. It was at this time that the Canadian Biotechnology Secretariat and the 1999 Canadian Biotechnology Advisory Committee were also established. The Biotechnology Assistant Deputy Minister Coordinating Committee (BACC) includes representatives from all federal departments receiving funds from CBS (including Fisheries and Oceans Canada or DFO) and is the management committee of the CBS. CBS priorities are set by the Biotechnology Ministerial Coordinating Committee (BMCC) under the leadership of the minister of Industry. Industry Canada chairs the Biotechnology Deputy Minister Coordinating Committee, which, among other things, plays an advisory role to the BMCC (Treasury Board of Canada Secretariat 2004). This horizontal management structure was a clear example of the difficulties posed by new technologies such as biotechnology, which demand multiple knowledge bases.

The CBS's three pillars are stewardship, citizen engagement, and innovation (Treasury Board of Canada Secretariat 2004). As such, one of the goals of the CBS is to maintain standards based on the Canadian Framework, including the commitment to public involvement. Central to this new strategy is a broad policy framework that addresses social, ethical, health, economic, environmental, and regulatory considerations, while fostering public information and participation (Natural Resources Canada 2005). As proof of their desire to remain inclusive, many agencies were involved in the National Biotechnology Strategy revision/CBS consultation process, including Natural Resources Canada, Industry Canada, Health Canada, Environment Canada, Agriculture and Agri-Food Canada, the DFO, the Department of Foreign Affairs and International Trade, provincial officials, industry stakeholders, NGOs, scientific and academic communities, and others (Natural Resources Canada 2005).

Transgenic Salmon

In May 2004, following the Canadian Framework's directive to avoid regulatory overlap, the DFO, Environment Canada, and Health Canada created a memorandum of understanding that stated how the departments would regulate transgenic aquatic organisms until new regulations concerning transgenic fish became established under the Fisheries Act (Health Canada 2005). Until these new regulations are in place, transgenic fish intended for human consumption will fall under Health Canada's novel food regulations. Environment Canada will conduct assessments of transgenic animals, and the DFO will take the lead in drafting new regulations pertaining to transgenic fish (Fisheries and Oceans Canada 2004).[13] These regulations will cover manufacturing and research aspects of transgenic fish, among other things. Assessments of indirect human health impacts, as well as environmental impacts, will continue in the interim to be authorized under the Canadian Environmental Protection Act, 1999 (CEPA 1999), the key authority for ensuring the safety of all new substances.[14] The Canadian Framework therefore results in a situation in which a number of federal government departments and agencies are granted authority over Canadian biotechnology products, including transgenic fish, though no agency has taken or been given a leading role.

Yet, the regulatory realities are actually much more complex than suggested in the 2004 memorandum of understanding (see Table 5.2). For example, under the current regulatory system, multiple bodies have oversight over closely related or overlapping issues while other agencies are, in the views of some, stepping outside their mandated authority. Environmental aspects, in particular, are subject to multiple and distinct layers of regulation. For example, Health Canada must assess the various human health risks associated with biotechnology food products. However, the department is also one among many bodies that will help address the environmental risks posed by transgenic fish. The department must conduct environmental assessments of transgenic, and other, food products regulated under the Food and Drugs Act thanks to responsibilities established under the New Substance Notification Regulations of CEPA 1999.[15] Environment Canada will conduct assessments of transgenic animals with guidance from the Canadian Food Inspection Agency's (CFIA's) Animal Biotechnology Unit through requirements under CEPA 1999 and the New Substances Notification Regulations. The CFIA regulates biotech products such as animal feeds and fertilizers while also examining the environmental risks associated with transgenic plants. In addition, this body monitors trials, import permits, and others issues regarding the registration of these products. Lastly, the DFO will also share responsibility for environmental assessments of transgenic fish (Fisheries and Oceans Canada 2004).

The role of the DFO is further confused. The original mandate of the DFO was the management of wild Canadian fisheries. As such, many view its position as a regulator of aquaculture development as a conflict of interest with its original mandate (Auditor General of Canada 2001).[16] Similar objections can also be directed at the Fisheries Act. As reported by the commissioner for Aquaculture Development and referred to by Melanie Power in her paper on salmon aquaculture, genomics, and ethics, "many of the regulations under the Fisheries Act are not well adapted or directly

Table 5.2

Overview of Canadian regulatory bodies responsible for transgenic salmon

Regulatory body	Responsibility
Health Canada	• Assessing the various human health risks associated with biotechnology food products • Helping address the environmental risks posed by transgenic fish by conducting environmental assessments of transgenic, and other, food products regulated under the Food and Drugs Act because of responsibilities established under the New Substance Notification Regulations of CEPA 1999
Environment Canada	• Conducting assessments of transgenic animals with guidance from the Canadian Food Inspection Agency's Animal Biotechnology Unit
Fisheries and Oceans (DFO)	• Regulating wild Canadian fisheries (original mandate) • Conducting research to determine and minimize any risks that transgenic fish may pose to wild stocks and the natural environment • Helping to draft new regulations pertaining to transgenic fish • Sharing responsibility for environmental assessments of transgenic fish
Canadian Food Inspection Agency	• Helping Environment Canada conduct assessments of transgenic animals through requirements under CEPA 1999 and the New Substances Notification Regulations

relevant to aquaculture–a situation that results in the aquaculture industry being managed as a subset of the traditional fisheries. This is analogous to equating traditional livestock and crop agriculture to the hunting and gathering of animals and plants" (Power 2003).

With the appearance of transgenic fish, as with aquaculture generally, the Fisheries Act will be required to stretch its regulatory power to cover issues that could not possibly have been conceived of when it was originally drafted. This has been widely recognized in government as evidenced, for example, by Health Canada's July 2003 consultation report's focusing specifically on evaluating regulatory possibilities for the environmental assessments of products falling under the Food and Drugs Act.[17] Even more revealing is that Health Canada, the CFIA, and Environment Canada asked the Royal Society of Canada to convene an expert panel to study the regulation of food biotechnology in Canada, an investigation that included presentations by government agencies, NGOs, stakeholders, and the public. In its report, the panel states that philosophical and metaphysical issues are outside its mandate, but it goes on to say this:

> Although the focus of the Expert Panel's enquiry was on the scientific aspects of the new technologies and their effective regulation, the Panel would need to address many peripheral issues that touch on the question of the appropriate use of science in the regulation of risks . . . It is important to understand that answers to questions not specifically within our mandate are often relevant to, and influence answers to questions that are within it. The health and environmental safety issues posed to the Panel in the Terms of Reference, though largely scientific in nature, *often cannot be addressed fully without reference to broader ethical, political and social issues and assumptions.* (Royal Society of Canada 2001, 3, emphasis added)

In both Canada and the United States, there have been assorted calls for moratoria on GM salmon. These moves likely reflect a hierarchy in the public acceptance of GM food products (Royal Society of Canada 2001; Epstein 2002; Office of the Auditor General of Canada 2001). Opinion polls indicate that citizens of the European Union find GM animals the most ethically unacceptable of all forms of GM (Mepham 2000). Canadian Biotechnology Advisory Committee consultations appear to confirm this finding in Canada. The 2001 report on the Saskatoon Roundtable on GM Food finds that plant-to-plant initiatives are associated with the least concern. Concerns become greater as we move to transgenic initiatives on animals and are greater still for any initiatives involving human genes (CBAC 2001). Recognition of this hierarchy and the case-by-case judgments made by publics led the Canadian Biotechnology Advisory Committee to initiate discussions of an "acceptability spectrum" for GM foods, first mentioned in April 2001. This tool, later renamed the GMFF (genetically modified food and feed) Dialogue Tool in March 2003 includes four categories: (1) "acceptable; (2) acceptable with certain conditions; (3) unacceptable at the present time and until more is known (i.e., moratorium) or a given standard is met; or (4) not acceptable under any circumstances" (i.e., a ban) (CBAC 2003a). The framework uses a broad range of considerations to examine groups of foods or individual food products, including health, environmental, socio-economic, ethical, and broader societal interest and international considerations (CBAC 2003b).

Conclusion

This chapter provides a window on the state of the regulatory regimes that both reflect and shape the development and integration of transgenic salmon technology. In the United States, we observe a governing authority that does not wish to unduly slow technology and a cultural context of narrow examination of biotechnology. Canadian biotechnology regulations, on the other hand, have evolved into broad frameworks that are said to be founded on pillars of stewardship and citizen engagement, as well as innovation. Yet, in both contexts, the regulatory structure does not cover all significant aspects of evaluating transgenic salmon.

We began this chapter by raising the question of whether the social implications of *anticipated* technologies are addressed in a more systematic and constructive manner by virtue of their being subject to extensive regulatory consideration. The US and Canadian experience with agricultural biotechnology provides an interesting counterpoint to this discussion of transgenic salmon. Agricultural biotechnology was anticipated and, yet, the US and Canadian governments emphatically chose not to view agricultural biotechnology as a novel regulated category. Thus, there was no anticipatory regulation per se (Marden 2003). There have been numerous high-profile disputes over surrounding agricultural biotechnology that could potentially be linked to this lack of anticipatory regulation. Further research would be needed, however, to make this definitive cause-effect link.

This case study on transgenic salmon suggests that enacting regulation in anticipation of a transformative technology may not adequately address the range of issues raised. We offer the following conjectures and hypotheses.

First, it appears to us that it is the potentially transformative–and even disruptive–character of transgenic salmon that is the primary cause of the consternation we see within the regulatory systems. This is based not only on the example of transgenic salmon but also on the authors' experiences–in some cases as participants–with regulatory systems. We know of any number of new technologies that the FDA and Health Canada have dealt with in a timely and comprehensive manner.[18] However, when the emerging technology appears (to a sizable number of citizens) to challenge some basic values, as in the case of transgenic salmon, regulatory systems may be facing issues that their mandates and processes are ill-equipped to handle.

Second, it appears that the problems caused by gaps and overlaps within regulatory systems are exacerbated when agencies try to deal with transformative technologies within existing regulatory frameworks. This seems to be the case even where regulatory foresight is applied, as has been the case with transgenic salmon. In both the United States and Canada, regulatory agencies have been twisted into knots trying to decide which bodies will regulate what, and, after many years, have not come close to establishing a clear, accountable, transparent, and effective approach. Without a mandate from legislative bodies, regulatory agencies are placed in a reactive rather than proactive role, struggling to manage this product despite (and perhaps due in part to) initiatives to avoid regulatory duplication (i.e., the 1993 Canadian Framework and the US Coordinated Framework). We project that other transformative technologies, such as nanotechnology, that are anticipated may encounter similar difficulties.

Third, we have observed that when a potentially transformative technology confronts issues beyond the mandates of the regulatory authorities, those authorities often call upon external bodies having no decision-making authority to provide advice and guidance. In the case of transgenic salmon, the agencies called upon the National Academy of Sciences and the Royal Society of Canada. Although both of those agencies are centred squarely within the sciences, they both recognized and took into account a range of considerations well beyond science-based ones–an illustration of the challenges posed by strict adherence to a solely science-based framework. The transgenic salmon case does demonstrate that being subject to regulation is not always a solution for addressing the full range of social issues raised by the technology.

Finally, we suggest that merely expanding the mandates of regulatory agencies to include a broader range of social issues is only one way to handle the regulation of potentially transformative technologies. Furthermore, new approaches are urgently needed to address social implications of new technologies. Such approaches could include regulatory agencies learning how to better integrate public and stakeholder views and/or fuller legislative involvement. Processes of horizontal regulation, although already implemented, need to be honed further to include the embedding of institutional learning mechanisms in regulatory and policy development.

Notes

The authors wish to thank Renita Sherma and Daisy Laforce for their contributions to this article. An earlier version of this chapter, which addressed public consultation issues raised by transgenic salmon, was published in the *Journal of Integrated Assessment* (2006).

1 In our effort to characterize how transgenic salmon was addressed systematically by regulatory authorities, we do not attempt to gauge the impact of other factors, such as public concern, media attention, and public consultations, on regulatory responses. We do address these factors in more detail in a companion paper on transgenic salmon that was published in the *Journal of Integrative Assessment* (Marden, Longstaff, and Levy 2006).

2 In both countries there are municipal and regional bodies as well provincial or state bodies that can affect policy within their jurisdictions. Nevertheless, in both cases the federal level is the primary policy setter; the main actions open to other jurisdictions are to opt out of, or otherwise modify, federal policy. For example, in Canada, even if transgenic salmon were approved federally, a provincial government might be able to severely restrict the location of aquaculture.

3 In the realm of basic scientific research, genomics is the study of genes and their function. Various branches of genomics concentrate on gene mapping, genetic evolution, and gene function and expression. As such, discoveries in salmon genomics could be applied to understanding, diagnosing, and treating fish pathologies; identifying and tracking subpopulations; and modifying the salmon genome to create an organism more suitable

for aquaculture. We fully recognize that the latter is only one of the applications and that the focus of basic scientists is less on applications than on understanding.

4 When a foreign gene (the transgene) is transplanted into the germ line of an organism, the transgenic organism will then express the gene product of the inserted DNA. Our definition of a genetically modified organism is sourced from the *Oxford Dictionary of Biology*.

5 To date, on-land salmon farms have not been widely pursued, largely because they are so expensive to run. Many experts, however, feel on-land systems would provide a good alternative to netpen salmon farms because the environmental risks are not as great.

6 This is certainly the case with respect to transgenic crops under the Coordinated Framework for Biotechnology in the United States. For a good discussion along these lines, see (Mandel 2004).

7 For the purposes of this chapter, we define *transgenic* in accordance with the RSC Expert Panel Report on the Future of Food Biotechnology in Canada as a process in which genes are "altered and transposed between organisms by processes that would not occur 'naturally,' crossing species and kingdom barriers and producing life forms (transgenic plants and animals) that would not be produced by the 'natural' processes of evolution" (Royal Society of Canada 2001). In the media, the transgenic terminology is mostly used, but not necessarily technically correct.

8 The conceptual framework of the Coordinated Framework can apply to any product of biotechnology, whether intended for use as a food or drug. In reality, however, the Center for Drug Evaluation and Research at the FDA undertook its own review of drugs developed through biotechnology and, in discussion with industry, ensured existing safety and efficacy standards were safeguarded. Thus, the Coordinated Framework was intended to apply only to other FDA-regulated products, including food, feed, food additives, and veterinary drugs, and thus focuses almost exclusively on risks–that is, assessments of benefits are not considered.

9 Some interested parties have argued that the EPA could assert regulatory authority over GM fish by defining the products of the inserted genes as "new chemical substances" pursuant to the Toxic Substances Control Act (TSCA). The EPA, however, has not exercised this authority, and there is some question of whether such an interpretation would be upheld (Pew Initiative on Food and Biotechnology 2003).

10 It is interesting to observe that the regulatory attention to the first GM fish is analogous to the regulatory attention given to the Calgene Flavr Savr tomato, which was introduced in 1992 as the first GM food. At that time, Calgene and the FDA agreed that the genetic modification in the tomato would be reviewed as a food additive, affording the FDA the most stringent pre-market review available for food products. Subsequent entries to the GM food category have been subject only to voluntary pre-market consultations with the agency under the presumption that most modifications result in foods that are "substantially equivalent" to existing products and therefore are technically exempt from any mandatory agency review (Marden 2003).

11 The FFDCA defines *drug* to include "articles . . . intended to affect the structure or any function of the body of man or other animals" (21 U.S. Code section 321[s]). Under the FFDCA, a new animal drug's safety is determined with "reference to the health of man or animal" (21 U.S. Code section 321[u]). FDA interprets this statutory language to include "environmental effects that directly or indirectly affect the health of humans or animals" (CEQ and OSTP 2001).

12 In recognition of the absence of clarity on regulatory authority over the fish, the Center for Food Safety also filed citizen petitions with the USDA and the Commerce Department, asking those entities to enforce regulations–such as the Endangered Species Act–that could have a bearing on the presence of GM salmon on the market.

13 It appears that DFO is still in the process of developing these new regulations, and it is not clear when these regulations will be established under the Fisheries Act.

14 Although CEPA 1999 is the key authority for ensuring the safety (in terms of both human and environmental health) of new substances, substances regulated by other acts are exempt from the New Substance Notification requirements in order to reduce regulatory overlap.

15 Health Canada may create new assessment regulations with Environment Canada concerning the impact that new products may have on human health and the environment.
16 For example, scientists at the DFO are conducting research to determine and minimize any risks that transgenic fish may pose to wild stocks and the natural environment. This activity is taking place in spite of there being a tension in the Royal Society of Canada's report between the panel's call for a moratorium of GM fish in ocean netpens and their call for more research on GM fish.
17 The document was posted on the Environmental Impact Initiative website in order to allow for sixty days of public commentary and additional meetings or sessions that followed.
18 Examples from the area of health include the introduction of lasers and many other new diagnostic and therapeutic devices over the last thirty years. As well, the discovery or identification of every new class of biological targets–such as proteases, cytokines, seratonin transporters—spawns the discovery and development of new technologies. Of course, this is not to say that the regulatory systems' handling of these emerging technologies cannot be improved.

References

Auditor General of Canada. 2001. *Aquaculture in Canada's Atlantic and Pacific regions: The Standing Senate Committee on Fisheries interim report.* http://www.parl.gc.ca/37/1/parlbus/commbus/senate/com-e/fish-e/rep-e/repintjun01-e.htm.

CBAC (Canadian Biotechnology Advisory Committee). 2001. *Highlights of Saskatoon Roundtable on GM Food.* http://www.cbac-cccb.ca/epic/internet/incbac-cccb.nsf/en/ah00392e.html.

–. 2003a. *Genetically modified food and feed dialogue tool.* http://www.cbac.gc.ca/epic/internet/incbac-cccb.nsf/en/ah00351e.html (accessed May 14, 2006).

–. 2003b. *Report of the Exploratory Committee submitted to CBAC.* http://www.cbac-cccb.ca/epic/internet/incbac-cccb.nsf/en/ah00343e.html.

Center for Food Safety. 2001. Petition seeking a moratorium on the domestic marketing and importation of transgenic fish. http://www.centerforfoodsafety.org/legal_acti2.cfm.

CEQ and OSTP (Council on Environmental Quality and Office of Science and Technology Policy). 2001. Case study no. 1: Growth enhanced salmon. http://ostp.gov/html/012201.html.

CFIA (Canadian Food Inspection Agency). 2003. Federal government agrees on new regulatory framework for biotechnology. http://www.inspection.gc.ca/english/sci/biotech/reg/fracade.shtml.

Environment Canada. 2006. A glowing success: The case of the GloFish™. *EnviroZine: Environment Canada's Online Newsmagazine* 6 (August 10), http://www.ec.gc.ca/envirozine/english/issues/67/feature3_e.cfm.

Epstein, Ron. 2002. Redesigning the world: Ethical questions about genetic engineering ethical issues in biotechnology. In *Ethical issues in biotechnology,* ed. Richard Sherlock and John D. Morrey, 47-70. New York: Rowman and Littlefield.

FDA (US Food and Drug Administration). 2003. FDA statement regarding Glofish™. December 9. http://www.fda.gov/bbs/topics/NEWS/2003/NEW00994.html.

Fisheries and Oceans Canada. 2004. Fisheries and Oceans Canada response to the interim report of the Standing Senate Committee on Fisheries entitled "Aquaculture in Canada's Atlantic and Pacific regions." http://www.dfo-mpo.gc.ca/communic/reports/aquaculture/response-response_e.htm.

–. 2005. Response of the federal departments and agencies to the petition filed November 21, 2001, by Greenpeace Canada under the Auditor General ACT. http://www.dfo-mpo.gc.ca/science/aquaculture/biotech/greenpeace_e.htm#framework.

Government of Canada Biostrategy. 2005. The 1998 Canadian Biotechnology Strategy: An ongoing renewal process. http://biostrategy.gc.ca/english/view.asp?x=535&mid=15

Health Canada. 2005. Action plan of the Government of Canada progress report. http://www.hc-sc.gc.ca/sr-sr/pubs/gmf-agm/pr6_02_2005_e.html.

–. 2006. Food and nutrition: Frequently asked questions. http://www.hc-sc.gc.ca/fn-an/gmf-agm/fs-if/faq_1_e.html (accessed May 22, 2008).
Krimsky, Sheldon. 1991. *Biotechnics and society: The rise of industrial genetics*. Westport, CT: Praeger Publishers.
Mandel, Gregory N. 2004. Gaps, inexperience, inconsistencies, and overlaps: Crisis in the regulation of genetically modified plants and animals. *William and Mary Law Review* 45 (5): 2167-2259.
Marden, Emily. 2003. Risk and regulation: US regulatory policy on genetically modified food and agriculture. *BC Law Review* 44 (3): 733-88.
Marden, Emily, Holly Longstaff, and Ed Levy. 2006. The policy context and public consultation: A consideration of transgenic salmon. *Journal of Integrated Assessment* 6 (2): 73-97.
McDaniels, Timothy L., Hadi Dowlatabadi, and Sara Stevens. 2005. Multiple scales and regulatory gaps in environmental change: The case of salmon aquaculture. *Global Environmental Change Part A* 15 (1): 9-21.
McGarity, Thomas O. 1987. Federal regulation of agricultural biotechnologies. *University of Michigan Journal of Law Reform* 20: 1089-1155.
Mepham, Ben. 2000. Comments on Matthew Freeman's paper: Genetic modification of animals. Agriculture and Environment Biotechnology Commission. http://www.aebc.gov.uk/aebc/pdf/comments.pdf.
National Academy of Sciences. 2002. *Animal biotechnology: Science-based concerns*. Washington, DC: National Research Council.
Natural Resources Canada. 2005. Canadian biotechnology strategy. http://www.nrcan-rncan.gc.ca/cfs-scf/science/biotechnology/canadi_e.html.
Nelkin, Dorothy. 1978. Threats and promises: Negotiating the control of research. *Daedalus* 107 (Spring): 191-209.
—, ed. 1992. *Controversy*. London: Sage Publications.
Office of the Auditor General of Canada. 2001. Petition no. 38A: Genetically engineered fish (Greenpeace Canada as petitioner). http://www.oag-bvg.gc.ca/domino/petitions.nsf/viewe1.0/FCB25273AD52B41785256DC7004C54C9.
Pew Initiative on Food and Biotechnology. 2002. Biotech in the barnyard: Implications of genetically engineered animals. http://pewagbiotech.org/events/0924/proceedings1.pdf.
–. 2003. Future fish: Issues in science and regulation of transgenic fish. Washington, DC: Pew Initiative on Biotechnology.
Power, Melanie D. 2003. Lots of fish in the sea: Salmon aquaculture, genomics, and ethics. Electronic Working Papers Series, W. Maurice Young Centre for Applied Ethics, University of British Columbia. http://www.ethics.ubc.ca (accessed May 22, 2008).
Rogers, Michael. 1977. *Biohazard*. New York: Alfred A. Knopf.
Royal Society of Canada. 2001. Elements of precaution: Recommendations for the regulation of food biotechnology in Canada. http://www.rsc.ca//files/publications/expert_panels/foodbiotechnology/GMreportEN.pdf.
Treasury Board of Canada Secretariat. 2004. Canadian biotechnology strategy (CBS). http://www.tbs-sct.gc.ca/rma/eppi-ibdrp/hrdb-rhbd/dep-min/ic/cbs-scb/2004-2005_e.asp (accessed May 22, 2008).
US Congress. 1983. Environmental implications of genetic engineering: Hearing before the Subcomm. on Investigations and Oversight and the Subcomm. on Science, Research and Technology of the House Comm. on Science and Technology, 98th Cong. 2-3.
–. 1984. Biotechnology regulation: Hearing before the Subcomm. on Oversight and Investigations of the House Comm. on Energy and Commerce, 98th Cong. 98-193.
–. 1985. Planned releases of genetically-altered organisms: The status of government research and regulation; Hearing before the Subcomm. on Investigations and Oversight of the House Comm. on Science and Technology, 99th Cong. 72.
US *Federal Register*. 1986. Coordinated framework for the regulation of biotechnology. *Administrative Publication* 51 (June 26): 23302-50.
Volz, Matt. 2005. FDA may clear genetically enhanced salmon. Checkbiotech. http://www.checkbiotech.org (accessed September 3, 2005).
Wade, Nicholas. 1977. *The ultimate experiment: Man-made evolution*. New York: Walker and Company.

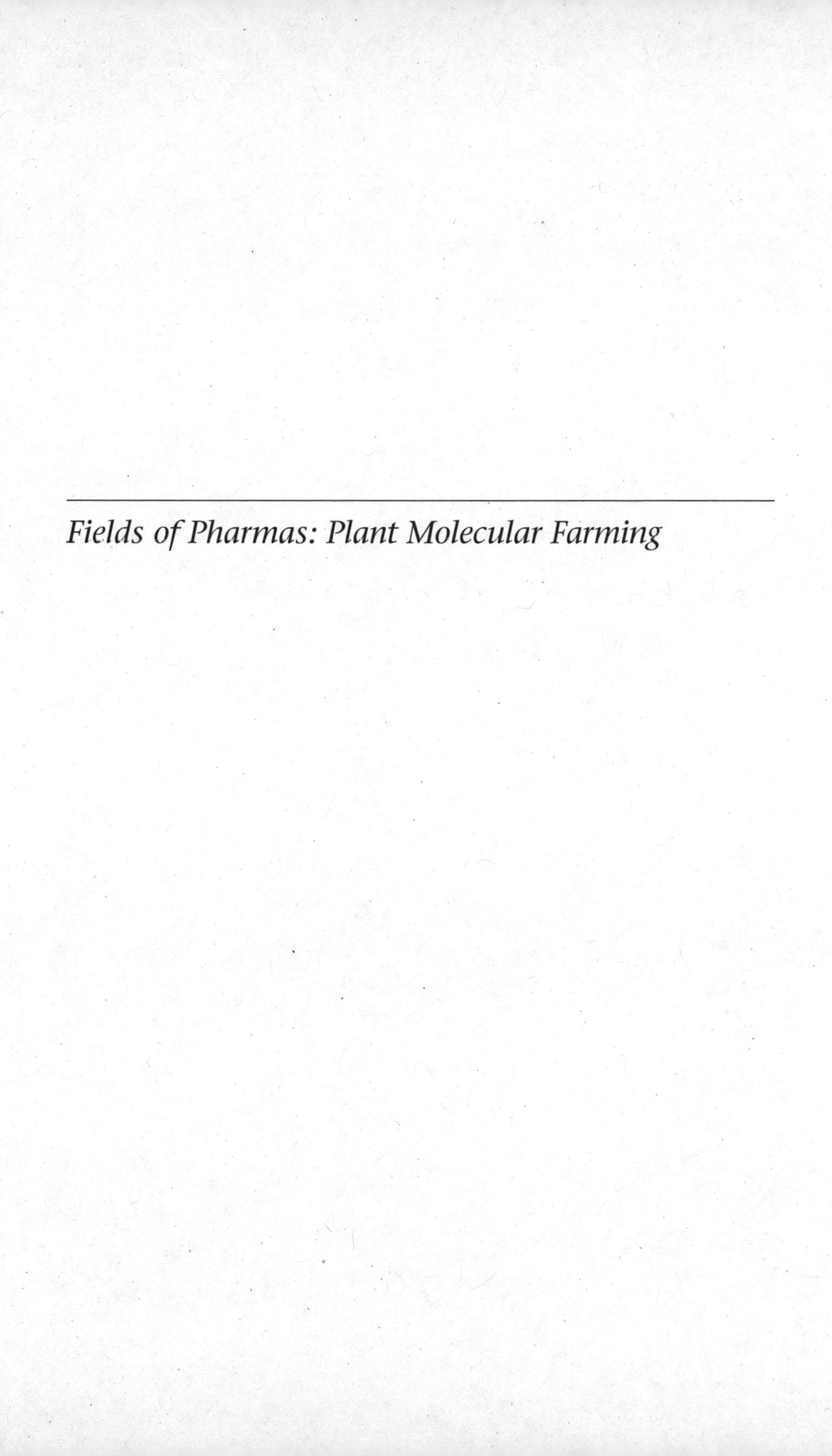

Fields of Pharmas: Plant Molecular Farming

6
The Emerging Technology of Plant Molecular Farming

Michele Veeman

In the area of biotechnology research known as plant molecular farming (PMF), the technologies of genetic engineering are being applied to the potential plant-based production of a variety of proteins and related compounds that have pharmaceutical, industrial, and consumer uses. Worldwide, few PMF products have actually been commercialized (none yet in Canada), but research to develop new plant-based pharmaceutical products is ongoing, and field trials of plants modified to express recombinant proteins that have potential pharmaceutical and industrial uses have been pursued for more than a decade in the United States, Canada, and other countries. Product testing, including initial clinical trials, has been occurring for a handful of plant-based vaccines and other pharmaceuticals (PROVACS 2005).

The first plant-based recombinant product to be commercialized (TrypZean, a plant-produced bovine-sequence trypsin), an enzyme with potential industrial and pharmaceutical uses that is mainly used for research, became available in 2002 from the United States PMF firm ProdiGene, Inc. (2005). The first plant-based vaccine to receive regulatory approval, developed for veterinary use, was announced in January 2006 by Dow AgroSciences LLC (2006). Nonetheless, PMF technology is still emerging and faces numerous scientific and other challenges (see, e.g., Vermij 2004; Arntzen, Plotkin, and Dodet 2005; van der Laan et al. 2006).

Social scientists and others have only recently started to ask questions about the potential benefits and possible risks, costs, and social acceptability of these new technologies. As yet, there has been relatively little public discussion of appropriate policy regulating new products produced though PMF systems. A small number of regulated field trials involving PMF plantings occur in Canada each year under the supervision of the Canadian Food Inspection Agency (CFIA), the federal body charged with oversight of food safety. Regulations to govern possible commercial applications of these technologies have been discussed but not yet developed (CFIA 2005a,

2005b, 2005e). Similarly, public discussions of regulatory policy for PMF have not yet occurred in Canada beyond an early workshop consultation in 2001, two further technical workshops organized by the CFIA (CFIA 2001, 2004, 2005c, 2005e), and a subsequent more widely based internet consultation on potential draft regulations for release of such novel plants, undertaken in 2006 (CFIA 2006).

In the current context where the science of PMF is being developed and starting to be applied, where regulatory policy directed specifically at PMF technology is proposed but relatively undeveloped and there is as yet little independent assessment of possible social, economic, and policy implications of these new technologies, it is useful to look at the social science literature that does exist and to search for possible lessons from related issues and analyses. Thus, in addition to exploring apparent benefits and limitations of PMF technology, one objective for this chapter is to give an overview of the limited information on social science evaluations related to PMF technology. Another objective is to consider some of the questions and research relating to genetically modified (GM) food as a source of potential lessons that might apply to emerging applications of plant molecular farming. The chapter concludes with an overview of issues associated with the evolving regulations for release of PMF plants.

Possible PMF Benefits and Risks

What are the potential benefits and costs of the new PMF technologies? Scientists and industry spokespersons postulate many compelling potential benefits. In addition to potential production of industrial and consumer products, these include prospects of new, cheaper, and safer pharmaceutical proteins with diagnostic, therapeutic, and prophylactic uses. It is expected that by using plant-based production systems some, but not all (Ma et al. 2005), pharmaceutical proteins could be produced in much larger volumes and in much shorter periods than in current pharmaceutical production systems, which are typically based on microbial fermentation or mammalian cell culture techniques.[1] One advantage is that the possibility of contamination of PMF pharmaceutical systems by viruses or prions is lower than for production systems based on mammalian cell cultures or transgenic animals (Ma, Drake, and Christou 2003).

The possibility of producing basic materials for PMF products based, in effect, on agricultural practices of plant cultivation, rather than on the biological-reactor processes of mammalian cell culture that dominate current pharmaceutical production, also leads to expectations that PMF systems could involve much lower pharmaceutical production costs (see, e.g., Pew Initiative on Food and Biotechnology 2002; Ma, Drake, and Christou 2003; Arcand and Arnison 2004; Markley et al. 2006). Initial suggestions of very optimistic commercial prospects for PMF production systems were extrapolated from observations of the expanding demand for biological pharmaceuticals. Subsequently and increasingly, projections of lower costs of PMF pharmaceuticals are based on evidence from field trials. A tenfold or greater difference between PMF and mammalian cell culture system costs has sometimes been suggested (e.g., Pew Initiative on Food and Biotechnology 2002; BIOTECanada 2004). However, cost differences could vary markedly, reflecting the costs of processing plant materials (required for purification and standardization), the particular type of PMF system (including the plant platform and its production scale), and whether production and processing occurs in an environment of relatively high or low labour costs, e.g., in a developed or developing country context (Pew Initiative on Food and Biotechnology 2002; PROVACS 2005, 2006). For example, it has been suggested that North American production costs for a specific antibody could be 25 percent less when produced in a PMF system, as opposed to current mammalian-cell-based systems (Gellene 2002). Much wider differences in costs are cited for potential developing country PMF production (PROVACS 2006).

Some indication of the possible cost advantages of PMF-based pharmaceuticals is given by the Center for Infectious Diseases and Vaccinology of the University of Arizona, a leading centre of research on vaccine and antigen development, which reports cost calculations for a delivered packaged dose of PMF-derived hepatitis B antigen to range between US$0.04 and US$0.16, depending on which of the three cited countries of production is considered (the lower cost estimates are for India and Korea, as compared to production in the United States) (PROVACS 2006). In contrast, this source also reports estimates of a packaged dose of yeast-based hepatitis B antigen (produced in Korea or India rather than in the United States) to cost US$0.42 (PROVACS 2006). However, it is not evident that these cost estimates include any allowance for return to investment, patent charges, or licensing fees, which could have appreciable impacts on companies' costs and on the costs to users.

The considerable potential importance of return on investment to costs per dose–as well as the impact of competition (more specifically, lack of competition) on users' costs for particular pharmaceutical products–is illustrated in two case studies of particular potential PMF products, one an "orphan" drug, the other a more widely demanded blood component, assessed by Kostandini and Mills (2005) and Kostandini, Mills, and Norton (2004).[2] The economic modelling analysis of these authors highlights expectations that benefits for consumers and developers will be limited by the actual cost advantage of the PMF pharmaceutical and that the patenting system will enable capture of much of that cost difference for developers as their return to innovation. Similarly, accurate estimates of the costs for PMF systems need to recognize the costs of containment, testing, and monitoring that will be necessary to avoid contamination of other plants and wildlife.

Nonetheless, despite these significant qualifications, it is generally expected that costs for PMF-based vaccines that could be produced and delivered in a developing country could be appreciably less than they are currently. A related major benefit expected for PMF pharmaceuticals is the prospect of overcoming the financial and practical constraints of scaling up current production systems to meet the rapidly increasing demand for newly developed therapeutics (Elbehri 2005).

Potential PMF pharmaceuticals include blood components and products, hormones, antibodies, vaccines, and other medically useful proteins (Ma, Drake, and Christou 2003). It is envisaged that new pharmaceuticals expressed in seeds or other plant components that could be freeze-dried would be heat-stable and thus not require cold storage. Consequently, these would escape cold-chain transportation and storage requirements that increase costs and limit use, for example, of current human vaccines, particularly in rural areas of the developing world, where cost and lack of availability prevent routine vaccination of much of the population against common diseases. Another objective for most PMF-based vaccines is that these be effective orally, with the benefit of avoiding the considerable costs and risks of injection administration of vaccines and other pharmaceuticals. (Reuse of needles, a procedure stimulated by cost constraints, is recognized as a significant source of infection in poor nations.)

The potential for plant-based pharmaceuticals was first demonstrated through the expression of human growth hormone in transgenic tobacco in 1986, while the first such antibody was expressed, also in tobacco, in 1989 (Ma, Drake, and Christou 2003). Since then, researchers have sought to develop a variety of PMF products in numerous plant platforms. By 2004, US researchers were reported to have successfully applied small-scale safety trials for oral vaccines, produced in various plants, against diseases such as enterotoxic E. coli, Norwalk virus, hepatitis B, and rabies (Ma, Drake, and Christou 2003; Vermij 2004). Researchers have also sought PMF-systems of producing vaccines, antigens, and other proteins for measles, diabetes, cystic fibrosis, tuberculosis, HIV/AIDS, and SARS, among others (Amos 2004). These efforts include attempts to develop a contraceptive for non-indigenous animals that have become pests (Polkinghorne et al. 2005). New vaccines that might protect against diseases that could be associated with bioterrorism, such as anthrax or plague, are yet another recent focus of PMF research.

Other focal areas of PMF research involve plant-based systems directed to production of industrial products such as industrial enzymes, biofuels, and bioplastics. Biofuels are renewable energy resources produced from modified crops, among other feedstocks, providing substitutes for depleting stocks of oil and natural gas. Bioplastics, which have additional advantages of rapid biodegradability, are another example of PMF-based industrial products.

A third PMF focus involves potential new plant-based consumer products, such as food crops that would contain necessary micronutrients (essential minerals, such as iron or zinc, and vitamins that are required for human health but lacking in the diets of many of the poor in developing nations; "golden rice" is one often-cited example), or oilseed crops modified to reduce current unhealthy components (such as triglycerides) or to express desirable combinations of dietary fat profiles (some oilseed crops along these lines are already being marketed). New oilseeds with healthy components (enhanced omega-3 fatty acids) are being pursued.

The development of new PMF plant varieties has also been suggested to hold the promise of benefiting the economic situation of farmers by increasing their alternative crop production choices and market opportunities (e.g., Swoboda 2002), in addition to contributing to the potential for development of new industrial and pharmaceutical processing industries, with consequent benefits to national employment and economic growth (BIOTECanada 2004).

Nonetheless, in the case of PMF pharmaceuticals, it is misleading to suggest that these are likely to present an appreciable "added value" production option for typical farmers. Even with the solution of existing scientific, regulatory, and commercial challenges to commercialization of PMF pharmaceutical products, these are likely to be based on relatively small areas of land, be highly regulated, and involve production in specialized conditions that are either highly confined or fully contained.

With regard to the likely land areas that could be involved in PMF-based pharmaceuticals, it has been suggested that because of the efficiency of plant-based systems, potential PMF-based production of insulin grown in safflower sufficient to supply world requirements would need only ten to twenty thousand acres (Baum 2006), currently represented by two medium-sized farms (Crowley 2005). Similarly, it is suggested that sufficient hepatitis B antigen for global vaccination of all newborns could be produced from plants grown on two hundred acres of land, while sufficient HBV vaccine for all annual vaccinations required in China could be produced in plants grown on forty acres of land (PROVACS 2005, 7).

The small land base that would be required, coupled with the relatively concentrated structure of the pharmaceutical industry, lead to the conclusion that little if any economic benefit would accrue to typical farmers from PMF pharmaceutical production systems. This conclusion does not necessarily apply to production of other PMF products: PMF systems for higher valued food crops or safe alternative non-food crops could potentially involve much larger planted areas and represent additional crop and market alternatives that would be welcomed by many farmers.

Possible Limitations and Risks of PMF Technology

In addition to anticipated potential benefits, new technologies invariably raise new risks and concerns. A major concern about PMF technology is that its products or waste might accidentally contaminate food supplies. A second concern is possible adverse impacts on the environment that could occur from PMF-based production systems. A related issue is the harm to wildlife that might arise from possible accidental ingestion of PMF materials by birds, deer, insects, and other animals and organisms.

The consequences of accidental contamination of the food chain by plant-based pharmaceutical and industrial products could be serious. The possibility of food-crop contamination is not far-fetched, as two incidents in Nebraska and Iowa in 2001 illustrate. The PMF company ProdiGene, which subsequently developed a PMF-based trypsin enzyme and had also been researching PMF-based veterinary diarrhea treatments, was fined in December 2002 following investigations by officials of the US Department of Agriculture, in consultation with the US Food and Drug Administration, of charges that there had been violations of the conditions of containment permits for experimental corn-based PMF cultivation (USDA 2002).

These incidents each involved PMF-modified corn plants grown by local farmers under contract to ProdiGene. In one instance, volunteer modified plants were evident in a subsequent soybean crop in Nebraska but not removed and destroyed before harvest of that crop. The second incident, in Iowa, involved possible cross-pollination from PMF corn plants for which tassels had not been removed, raising the possibility of pollen drift to an adjacent cornfield. ProdiGene was fined US$250,000 and required to cover costs of product destruction and disposal of some 500,000 bushels of warehouse-stored soybeans and 155 acres of corn, at a cost of about US$3.5 million. A US$1 million bond was required to be posted for continued field operations.[3] The US Department of Agriculture tightened its PMF field-trial permit requirements in 2003, but not to the point of disallowing trials involving pharmaceutical expression in food crops such as corn (USDA 2002; Reuters 2002; Fox 2003).

There are also scientific, regulatory, and commercial limitations to the commercial development of PMF pharmaceuticals and industrial products. For example, work to date on developing human vaccines through plant-based systems has largely been undertaken by university researchers and has been slow to move into commercial production. By 2004, more than forty-five different antigens had been produced in a variety of plants, but only five small phase 1 tests (the first stage of human clinical testing) and a successful phase 2a test were reported to have been completed . To date, no PMF vaccine has yet completed the testing required for human use (Vermij 2004; Arntzen, Plotkin, and Dodet 2005; van der Laan et al. 2006; PROVACS 2005, 2006).

Although initial trials are reported to indicate the safety of a small number of plant-based vaccines and the handful of clinical trials that have occurred are reported to be very promising, there are scientific challenges to the demonstration of the levels of immune responses needed for successful oral vaccine development and to ensuring the dose-by-dose consistency, uniformity, and purity needed for regulatory approval and commercialization (Vermij 2004; Arntzen, Plotkin, and Dodet 2005; PROVACS 2005). Vaccines generally are not the most profitable products for pharmaceutical companies, and it has been difficult for researchers to find manufacturers willing to finance the large-scale trials necessary to demonstrate efficacy of new vaccines (Vermij 2004).

Concerns about inadequate returns to investment and lack of a regulatory framework are viewed to underlie a relative lack of interest of most research-based international companies in PMF-based human vaccines (PROVACS 2005). Overall, it seems that despite considerable promise, there are still many challenges to approved production of PMF-based human vaccines. Judged on the basis of the numbers of confined PMF field trials conducted in Canada, it may be that there is relatively more interest in the development of PMF industrial products than pharmaceutical products in this country. CFIA reports on confined research trials of plants with novel traits indicate eight such trials in 2005, seven of which were for industrial products (CFIA 2005d).[4]

Social Acceptability of PMF Systems

Some very optimistic views have been expressed of the commercial potential for PMF-based pharmaceutical industries, largely based on projections of growing markets for biologic-based pharmaceuticals relative to high investment costs and lengthy periods for conventional systems to grow beyond current capacity levels (Elbehri 2005). However, social acceptability is also necessary for any new technology to be successful. Thus, Castle and Dalgleish (2005) indicate the necessity for proactive anticipation and evaluation of social determinants of the success of plant-based vaccines for these to be an effective means of reducing the significant inequities in health that are common in developing countries.

Some assessments of public reactions, in the United States and Canada, to the prospect of PMF technologies have been reported. Nevitt et al. (2003) engaged in interviews and small group discussions with various regional stakeholder representatives (farmers, policy makers, and regulators) and also undertook a 2003 US-wide telephone survey to assess the public's perceptions and attitudes toward two applications of agricultural biotechnology: possible tobacco plant-based pharmaceuticals and genetic modification of rice to impart insect resistance. These authors report that more concerns tended to be expressed by the group of respondents queried about agricultural biotechnology directed to GM food than were expressed by respondents queried about the pharmaceutical product. Even so, nearly one-fifth of respondents to the telephone survey indicated that they would not purchase a PMF pharmaceutical product (Nevitt et al. 2006).

A second assessment of US public opinion is provided by Stewart and McLean (2005) based on a 2004 telephone survey of 680 respondents in five mid-south US states. These authors report that relative to the application of agricultural biotechnology to produce plants modified for chemical or pest resistance, plants modified to improve food quality and to produce industrial or pharmaceutical products were seen by the respondents to have increasing levels of perceived benefits. However, nearly two-thirds of respondents assessed industrial and pharmaceutical plants to be likely to accidentally enter the US food supply. Queried on their anger if this were to occur, a slightly higher proportion of respondents indicated more anger if they were to consume industrial PMF products than pharmaceutical PMF products.

Assessments of public opinions on PMF applications have also been undertaken in Canada. Einsiedel and Medlock (2005) used a focus-group format in several Canadian cities, providing the forty-eight participants with advance briefing materials as a basis for discussions on five industrial and pharmaceutical PMF products expressed in different types of plants. These authors observed that views on the acceptability of the cited PMF applications varied on a case-by-case basis, suggesting more support for applications that provided human health or environmental benefits and less support for industrial applications. The potential for cross-pollination and contamination of food crops was a dominant concern. This concern was also expressed in a series of interviews with individual stakeholders reported by Medlock and Einsiedel (2005). However, these individuals had more prior knowledge of PMF technologies and did not assess industrial PMF applications to be more of a concern than pharmaceutical applications.

Building on initial focus-group assessments, Veeman and Adamowicz (2006) undertook a 2005 Canada-wide survey of 1,574 respondents, drawn from a national representative panel with the aid of a market research company. Respondents were queried on their assessments of risk, benefit, and priority rankings for various types of PMF applications. Although relatively few respondents expressed familiarity with PMF technologies, the perceived benefits of four cited types of GM applications clearly influenced respondents' risk rankings. Thus, the use of genetic modification or engineering to produce medicines, industrial products (like plastics, fuel, or industrial enzymes), or increased nutritional qualities of food were all rated to be less risky than the use of genetic modification or engineering to increase crop production. Contamination of food supplies and the environment were nonetheless seen as major risks of PMF technologies.

Asked to vote on desired levels of restrictions on PMF research plantings, a majority of respondents to this survey chose containment in a greenhouse or more rigorous restrictions over confined field-planting options. Among other issues, respondents were queried on opinions relating to liability responsibilities. More than two-thirds agreed with a statement that "industry and government should share the liabilities and costs if there is a problem" (Veeman and Adamowicz 2006). Respondents' preferences for research on five areas of PMF research were also assessed through a stated choice experiment on alternative research-funding levels and allocations. Analysis of the PMF research-funding choices made by respondents indicates a preference for funding to be directed to health applications of PMF research; environmental applications were also favoured, while industrial applications were the least preferred (Tao, Veeman, and Adamowicz 2008).

A somewhat different aspect of social acceptability of PMF-based pharmaceuticals is raised by Kirk and McIntosh (2005), who focus on the willingness of 706 residents of Phoenix, Arizona, to use vaccines produced from modified plants. Respondents were queried on their preferences for an oral-based versus an injection-administered vaccine, on their beliefs about the extent to which current vaccines are the product of biotechnology, and on how likely they would be to accept a vaccine produced in a GM plant. Slightly more than two-thirds of respondents indicated that they would be likely to use a PMF-based vaccine.

In this context, it is interesting to note the focus of presentations given by representatives of health-patient interest groups at two recent stakeholder conferences, one being the Pew Initiative on Food and Biotechnology workshop on PMF held in Washington, DC, in 2002, the other being the workshop on bio-based molecular production systems co-sponsored by the Canadian Agri-Food Research Council, the Government of Canada, and the Genome Prairie GE3LS research group, held in Ottawa in April 2004 (Pew Initiative on Food and Biotechnology 2002; CARC 2004). In neither case was the issue that PMF pharmaceuticals would be products of genetic modification raised as a significant concern by the representatives of patient interest groups. Rather, the emphasis expressed on behalf of these patient interest groups was that pharmaceutical products should be "the best that science can offer," together with assurances that "these products are safe and stringently regulated" (Macdonald, quoted in CARC 2004, 10). As expressed at the Pew Initiative on Food and Biotechnology workshop, "most patients will not be concerned about the source of the protein . . . or the pharmaceutical product" but are in general "always on the side of searches for the cure . . . that can provide us with safer, cleaner, more effective medicines" (Loving, quoted in Pew Initiative on Science and Biotechnology 2002, 10).

In summary, despite a high level of skepticism about the merits of PMF by environmentally focused nongovernmental organizations (see, e.g., Freese 2002), the assessments of current public perceptions of PMF indicate a fairly positive public view of the potential benefits of PMF production systems, particularly for applications that have clear medical and health benefits, conditional on assurance of strict regulations that ensure food and environmental safety. However, past experience of incidents and situations where food safety has been compromised suggests that generally positive views of PMF-based production systems could erode rapidly if food-safety incidents were to occur from accidental contamination of the food chain by PMF materials or if environmental safety was seen to be harmed.

Lessons Learned from GM Food Experience and GE3LS Research

Any incidents of accidental contamination of food systems by PMF pharmaceutical and industrial materials could not only cause considerable food-safety risk but also involve major costs and market losses throughout the national food chain and the supply chains for exported farm products. Losses would arise from export-market closures and other market disruption, including product recall and destruction, costs of thorough cleaning or decontamination of equipment and facilities, subsequent added testing and monitoring costs, associated litigation, and the loss of reputation as a reliable national source of high-quality food products. Because much of Canada's farming industry is highly export dependent, reports of possible food chain contamination that led to closure of export markets would have major adverse impacts on the livelihoods of affected farmers and other members of the food-supply chain. The possibility of food chain contamination can be considered to be more likely where there is field-based cultivation of PMF materials (as opposed to closed-system greenhouses or production in other fully contained structures). The possibility of food chain contamination also appears to be more likely where PMF-products are expressed in food-crop plants such as corn or oilseeds, rather than in non-food plants such as tobacco. Consequently, food retailers, food manufacturers, and consumer interest groups have opposed the use of PMF-systems in food crop plants.[5]

As noted earlier, the possibility of accidental commingling of non-food PMF materials or wastes with food crops is not far-fetched, as was seen in 2002 in the ProdiGene incidents, though the contamination by PMF materials did not extend through the food chain in those instances. However, one incident of widespread accidental contamination of food systems by a non-approved GM crop occurred in 2000, with StarLink corn. This product, which was grown on less than 0.5 percent of the total US area planted with corn, had been approved for animal-feed use but not for human use because of suspicion that it could cause a relatively uncommon allergic response in some people. Nevertheless, evidence of StarLink corn content was found at the retail level in numerous corn-based food products in North America and in export shipments of corn to Japan, causing widespread market disruption and hundreds of millions of dollars in losses and costs (Uchtmann 2002; Harl et al. 2003; Wisner 2005). The instances of accidental commingling with food crops that have occurred and the potentially drastic food-safety and economic implications that could result if high-risk PMF materials were to contaminate food supplies indicate the need for a high level of caution to be incorporated in regulations governing PMF production systems.

In this context it is of interest that the potential production of the first approved plant-based vaccine, developed for veterinary use by Dow AgroSciences, does not depend on field-based plant production but on a plant-cell-based production system (Dow AgroSciences 2006). This type of system is entirely contained and thus substantially reduces the possibility of accidental contamination of food. Reflecting the high level of containment provided by plant-cell-based methods of production, root culture methods are suggested as one of three approaches outlined in a blueprint of PMF vaccine production by PROVACS (2005), which proposes that costs of this approach be assessed. The second suggested approach is to use a non-food crop plant (tobacco), and the third involves physical isolation through containment (e.g., greenhouse conditions, potentially combined with male sterility systems). These seem to be useful approaches that are particularly appropriate for potential high-value PMF applications in view of the problems of potential food and environmental contamination from field-based PMF systems.

It may be relevant to consider lessons for potential PMF commercialization and regulation from previous social science research on GM food, as from the Genome Prairie GE3LS research program.[6] One component of this is directed to assessment of the impact of information on individuals' views of GM food. Based on findings from a representative Canada-wide survey of some 880 respondents, it was observed that government and industry are not highly trusted sources of information on GM issues (in contrast to researchers and consumer groups, which are more highly trusted sources of information). This survey also demonstrated that many people do not actually access information that has been made available but nonetheless expect that information be provided and that regulatory processes be transparent and participatory (Gao, Veeman, and Adamowicz 2005; Hu, Veeman, and Adamowicz 2005; Veeman et al. 2005).

Considerable differentiation in Canadians' attitudes to GM food was also shown by a Canada-wide experimental survey of some 450 people, where respondents chose between hypothesized alternatives of a basic food product, bread, in which prices and other bread features, including the presence or absence of GM ingredients, a health attribute ("healthy vitamins"), and an environmental attribute ("environmentally friendly" production) were systematically varied. Analysis of respondents' product choices indicates four distinct groups: 51 percent ("value seekers") were most interested in the price of the product and not significantly influenced by its GM content; a small group, accounting for 4 percent of respondents ("fringe consumers"), strongly valued the health attribute; 14 percent ("traditional consumers"), strongly preferred their standard bread product and avoided any changes to it, including GM ingredients; and a significant number (the "anti-GM" group), consisting of 32 percent of respondents, strongly opposed GM ingredients irrespective of introduced attributes (Hu et al. 2004).

The conclusion that a small majority of Canadians (value seekers plus fringe consumers) appears to perceive little or no risk from a specific GM food, while this food is strongly avoided (and highly opposed) by a large minority (46 percent, representing the anti-GM group plus traditionalists) indicates a dichotomy in Canadians' attitudes to GM food, a conclusion supported by other analyses (e.g., Veeman et al. 2005). This dichotomy and the evident concerns of possible food-safety incidents held by the food industry contribute to a considerable challenge for Canadian PMF policy to balance the rather different interests of PMF industry stakeholders with the interests of the food industry and consumers.[7]

Developing a Canadian Regulatory System for PMF Products

Regulation of novel plants and novel food products directly involves three national Canadian agencies. In very general terms, Health Canada and Environment Canada have the authority to define standards for health and environmental safety, respectively. The Canadian Food Inspection Agency (CFIA) has the responsibility to apply and enforce these standards. Industry Canada and Agriculture and Agri-Food Canada also have strong interests in issues associated with PMF activities and potential products but are not directly involved in the health and safety regulation associated with this technology. Canada's approach to regulation of agricultural biotechnology does not focus directly on the process of applying the techniques of molecular biology to the development of transgenic plants. Instead, the focus is on novel plants and their novel characteristics, that is, on "plants with novel traits" instead of on the process by which these are derived. In other words, all GM plants are considered to be plants with novel traits, but plants with novel traits that could be developed from conventional plant breeding practices are also subject to regulation. An outline of potential PMF regulatory roles and responsibilities based on current practices and regulations, extrapolated to PMF production that would occur in open fields (i.e., that would not be "contained"), is given by Macdonald (2005).[8]

Statements by the CFIA on the process of establishing regulations for commercial production of PMF products cite the necessity for caution and for public engagement and public acceptance (CFIA 2005a), though public engagement has not yet occurred to any significant extent. To this point, public discussion on PMF initiated at the federal level has been confined to an Ottawa-based public meeting on the evening of October 30, 2001, which preceded a two-and-a-half-day invited multi-stakeholder consultation on PMF issues attended by fifty-five people (CFIA 2001). This event has been credited with contributing to the tightening of scrutiny of experimental field trials of plants with novel traits and to the interim amendments of the conditions for research field trials that were subsequently posted (CFIA 2003). Even so, the ProdiGene and StarLink contamination incidents that had occurred in the United States and the 2003 tightening of US field trial requirements evidently also influenced the 2003 reconsideration of the rules for Canadian research field trials.[9] Since 2001, two CFIA technical workshops of invited stakeholders, primarily representing government and industry representatives, have been held in Ottawa. These included a March 2-4, 2004, workshop on the segregation and handling of potential commercial PMF products and by-products. Participants were reported to include representatives from the PMF industry, federal government organizations, agricultural and agribusiness associations, and experts in grain handling and identity preservation (CFIA 2004).

A subsequent CFIA technical workshop reported to involve representatives from the PMF industry, federal government organizations, agricultural and agribusiness associations, academia, and other stakeholders was held on March 2-3, 2005. This focused on discussion papers developed by the CFIA Plant Biosafety Office: "Developing a regulatory framework for the environmental release of plants with novel traits intended for commercial plant molecular farming in Canada" and "Draft assessment criteria for the evaluation of environmental safety of plants with novel traits intended for commercial plant molecular farming" (CFIA 2005a, 2005b, 2005e).

The proposed draft guidelines are modelled on procedures for confined research field trials. They include a proposed categorization of PMF risk relative to (1) environmental risk and (2) human and animal health risk. The executive summary of this workshop states, as the view of CFIA, that "As CFIA regulation is science-based, market risk will not be considered in the development of risk categories. However this risk will be assessed by those government departments that have the mandate to do so." (CFIA 2005e, 2). The exclusion of market-based risk considerations by the agency leading development of the draft PMF regulations in favour of the claim that the regulatory role in release of novel plants is entirely science-based is a highly debatable position for two reasons. First, science-based risk guidelines cannot be entirely divorced from social values. Second, these regulations should surely reflect the significant risk of economic harm to the agricultural and food industries, first, if PMF guidelines and related regulations are insufficient to ensure food and environmental safety, and second, if approval of a particular food-crop platform or particular PMF material for production in field conditions leads to effective market boycotts or export-market closures for Canadian food crops. Consideration of these issues in risk categorization may make this task more difficult, but it is, nonetheless, extremely important.

Translation of the current field-trial rules and procedures to a policy for commercial PMF production needs to be informed by general conclusions from the emerging social science research that assesses social acceptability of PMF systems. This indicates a public expectation of very high levels of precaution in production and segregation directed at food and environmental safety; opposition to food-crop use for PMF production; expectations of transparent policy and regulatory processes and expectations that in the event of any problems, liabilities, and costs will be shared by industry and government.[10]

Consequently, development of policy for commercial PMF production faces several questions and challenges. These include questions of whether field-based systems for specific PMF products that may be expressed in particular food and feed plant platforms create risks that are commercially and socially acceptable; the nature of socially acceptable confinement and containment and segregation protocols for PMF production systems that represent different risk situations and the likely costs of these protocols; and the nature and costs of planning requirements for various possible types of contamination incidents. Procedures and requirements for testing non-PMF food and feed crops for accidental contamination also need to be specified. Other salient issues requiring consideration and public debate are the responsibilities and mechanisms to recover costs associated with these activities and to cover associated liabilities that would be created by accidental contamination. Despite the length of time taken in considering PMF regulations to date, there appears to have been little research on these types of questions. It seems that there is considerable need for participatory social science assessment and related research on the issues noted here.[11]

Other related broad questions that require discussion, particularly in the farming community, are the conditions under which farmers and others might consider designating particular types of PMF systems to specific farming regions that are not highly dependent on food-crop production and exports. This could be the case, for example, for those industrial or consumer PMF products where large-scale production and relatively lower valued output (compared with higher-valued PMF pharmaceutical systems) would preclude containment production systems. A related issue concerns the public information requirements that would be necessary for PMF cropping systems to be socially accepted, including, for example, potential requirements for information to be made available to surrounding landowners and residents, as well as to the general public, about the details of approved confined PMF production systems. All these questions would benefit from public discussion–both within and beyond the agricultural sector. Such discussion and related research could contribute to an effective Canadian policy for PMF-based production of pharmaceutical, industrial, and consumer products.

Notes

Research funding from the Alberta Agricultural Research Institute, the Alberta Crop Industry Development Fund, and the GE3LS research program of Genome Alberta and Genome Canada relating to work reported here is acknowledged.

1 Another proposed production system, already used in a few instances, involves expression of novel proteins in the milk of transgenic goats or other animals; expression of pharmaceutical proteins in eggs from transgenic chickens is also proposed (for example, Martin 2005).

2 Some discussion of potential PMF pharmaceutical targets has focused on major diseases, for which there is large potential demand for effective treatments; other discussion suggests targeting orphan drugs (drugs that treat illnesses affecting relatively small proportions of the population–fewer than 200,000 people according to the US Food and Drug Agency definition) because these are not generally targets for commercial drug development by major pharmaceutical companies.

3 The US Environmental Protection Agency website also shows reported incidents in the state of Hawaii in 2002 and 2003 involving smaller fines relating to experimental permit violations for transgenic varieties by Dow AgroSciences and Pioneer Hi-Bred. These included a Dow AgroSciences fine of US$8,800 for lack of a windbreak and incorrectly described crop grown in the bordering rows; Pioneer was fined US$9,000 for planting in an unapproved location insufficiently distant from neighbouring field trials; another incident in 2003 led to a fine of US$72,000 for an apparent cross-pollination from an experimental plot.

4 However, these numbers do not reflect PMF field trials by Canadian researchers and Canadian PMF companies that are conducted in other nations (plant crop trials commonly occur in more than one country at least in part to take advantage of differences in climatic conditions). Nor do these numbers reflect PMF trials that occur in containment (i.e., contained structures). Both these practices are believed to be followed by Canadian PMF companies.

5 Some scientists and PMF industry representatives note ease and cost advantages of corn- and oilseed-plant-based PMF platforms. For example, tobacco-based vaccine systems must purify this PMF material to remove toxic alkaloids.

6 GE3LS stands for "Genomics, ethics, environment, economics, law and society." This

encompasses research in the humanities and social sciences relating to applications of genomics science.

7 US food manufacturers (unsuccessfully) argued against the use of food-crop platforms for US-regulated field-based PMF production systems. Canadian concerns about this issue are implied in the summaries of the workshops noted here. Use of food-plant platforms for PMF field trials is discouraged but not prohibited by Canada's current regulations (CFIA 2003).

8 Regulatory terminology refers to production within closed-system greenhouses or similar fully enclosed structures as being "contained," versus production in open-field systems, but with precautions against pollen or other material being transmitted to other crops or the environment referred to as "confined." This is a significant distinction in terms of current regulatory roles and responsibilities. Most discussion, questions, and concerns about regulation for commercial PMF production focus on field-based (confinement) issues.

9 The 2003 interim amendments to the Canadian provisions governing field trials for plants with novel traits specify that field trial applications be considered on a case-by-case basis. Recommendations are made against the use of major food- or feed-crop species for PMF purposes; use of crop species that are pollinated by bees and that contribute to commercial honey production is also discouraged; emphasis on confinement characteristics in plant selection is encouraged. (Issues noted here are the levels of outcrossing, mode of pollination, weediness, seed dormancy and dispersal, harvest efficiency, volunteer tendency, and reproductive control mechanisms, including genetic mechanisms to mitigate environmental exposures.) Provisions relating to records, reporting, isolation distances, inspection of residual materials, contingency plans for accidental security breaches, post-harvest land use, and seed logs were tightened. The amendments also specify notification of potential exposure and hazard information for PMF materials where these are expressed in a food or feed crop (for assessment by Health Canada and CFIA, respectively, during a sixty-day period) (CFIA 2003).

10 Transparency expectations seem somewhat at odds with current procedures, which provide limited information on field trials (i.e., no information is provided on the nature of the PMF product or planting location).

11 A study by Huygen, Veeman, and Lerohl (2004) on the costs of alternative tolerance levels relative to possible GM wheat approval is one example of the type of social science research that could inform PMF regulatory policy.

References

Amos, Jonathan. 2004. EU funding for GM plant vaccines. BBC News Science/Nature world edition. http://news.bbc.co.uk/2/hi/science/nature/3887517.stm.

Arcand, François, and Paul G. Arnison. 2004. *Development of novel protein production systems in Canada*. Report prepared for Life Sciences Branch, Industry Canada. Government of Canada, mimeograph.

Arntzen, Charles, Stanley Plotkin, and Betty Dodet. 2005. Plant-derived vaccines and antibodies: Potential and limitations. Preface to special conference issue. *Vaccine* 23 (15): 1752-56.

Baum, Nicholas. 2006. Canadian biotech company achieves commercially viable level of insulin in safflower (press release cited by Checkbiotech.org, July 18). PR Newswire. http://www.checkbiotech.org/blocks/dsp_docToPrint.cfm?doc_id=13157.

BIOTECanada. 2004. BIOTECanada Molecular Farming Committee white paper. Paper submitted to Industry Canada, October 28, mimeograph.

CARC (Canadian Agri-Food Research Council). 2004. *Bio-based Molecular Production Systems Workshop report 2004*. Report of workshop sponsored by CARC, Government of Canada, and the Genome Prairie GE3LS Research program, April 26-27. Ottawa: CARC.

Castle, David, and Jean Dalgleish. 2005. Cultivating fertile ground for the introduction of plant-derived vaccines in developing countries. *Vaccine* 23 (15): 1881-85.

CFIA (Canadian Food Inspection Agency). 2001. *CFIA multi-stakeholder consultation of plant molecular farming.* Report of proceedings. http://www.inspection.gc.ca/english/plaveg/bio/mf/mf_cnsle.shtml.

–. 2003. *Interim amendment to DIR2000-07 for confined research field trials of PNTs for plant molecular farming.* Plant Products Directorate, Plant Biosafety Office. http://www.inspection.gc.ca/english/plaveg/bio/dir/dir0007ie.shtml.

–. 2004. CFIA Technical Workshop on the Segregation and Handling of Potentially Commercial Plant Molecular Farming Products and By-products. http://www.inspection.gc.ca/english/plaveg/bio/mf/worate/woratee.shtml.

–. 2005a. Developing a regulatory framework for the environmental release of plants with novel traits intended for commercial plant molecular farming in Canada. Plant Biosafety Office. http://www.inspection.gc.ca/english/plaveg/bio/mf/fracad/fracade.shtml.

–. 2005b. Draft assessment criteria for the evaluation of environmental safety of plants with novel traits intended for commercial plant molecular farming. Plant Biosafety Office. Directive Dir200X-YZ. http://www.inspection.gc.ca/english/plaveg/bio/mf/fracad/evaluae.shtml.

–. 2005c. *CFIA Technical Workshop on the Segregation and Handling of Potentially Commercial Plant Molecular Farming Products and By-products.* http://www.inspection.gc.ca/english/plaveg/bio/mf/worate/woratee.shtml.

–. 2005d. Summary of confined research field trials of plants with novel traits (PNTs) intended for plant molecular farming (PMF) in Canada. Plant Biosafety Office. http://www.inspection.gc.ca/english/plaveg/bio/mf/sumpnte.shtml.

–. 2005e. Executive summary (summary of proceedings of Technical Workshop on Developing a Regulatory Framework for the Environmental Release of Plants with Novel Traits Intended for Commercial Plant Molecular Farming). Plant Biosafety Office. http://www.inspection.gc.ca/english/plaveg/bio/mf/fracad/techenviroe.shtml#1.

–. 2006. The Canadian Food Inspection Agency's online consultation on plant molecular farming: Summary of feedback (April 26-May 23, 2006). Plant Biosafety Office. http://www.inspection.gc.ca/english/plaveg/bio/conlige.shtml

Crowley, Catherine. 2005. Fields of insulin. *Food for Thought* (Summer). http://206.191.61.186/foodforthought/fft-summer-2005-fields-of-insulin.asp.

Dow AgroSciences. 2006. Dow AgroSciences achieves world's first registration for plant-made vaccines. Corporate news release, January 31. http://www.dowagro.com/newroom/corporatenews/2006/20060131b.htm.

Einsiedel, E.F., and J. Medlock. 2005. A public consultation on plant molecular farming. *AgBioForum* 8 (1): 26-32. http://www.agbioforum.org.

Elbehri, Aziz. 2005. Biopharming and the food system: Examining the potential benefits and risks. *AgBioForum* 8 (1): 18-25. http://www.agbioforum.org.

Fox, Jeffrey L. 2003. Puzzling industry response to ProdiGene fiasco. *Nature Biotechnology* 21 (1): 3-4.

Freese, B. 2002. Manufacturing drugs and chemicals in crops: Biopharming poses new threats to consumers, farmers, food companies and the environment. Friends of the Earth for Genetically Engineered Food Alert, July. http://www.foe.org/camps/comm/safefood/biopharm/BIOPHARM_REPORT.doc.

Gao, Ge, Michele Veeman, and Wiktor Adamowicz. 2005. Consumers search behaviour for GM food information. *Journal of Public Affairs* 5 (3 and 4): 1-9.

Gellene, Denise. 2002. Antibody creation on corn patented. Checkbiotech. http://www.checkbiotech.org/bocks/dsp_document.cfm?doc_id=3444.

Harl, N., R. Ginder, C. Hurburgh, and S. Moline. 2003. The StarLink situation. *Iowa State University Extension,* November 18. http://www.extension.iastate.edu/grain/pages/grain/publications/buspub/0010star.PDF.

Hu, W., A. Hünnemeyer, M.M. Veeman, W.L. Adamowicz, and L. Srivastava. 2004. Trading off health, environmental and genetic modification attributes in foods. *European Review of Agricultural Economics* 31 (3): 389-401.

Hu, Wuyang, Michele M. Veeman, and Wiktor L. Adamowicz. 2005. Labelling genetically modified food: Heterogeneous consumer preferences and the value of information. *Canadian Journal of Agricultural Economics* 53 (1): 83-102.

Huygen, I., M. Veeman, and M. Lerohl. 2004. Cost implications of alternative GM tolerance levels: Non-genetically modified wheat in western Canada. *AgBioForum* 6 (4): 169-77. http://www.agbioforum.org.

Kirk, Dwayne D., and Kim McIntosh. 2005. Social acceptance of plant-made vaccines: Indications from a public survey. *AgBioForum* 8 (4): 228-34. http://www.agbioforum.org.

Kostandini, Genti, and Bradford F. Mills. 2005. Market strategies for a tobacco biopharming application: The case of Gaucher's disease treatment. Paper presented at the annual meeting of the American Agricultural Economics Association, Providence, Rhode Island, July 24-27. http://www.agecon.vt.edu/biotechimpact/extension/papers.htm.

Kostandini, Genti, Bradford F. Mills, and George Norton. 2004. Potential impacts of pharmaceutical uses of transgenic tobacco: The case of human serum albumin (HSA). Paper presented at the annual meeting of the American Agricultural Economics Association, Denver, Colorado, August 1-4. http://www.agecon.vt.edu/biotechimpact/extension/papers.htm.

Ma, Julian K.-C., Pascal M.W. Drake, Daniel Chargelegue, Patricia Obregon, and Alessandra Prada. 2005. Antibody processing and engineering in plants, and new strategies for vaccine production. *Vaccine* 23 (15): 1814-18.

Ma, Julian K.-C., Pascal M.W. Drake, and Paul Christou. 2003. The production of recombinant pharmaceutical proteins in plants. *Nature Reviews Genetics* 4 (10): 794-805.

Macdonald, Phil. 2005. Technical Workshop on Developing a Directive for Plant Molecular Farming in Canada. Presentation by Biosafety Office, Plant Products Directorate, CFIA, mimeograph.

Markley, Nancy, Cory Nykiforuk, Joe Boothe, and Maurice Moloney. 2006. Producing proteins using transgenic oilbody-oleosin technology. *BioPharm International*, June 1. http://biopharminternational.findpharma.com/biopharm/Article/Producing-Proteins-Using-Transgenic-Oilbody-Oleosi/ArticleStandard/Article/detail/329358.

Martin, Ellen M. 2005. First production of human monoclonal antibodies in chicken eggs published in *Nature Biotechnology*. Public announcement, August 28. Eurekalert. http://www.eurekalert.org/pub_releases/2005-08/ka-fpo082505.php.

Medlock, Jennifer, and Edna Einsiedel. 2005. It's not just what's for dinner anymore but the future contents of our medicine cabinets. *Choices: The Magazine of Food, Farm and Resource Issues* 20 (4): 253-56. www.choicesmagazine.org/2005-4-08.htm.

Nevitt, J., G. Norton, B. Mills, M.E. Jones, M. Ellerbrock, D. Reeves, K. Tiller, G. Bullen. 2003. *Participatory assessment of social and economic effects of using transgenic tobacco to produce pharmaceuticals*. Department of Agricultural and Applied Economics, Virginia Tech. http://www.agecon.vt.edu/biotechimpact.

Nevitt, J., B.F. Mills, D.W. Reaves and G. Norton. 2006. Public Perceptions of Tobacco Biopharming. *AgBioForum* 9 (2): 104-10. http://www.agbioforum.org.

Pew Initiative on Food and Biotechnology. 2002. *Pharming the field: A Look at the benefits and risks of bioengineering plants to produce pharmaceuticals*. Washington DC, July 18. http://www.pewtrusts.org/our_work_detail.aspx?id=442.

Polkinghorne, Ian, Denes Hamerli, Phil Cowan, and Janine Duckworth. 2005. Plant-based immunocontraceptive control of wildlife: Potentials, limitations and possums. *Vaccine* 23 (15): 1847-50.

ProdiGene. 2005. Press announcement. http://www.prodigene.org (accessed January 2006).

PROVACS (Program for Production of Vaccines from Applied Crop Sciences, Center for Infectious Diseases and Vaccinology, University of Arizona). 2005. *Blueprint for the development of plant-derived vaccines for diseases of the poor in developing countries: Intellectual property considerations*. The Biodesign Institute, Arizona State University. http://www.biodesign.asu.edu/centers/idv/projects/provacs.

–. 2006. Plant-derived vaccines: Cost of production. http://www.biodesign.asu.edu/centers/idv/projects/provacs.

Reuters News Service. 2002. ProdiGene to spend millions on bio-corn tainting. Reuters, December 9. http://www.planetark.com/avantgo/dailynewsstory.cfm?newsid=18935.

Stewart, P.A., and W. McLean. 2005. Public opinion toward the first, second, and third generations of plant biotechnology. *In Vitro Cellular and Developmental Biology—Plant* 41 (6): 718-24.

Swoboda, Rod. 2002. Biopharming can add value to Midwest agriculture. Checkbiotech. http://www.checkbiotech.org/bocks/dsp_document.cfm?doc_id=3427.

Tao, Shiyi, Michele Veeman, and Wiktor Adamowicz. 2008. Preferences for research funding allocations to plant molecular farming. Department of Rural Economy, University of Alberta, mimeograph.

Uchtmann, D.L. 2002. StarLink: A case study of agricultural biotechnology regulation. *Drake Journal of Agricultural Law* 7 (1): 159-211.

USDA (US Department of Agriculture). 2002. USDA investigates biotech company for possible permit violations. Press release, November 13. http://www.aphis.usda.gov/lpa/news/2002/11/prodigene.html.

van der Laan, J.W., P. Minor, R. Mahoney, C. Arntzen, J. Shin, and D. Wood. 2006. Meeting report: WHO informal consultation on scientific basis for regulatory evaluation of candidate human vaccines from plants, Geneva, Switzerland, 24-25 January 2005. *Vaccine* 24 (20): 4271-78.

Veeman, Michele, and Wiktor Adamowicz. 2006. Canadians' assessments of food safety and environmental safety relative to plant molecular farming. Department of Rural Economy, University of Alberta, mimeograph.

Veeman, Michele, Wiktor Adamowicz, Wuyang Hu, and Anne Hünnemeyer. 2005. Canadian attitudes to genetically modified food. In *Crossing over: Economics in the public sphere,* ed. E. Einsiedel and F. Timmermans, 99-113. Calgary: University of Calgary Press.

Vermij, Peter. 2004. Edible vaccines not ready for main course. *Nature Medicine* 10 (9): 881.

Wisner, Robert. 2005. The economics of pharmaceutical crops: Potential benefits and risks for farmers and rural communities. Report commissioned by the Union of Concerned Scientists. Cambridge, MA: Union of Concerned Scientists.

7
Policy and Regulatory Challenges for Plant-Made Pharmaceuticals in the United States

Patrick A. Stewart

Of all genetically modified (GM) plant products, plant-made pharmaceuticals (PMPs) offer the most tantalizing promise. Although previous products derived from GM plants have tended to have gone relatively unheralded, and hence unheeded by the public, plants genetically modified to produce pharmaceutical drugs have the potential to revolutionize drug production while expanding the range of drugs available. Given the aging populations in industrial countries, and the health needs of developing nations throughout the world, PMPs promise a cornucopia of pharmaceutical drugs. However, if the current generation of policy makers and shapers do not address the perennial "questions of safety, accountability, risk management, patents, university-industry relations . . . environmental hazards and so on, raised by biotechnology and its applications" (Caton 2002, 206), future generations may see PMPs as a Pandora's box that should not have been opened.

Increased public fears may stifle the development of this promising new technology because of the perceived potential for unique environmental and health threats from unrestrained PMPs that are increasingly likely as more acreage is grown with a greater range of these plants. Because public perceptions of a newly exposed threat, not the actual risks posed by a technology, often drive legislation and regulation (Kingdon 1984), effective regulation that preserves the integrity of the agricultural system while not impeding the research and development of these PMPs may alleviate public fears and assure continuing development of PMPs. This chapter analyzes current US federal regulations as they pertain to PMPs and considers public perceptions of the risks posed by and benefits from PMPs before considering the linkages, or more accurately, the lack thereof, between current regulations and public opinion, before drawing conclusions as to the potential public response to PMPs.

Regulation of Plant-Made Pharmaceuticals

The regulation of PMPs, a subset of agricultural biotechnology, is a complex affair marked by changes responding to both greater comprehension of the science and technology behind the products and apprehension over the potential risks they pose to individuals and society. The original effort to regulate biotechnology throughout the 1970s and the early 1980s asserted that the new biotechnology was continuous with pre-existing technology (such as crossbreeding and fermentation), despite the revolutionary jump genetic modification made in precision and capacity to alter plants in, to this point, unexpected ways. As a result, existing federal government agencies and regulations were seen as sufficient to address any problems raised by the new products.

The Reagan administration's Office of Science and Technology Policy proposed the Coordinated Framework for Regulation of Biotechnology in 1985. In 1986, the Coordinated Framework was promulgated and, in so doing, coordinated regulation between the research agencies of the National Institutes of Health and the National Science Foundation and the regulatory agencies the Environmental Protection Agency (EPA), the Food and Drug Administration (FDA), and the US Department of Agriculture (USDA). In regulating new GM products, the Coordinated Framework is to deal with the jurisdictional overlap on a "pragmatic" case-by-case basis.

In the field, the regulation of GM agricultural products begins with the USDA-Animal and Plant Health Inspection Service (APHIS) as provided for by the Plant Quarantine Act and the Federal Plant Pest Act, both of which have since been replaced with the Plant Protection Act of 2000. The USDA also has regulatory authority through the Federal Meat Inspection Act, the Poultry Products Inspection Act, the Viruses, Serums, Toxins, Antitoxins, and Analogous Products Act, and the Federal Seed Act. The EPA has regulatory authority under the Federal Insecticide, Fungicide, and Rodenticide Act and the Toxic Substances Control Act. Finally, the FDA regulates GM products on the basis of the Federal Food, Drug, and Cosmetic Act and the Food Quality Protection Act–which also implicates the EPA (Stewart and Sorensen 2002; Krimsky and Wrubel 1996; Wright 1994).

USDA-APHIS' responsibility for the regulation of GM plant products in the fields is based on the potential for these products to be plant pests or to harbour plant pests. Here, based on an explicit statement by the Coordinated Framework, there is a focus on the characteristics of risk posed by the plants. The regulatory trigger is the use of recombinant DNA inserted through *Agrobacterium tumefaciens*–a bacterium that transfers genetic sequences into the targeted plant. In other words, according to the Coordinated Framework, only plants altered through the use of *Agrobacterium* trigger regulatory oversight. Although those plants using other methods of gene insertion, such as gene guns, use the current regulatory system, they do so voluntarily (NRC 2002).

Three regulatory paths for the field release of GM plants, as overseen by USDA-APHIS, currently are in place: permitting, notification, and petitioning. The permitting track, put in place in 1987, is the most stringent approach because of the uncertainty of the new GM products being put into the field. As such, there was a high level of oversight, extensive paperwork requirements, a relatively high level of expense, and a longer wait for approval. The notification track was put in place in 1993 to simplify the field release and transport process of six plant species (corn, cotton, potato, soybean, tobacco, and tomato) that were considered genetically well characterized and with limited threat of transmission to wild relatives, cutting time, effort, and expense. Also in 1993, the petition process was put in place to allow for the determination that certain plants were no longer regulated. In other words, this allowed the large-scale release of GM plants and their further breeding without oversight, if done through conventional crossbreeding. In 1997, the notification track was further expanded to include most GM plants (Stewart and Knight 2005). This decision was recently revisited by USDA-APHIS in response to a series of events that have attracted public awareness and greater involvement by pressure groups and interested citizens (Stewart and McLean 2005a).

Recent Regulatory Changes

The regulation of agricultural biotechnology has attracted little public involvement until recently. This is most likely because of the highly technical nature of its products. In addition, most of its products are concerned with replacing or reducing production inputs such as herbicides and pesticides, which has gone unheralded and hence relatively unnoted by the public. However, recent events have publicized and politicized the regulation of agricultural biotechnology as an expanding range of parties with economic and political clout have become involved (Stewart and McLean 2005a). A series of misadventures in which GM animal feed (StarLink corn) entered the human food supply in 1999-2000; the publication of an article in a prominent science journal in 1999 suggested long-term deleterious effects of corn genetically modified to express *bacillus thuringiensis* (Bt), a pesticide, on Monarch butterflies; and corn plants genetically modified to produce pig vaccine nearly entered the food supply on two occasions in 2002; as well as debate over GM wheat, which was ultimately pulled from the market (Taylor, Tick, and Sherman 2004; Stewart and Knight 2005), have piqued public concern. More recently, a report suggested that, for the past four years, an experimental corn plant genetically modified to express the Bt pesticide and an antibiotic used for marker purposes has been entering the food supply without regulatory approval and the awareness of the company producing it, the farmers growing it, and consumers eating it (Rosenwald 2005). While this corn does not necessarily present a health risk, its presence in the food chain raised questions about the implementation of regulations intended to protect the public from GM plants not intended for consumption.

The Office of Science and Technology Policy (OSTP) published "Proposed federal actions to update field test requirements for biotechnology derived plants and to establish early food safety assessments for new proteins produced by such plants" in August 2002 to address decreasing trust in institutions producing and regulating food. The notice provides guidance to the USDA, EPA, and FDA, updating field testing requirements and establishing early food-safety assessments for new plant proteins, such as those produced by PMPs and plant-made industrial products (PMIPs). This document relies on three principles to update the 1986 Coordinated Framework: (1) the level of field test confinement should be consistent with the level of environmental, human, and animal health risk associated with the introduced proteins and trait(s); (2) if a trait or protein presents an unacceptable or undetermined risk, field test confinement requirements would be rigorous to restrict outcrossing or commingling of seed and, further, the occurrence of these gene or gene products from these field tests would be prohibited in commercial seed, commodities, and processed food and feed; and (3) even if these traits or proteins do not present a health or environmental risk, field test requirements should still minimize the occurrence of outcrossing and commingling of seed. In this last case, low levels of genes and gene products found in commercial seed, commodities, and processed food and feed may be acceptable based on meeting applicable regulatory standards (OSTP 2002).

In light of concerns raised by increased experimentation with PMPs and PMIPs, addressed by OSTP in its notice (OSTP 2002), USDA's Animal and Plant Health Inspection Service changed rules concerning its field testing of PMPs in March 2003 (USDA-APHIS 2003b). The resulting regulations incorporate significant changes in the regulation of PMPs and PMIPs (USDA-APHIS 2003a). Here, APHIS established seven conditions grouped into three categories considering field test siting, the dedication of production equipment and facilities, and, finally, procedural matters.

Field test siting regulations increased the perimeter fallow zone from twenty-five to fifty feet to prevent inadvertent commingling with food or feed plants and restricted the production of food and feed plants at the field test site and perimeter fallow zone for the following season to prevent inadvertent harvesting.[1] The dedication of farm equipment and facilities to crop production requires test-site-dedicated planters and harvesters for the duration of the tests. Even though tractors and tillage attachments do not have to be dedicated, they must be cleaned according to APHIS protocols, and equipment and regulated articles must be stored in dedicated facilities for the duration of the field tests. The procedural requirements are submission and approval of seed cleaning and drying procedures to ensure confinement of plant material and the minimizing of the risk of seed loss or spillage. Finally, permittees are required to implement an APHI-approved training program to successfully comply with these permit conditions.

The key issue with regulation is ensuring compliance. APHIS increased the number of PMP field site inspections "to correspond with critical times relevant to the confinement measures" (USDA-APHIS 2003a, 11338). In addition to maintaining records of activities related to meeting permitting conditions, and increasing the likelihood of audits, APHIS is to inspect permitted PMP field tests up to five times during the growing season: once at pre-planting to evaluate the site location, once at the planting stage to verify site coordinates and adequate cleaning of planting equipment, at mid-season to verify reproduction isolation protocols and distances, at harvest to verify cleaning of equipment and its appropriate storage, and again at post-harvest to verify cleanup of the field site. In addition, two post-harvest inspections to verify that the regulated articles do not persist in the environment are to occur. Finally, APHIS may inspect more frequently if deemed necessary (USDA-APHIS 2003a).

The administrative reorganization of how USDA-APHIS regulates biotechnology led to the creation of the Biotechnology Regulatory Services (BRS). This move addresses concerns raised by PMPs specifically and by GM organisms generally. According to USDA-APHIS, "Given the growing scope and complexity of biotechnology, now more than ever, APHIS recognizes the need for more safeguards and greater transparency of the regulatory process to ensure that all those involved in the field testing of GE crops understand and adhere to the regulations set forth by BRS" (USDA-APHIS 2003c). With the changes, BRS has overall control over inspection of GM plants; however, through a 2004 memorandum of understanding, BRS shares responsibility with the Plant Protection and Quarantine unit for field inspections (USDA-OIG 2005).

Changes instituted by BRS include new training for APHIS inspectors in auditing and inspections of field trials; the use of new technologies such as global positioning systems; and analysis of historical trends to inform monitoring and inspection. According to APHIS, the changes will serve six overarching goals, with these key components: (1) enhanced and increased inspections in which risk-based criteria, along with other factors, will assess field test sites. Higher risk sites are to be inspected at least once a year, with other sites being randomly selected for yearly inspections; (2) auditing and verification of records of businesses and organizations to verify accuracy and implementation; (3) remedial measures to protect "agriculture, the food supply, and the environment in the event of compliance infraction," with the establishment of a first-responder group to deal with serious infractions; (4) standardized infraction resolution in which criteria will be established to determine the extent of an infraction and the response, whether further investigation, the issuance of a guidance letter, the issuance of a written warning, or referral to APHIS' Investigative and Enforcement Services unit for further action; (5) documentation, in which a database will be set up to track field test inspections and resulting compliance infractions while enhancing transparency to keep stakeholders and the public informed on the regulatory decision-making process; (6) continuous process improvements where, as the science of biotechnology advances, so too will regulations and permit conditions to allow safe field testing; (7) an emergency response protocol, being developed with input from EPA and FDA, in which a quick response plan will be put in place "to counteract potential impacts on agriculture, the food supply, and the environment"; (8) training for field test inspectors in their dealings with PMP and PMIP field test sites, as well as the latest in auditing; and (9) certification for compliance with the highest level of auditing standards (USDA-APHIS 2003b, 11338-39).

A recent audit by the USDA's Office of Inspector General (USDA-OIG) suggests that while these changes have been put in place administratively, their implementation has proved problematic. A total of twenty-eight recommendations based on nine findings in four areas (overall assessment, application process, management and oversight of field tests, and enforcement actions) were made in this seventy-two-page audit report that was carried out between 2003 and 2005 (USDA-OIG 2005). Although many of the issues raised in the report have bearing on both GM plants generally and PMPs and PMIPs specifically, concerns raised by the potential of PMPs (and PMIPs) entering the environment and the food supply suggest this technology spurred greater oversight.

Specific problems with the regulation of PMPs and PMIPs identified by the audit were abundant and troubling, especially given USDA-APHIS' stated regulatory changes. Despite concerns over their potential accidental release, and requirements for GPS coordinates, only five of thirteen PMP or PMIP permit holders provided this information (USDA-OIG 2005, 15). The required scientific review of official PMP or PMIP permit files does not appear to have occurred, thanks to both the lack of a standardized process and incomplete information in the files (USDA-OIG 2005, 24-25). Actual monitoring of PMPs and PMIPs in the field does not appear to have occurred either, contrary to assurances. Only one of twelve of the audited field test sites were inspected five times during the 2003 growing season, with only eighteen of fifty-five inspections of the remaining eleven sites being carried out. Furthermore, permit requirements that restricted food and feed crops being planted in PMP and PMIP field test sites the following season was not followed in at least one audited case (USDA-OIG 2005, 29-30). Progress reports for PMPs and PMIPs were not submitted in a timely manner, according to the USDA-OIG report. Specifically, of twenty-two field test sites for one PMIP and twelve PMP permits, APHIS could not produce eleven of the twenty field test data reports, 36 percent of planting notices, 10 percent of the four-week/twenty-eight-day reports, and 45 percent of harvest notices (USDA-OIG 2005, 36).

The report also finds a lack of control by APHIS over the final disposition of PMP harvests, with two sites storing large amounts of PMP crops for over a year without APHIS' knowledge. In one case, half a ton of PMP crop was stored for fifteen months and in the other case 1.4 tons was stored for seventeen months without APHIS approval and without assessment of safety protocols (USDA-OIG 2005, 41). In three cases, because of lack of APHIS guidance, the fields themselves were not "devitalized" at the conclusion of field tests to ensure the plants did not persist in the environment, producing offspring that may lead to the problems of weediness, spread of genetic material to domesticated varieties of the same crop, or outcompeting important plants, as is discussed in the NRC 2004 report (USDA-OIG 2005, 44-45).

A final concern, and one having direct bearing on the ability to redress problems caused by grower negligence or malfeasance, is that applicants are not required to provide proof of financial responsibility under current laws and regulations. This is especially problematic because of the great majority of PMPs and PMIPs being field tested by small companies (with the remainder accounted for by university researchers). Although these small companies bring with them substantial entrepreneurial spirit, the inability to hold them financially responsible for not following regulations or, for that matter, after-the-fact legal liabilities (Kershen 2004), removes the teeth from regulatory implementation or legal redress. For example, in 2002, corn plants genetically modified by ProdiGene to produce pig vaccine nearly entered the food supply, focusing attention on flaws in the APHIS field test program. This led to a civil penalty of US$250,000 and the destruction of 500,000 bushels (US$2.7 million dollars' worth) of soybean in Nebraska and the incineration of 155 acres of corn in Iowa (Stewart and Knight 2005). Although ProdiGene was able to pay the civil penalty and reimburse the USDA for destruction of the soybean, without regulatory changes, financial responsibility might not be assured. Further, in 2002, a company field testing PMPs went out of business, leaving a cooperator with a field of PMPs but no legal responsibility for their monitoring. While the cooperator agreed to monitor the field, such assistance was voluntary (USDA-OIG 2005, 48).

What is apparent is that regulation of GM crops generally, and PMPs and PMIPs specifically, is a complex endeavour requiring precise oversight by extensive and well-trained professional staff. This, as seen in the USDA-OIG report, is not the case. In the absence of systematic implementation, USDA regulations are largely symbolic.

The difficulty of policing an invisible commons where infractions are not readily apparent to the naked eye, even when trained, is compounded by the reproducing and potentially cascading nature of biological pollution in which unwanted, yet successfully reproducing, plants may survive, indeed thrive, despite efforts at biological control. Regulations in place to date have reflected an expectation that GM plants are an extension of plant production methods long in use, and as such its regulation expects that the status quo approach, having worked so well in the past, will continue to be a successful strategy. This perception is based on an incomplete understanding of transgenic organisms and their interaction within varied environments. As difficulties in regulating GM plants have become apparent through various misadventures in the field and food supply, change in regulatory language has occurred, albeit without concurrent change in implementation capacity. The disjoint between symbolic policy changes and true regulatory implementation likely has come about because of the perceived potential for the expansion of the scope of conflict beyond the current constrained policy community of investors, researchers, regulators, and activists to include more powerful actors, such as large food corporations and their interest groups as well as the general public (Schattschneider 1975; Stewart and McLean 2005a) and combined with the lack of resources and political will to implement the newly promulgated regulatory policies. This suggests that regulatory change may occur only if a major misadventure involving a truly transgenic plant such as a PMP, which combines pharmaceuticals with food, focuses public attention on a potential threat to their health (Kingdon 1984). In this case, understanding what drives public attitudes toward GM plants, specifically PMPs, likely will lead to understanding how the public might react, and how different publics may be addressed in response.

Public Risk Perceptions of PMPs

Plant-made pharmaceuticals promise to radically change public perceptions by drawing attention to the changed character of the plants themselves. Although GM plants are currently a pervasive part of the agricultural landscape and consumed extensively, with the staple crops of corn and soybean grown in GM variants in the United States and comprising 45 percent and 85 percent of total acreage in 2004 (Pew Initiative on Food and Biotechnology 2004), there is little general public awareness of their presence in the food supply (Pew Initiative on Food and Biotechnology

2003; Hallman et al. 2004; Stewart and McLean 2005b). This is mainly because of these plants being genetically modified to express agronomic qualities, such as herbicide tolerance and pest resistance, which are typically not brought to the attention of the public.

PMPs, on the other hand, are likely to arouse public interest and concern, as they will not be part of the food chain and may be seen as a distinctly different technology. While plants have long been used for medicinal purposes and food can have healing properties, plants genetically modified to express pharmaceutical products, especially products using food plants as a delivery vehicle, will likely draw public attention by underscoring the transgenic nature of the products that span species and may even incorporate human DNA. Although it is highly unlikely that the same plant will be processed for both food and pharmaceutical purposes, it is quite likely that the public will perceive PMPs as involved in both, especially if food plants are used as production vehicles.

In terms of public response to date, there have been quite different paths in the reception of GM pharmaceuticals and medical products and that of GM plants (Fischhoff and Fischhoff 2001). There has been little controversy over GM pharmaceutical products, likely because people are willing to take risks to recover the health they have lost (Kahneman, Slovic, and Tversky 1982). On the other hand, people are likely to avoid risking change in their diet, even if there are tangible benefits, especially when they see nothing wrong with it in the first place. By combining the potential for health benefits from new pharmaceuticals with that of accidentally consuming them in food, PMPs are likely to arouse ambiguous public response.

Much of public risk perception research focuses on describing how familiar or unfamiliar technology is and how dreaded that technology is. This psychometric approach has been valuable in understanding the nature of risk perceptions (Slovic 1992) as it pertains to agricultural biotechnology (Fischhoff and Fischhoff 2001; Midden et al. 2002). However, recent advances in the neurosciences have led to researchers appreciating the role of emotion. Specifically, dread may be interpreted as reflecting the core emotion of fear-worry, which has a distinct physiological signature, recognizable communication characteristics, and predictable behavioural implications (Ekman and Davidson 1994; Lewis and Haviland-Jones 2004). Increase in fear-worry plays a key role in the collection of more information (Brader 2006; Marcus, Neuman, and MacKuen 2000), as well as driving regulatory policy making (Schubert, Stewart, and Curran 2002; Lerner et al. 2003). Worry may be seen as driving public concerns over the new agricultural biotechnology because many of its products are intended for personal consumption (i.e., eating) and thus become both public/environmental and personal/health threats if they inadvertently enter the food supply (Stewart and Sorensen 2002; Stewart and McLean 2005a, 2005b).

Although benefits from new technologies have often been considered the polar opposite of fears over them, as if they are end points on a single-dimension of risk perception, benefits may be seen as existing on a different, orthogonal dimension as a subset of how known or unknown a risk is (Slovic 1992). If a new technology has either a direct personal benefit or a societal benefit, it will likely be seen as more acceptable. In this case, risks and resultant fears may be accepted (or ignored) for the sake of the benefits if the trade-off is advantageous. In their study of European public perceptions, Midden et al. (2002) found that two factors emerged: the perception of risk, and a factor comprising judgments concerning its use, moral acceptability, and encouragement. Both these factors appear to map well onto the emotional response of worry and perceptions of benefits.

Research considering public perception of PMPs is relatively sparse, which is understandable given it is an emerging technology with unrealized benefits and risks. However, the research to date suggests that there is both public recognition of PMP benefits and awareness of the potential risks they pose to the environment and public health (Pew Initiative on Food and Biotechnology 2003; Einsiedel and Medlock 2005; Kirk and McIntosh 2005; Stewart and McLean 2005b; see also Chapter 6).

To further understand both perceived benefits and worries over PMPs, we analyzed telephone survey data collected in the summer of 2004 from a total of 680 respondents in the five-state EPA Region 6 of the United States (Arkansas, Louisiana, New Mexico, Oklahoma, and Texas).[2] Subjects were given a short description of PMPs and then asked to assess the technology. Specifically, PMPs were described as "plants genetically modified to produce pharmaceutical drugs. These are plants that are modified to produce compounds used in manufacturing vaccines for diarrhea, antibodies to fight cancer, and drugs to treat such illnesses as cystic fibrosis." Benefits were assessed by asking, "How much benefit do you believe you would personally get from this plant?" Responses on a four-point scale averaged three. In other words, respondents assessed PMP benefits relatively highly. Worry was assessed by asking, "How worried would you be if you ate food coming from this type of plant?" and was measured on a five-point scale. The mean score of three suggests moderately high levels of worry but with polarized response (see Table 7.1). Further analysis suggests that respondents viewed benefits from PMPs and worries over them as distinct.

Those factors driving perceptions of benefits from PMPs and worry over consuming these plants include familiarity with similar technologies, knowledge of science, level of education, trust in producers of the technology

Table 7.1

Perceptions of benefits and worries about PMPs
(2004 telephone survey in US EPA Region 6)

Question	Response
How much benefit do you believe you would personally get from this type of plant [PMPs]?	None at all – 13.3 2 – 11.8 3 – 33.6 A great deal – 41.3
How worried would you be if you ate food coming from this type of plant [PMPs]?	Not at all worried – 22.8 2 – 21.0 3 – 17.2 4 – 14.1 Very worried – 24.9
Have you seen, read, or heard about biotech/ GM food sold in grocery stores?	Nothing at all recently – 29.9 2 – 30.3 3 – 31.7 A great deal – 8.1
How likely would you be to choose to eat GM foods?	Not at all likely – 31.5 2 – 28.8 3 – 30.4 Very likely – 9.4
As far as you know, have you ever eaten GM foods?	No – 73.0 Yes – 27.0
How likely is it that these [PMP] plants might accidentally enter the US food supply?	Very unlikely – 14.8 2 – 13.3 3 – 4.6 4 – 43.3 Very likely – 23.9

and those regulating it, and background information such as age, sex, ethnicity, and religiosity.

Familiarity with Agricultural Biotechnology

Key in determining individual assessments of the benefits from PMPs and what causes people to worry is familiarity with and assessment of the products' risk and willingness to consume them if they accidentally enter the food supply (Slovic 1992). Perceptions and awareness of the products of agricultural biotechnology generally, and PMPs specifically, suggest a role is played, but this role varies based on explanation of perceived benefits or worries. Awareness of GM food sold in grocery stores is significantly correlated with perceived benefits, with the more seen, read, or heard about GM foods, the less the benefits are perceived as deriving from PMPs. As can be expected, worries increase the more likely respondents see themselves as eating GM foods, while awareness of having eaten GM foods leads to greater perceived benefits. Finally, that PMPs might accidentally enter the US food supply was significant and positively correlated with both benefits of and worries about PMPs. This finding suggests that respondents are circumspect in seeing increased exposure as being tied to both risks and benefits.

Science Knowledge and Education

Proponents and opponents of GM plants both believe that science knowledge and education play a role in perceptions of risks and benefits. A common belief of those in the scientific community, as well as in many policy circles, is that public scientific literacy leads to greater support for science and technology (Gaskell, Thompson, and Allum 2002; Hoban 1997). However, this has not been supported cross-culturally, with superior scientific literacy leading to greater acceptance in some cases and higher levels of rejection in other situations (Chess 1998; Gaskell, Thompson, and Allum 2002). In other words, interest and hence knowledge of science might be driven by proponents of progress in science and technology as well as by skepticism about its threat to society (Gaskell et al. 2001).

Specific objective science knowledge is the sum of four correctly identified science and technology knowledge statements (NSF 2004).[3] Highest level of education serves as a proxy measure of overall science knowledge and ranges from not finishing high school to obtaining a graduate or professional degree. Greater respondent science knowledge is correlated with less worry about PMPs, and those with greater education worry less. This suggests that greater science education both generally and specifically leads to greater support of PMPs (see Table 7.2).

Institutional Trust: Federal Government and Farmers

Trust in federal government agencies and their regulatory system and in farmers is also seen as important in the acceptance of GM products. In other words, governmental and farming institutions serve as buffers against the uncertainty of the outside world in the absence of direct knowledge (Gaskell, Thomson, and Allum 2002). Therefore, trust can be expected to reduce worry over new technologies such as GM plants (Gaskell et al. 2001), despite PMPs not being grown in the traditional farming situation (Stewart and Knight 2005; Taylor, Tick, and Sherman 2004). To assess trust we asked survey respondents their levels of trust in the institutions responsible for food safety in the United States. The first three questions consider government institutions responsible for regulating the new agricultural

Table 7.2

Determinants of perceptions of benefits of and worries about PMPs

Variable	Benefits of PMPs	Worries about PMPs
Have you seen, read, or heard about biotech/ GM food sold in grocery stores?	0.080*	0.022
How likely would you be to choose to eat GM foods?	0.051	–0.105**
As far as you know, have you ever eaten GM foods?	0.101*	–0.056
How likely is it that these (PMP) plants might accidentally enter the US food supply?	0.126**	0.132***
Science knowledge scale	–0.031	–0.110**
Highest level of education	0.057	–0.152***
Government trust factor score	0.077	–0.009
Farmer trust factor score	0.117**	–0.174***
Religiosity	–0.061	0.056
Family income	–0.025	–0.149***
Age	–0.049	–0.009
Sex	–0.030	0.066
Ethnicity	–0.009	–0.109**

Spearman's rho zero-order correlation coefficients; 2-tailed significance in parentheses.
* $p < .05$; ** $p < .01$; *** $p < .001$

biotechnology–the US Department of Agriculture, the US Environmental Protection Agency, and the US Food and Drug Administration. A second type of institution concerns corporate and family farms, which are responsible for the food supply and its safety.[4]

Trust in institutions has mixed effects on perceived benefits and worry. Trust in the federal government agencies of the USDA, EPA, and FDA is not significantly correlated with either benefits or worries. However, what seemed to matter most for both perceived benefits and worry is trust in corporate and family farms, as increased trust increases perceived benefits and decreases worry.

Religiosity

Religiosity often influences how new technologies are assessed, with genetic modification opposed as playing God (Fox 1992; Stewart and McLean 2005a) and supported by the belief that God gave humans dominion over creation and, hence, genetic modification. To assess religiosity, we asked participants how religious they considered themselves to be. Findings suggest that religion was not significantly related to perceptions of PMP benefits or worries.

Demographics

Finally, the impact of one's social grouping plays a role in public views. Of the demographic variables, increased income led to decreased worry, and being a part of the white majority likewise led to decreased worry. These demographic variables might reflect institutional trust in an indirect manner, as the socio-political system tends to benefit those who hold privileged positions within the system, engendering greater feelings of trust in and support for the system's institutions. Finally, age and sex did not play significant roles.

Discussion

Plant-made pharmaceuticals, by bridging the boundaries of both medical and agricultural biotechnology, as well as transgenic boundaries, have shifted the debate over genetically modified organisms. Until recently, debate was insulated from public opinion and carried out in a relatively small policy community that focused on technical regulatory issues not easily comprehended by the general public. This has recently changed, mainly because of problems with GM plants unintentionally entering the food supply and the potential for PMPs to do the same, yet with more deleterious consequences, alerting policy makers to potential public response. The findings of this chapter that the potential for PMPs entering into the food supply is positively correlated with both greater benefits and worry over eating its products suggests survey respondents see a relationship between both greater benefits and greater risks. Although the general public has yet to become involved with the debate over GM plants, especially PMPs, its response to potential benefits and threats is circumspect.

Within the policy community, the hard-line opposition suggesting plant genetic-modification risks far outweigh benefits has become more circumspect as well and are now concerned more about the production of PMPs using food crops and the need for better control of the production process (Stewart and McLean 2005a). Different actors have become involved in the debate, with food producers and processors contributing their considerable clout. This is likely a major driving force behind regulatory changes in the wake of the ProdiGene incidents in 2002 (Stewart and Knight 2005). Institutional trust, from farmers' fields to the supermarket, is an important driver of policy, with lost trust leading to lost profit margin. Indirectly and directly, this trust may be seen as a major factor in reducing worry, with greater trust in institutions leading to less worry. Further, institutional trust may help keep conflict contained, reducing the likelihood of conflict expansion and major changes to regulations.

The findings of this chapter that survey-respondent trust in farmers is strongly correlated with both perceptions of benefits and reduced worry over PMPs entering the food supply underscores the important roles farmers hold in the debate over PMPs. While family and corporate farmers likely will not be directly involved in the production of PMPs (Stewart and Knight 2005), the findings here suggest trust in this institution drives support for PMPs. As such, the potential for loss in trust in the institution of farms through accidental release of PMPs into the environment or the food supply is elevated and potentially devastating.

That trust in US federal governmental institutions does not play a role in either perceived benefits from PMPs or worries over PMPs being unintentionally consumed might presage regulatory problems. To date, regulatory policy implementation appears quite disappointing. The USDA's internal audit suggests that many of the policies put in place play a symbolic role, with little true oversight.

Conclusion

Given the nature of genetically modified organisms, which are not only difficult to identify once in the field but are also potentially capable of out-competing other plant species, there is the potential to turn the natural environment into a tragedy on the invisible commons. Although risks from PMPs accidentally entering into the environment may be vanishingly small, "a small experimental effect does not imply a small ecological effect" (National Research Council 2002, 194). Greater production likely will use greater acreage; this, combined with proliferation of novel pharmaceutical products produced in plants, increases potential risk to the environment, and likely to the food supply, if not controlled effectively.

Although it is hopeful that changes suggested by the USDA's Office of Inspector General, and put in place by APHIS, will address potential problems, thus allaying interested publics' concerns and enhancing public trust in government's regulatory institutions, the apparent lack of post-commercialization regulatory institutions and plans for dealing with accidental release into the environment or food supply is not hopeful. Even if APHIS exercises stringent oversight of PMP field experimentation, lack of long-term monitoring of the environment using a range of ecological indicators and covering greater acreage than is currently the case presents the potential for health and environmental risks that may prove to be unacceptable to the public (National Research Council 2002).

Notes

This research was made possible by a grant from the Arkansas Biosciences Institute.

1 Specific permit conditions for pharmaceutical corn have also been instituted, likely because corn was the organism of choice until the ProdiGene incidents drew public concern and regulatory changes were put in place in its wake.

2 The data analyzed involved ten-to-fifteen-minute phone interviews with a random sample of consumers. Interviews were conducted between May 9 and June 10, 2004, by paid, trained, and supervised interviewers of the Arkansas State University's Center for Social Research. This project used computer-assisted telephone interviewing technology. The response rate was very good, with a total of 680 interviews completed for a response rate of 61 percent and a margin of error of +/- 3.5 percent.

3 The first statement is "The oxygen we breathe comes from plants." (True; National Science Foundation survey findings = 87 percent; Arkansas Biosciences Institute survey findings = 89.7 percent.) The second statement is "It is the father's gene which decides whether a baby is a boy or a girl." (True; NSF = 65 percent; ABI = 68.5 percent.) The third statement is "Antibiotics kill viruses as well as bacteria." (False; NSF 51 percent; ABI = 54.6 percent.) And, finally, the fourth statement is "Radioactive milk can be made safe by boiling it." (False; NSF = 65 percent; ABI = 70.3 percent.)

4 To assess trust, we ran a principle components factor analysis with a varimax rotation on five trust variables. These variables assessed trust in federal regulatory agencies and farmers on a five-point scale and loaded on two variables. The first factor, trust in government institutions, has an eigenvalue of 2.838 and explains 56.75 percent of the variance with USDA (0.839), EPA (0.882), and FDA (0.890) loading strongly on the variable after rotation. The second factor, trust in farmers, has an eigenvalue of 1.049 and explains 20.982 percent of the variance, with corporate farmers loading at 0.752 and family farmers loading at 0.901.

References

Brader, T. 2006. *Campaigning for hearts and minds: How emotional appeals in political ads work*. Chicago: University of Chicago Press.

Caton, H. 2002. An ecological approach to agricultural biotechnology policy. In *Human nature and public policy: An evolutionary approach,* ed. S. Peterson and A. Somit, 205-24. New York: Palgrave Macmillan.

Chess, C. 1998. Fearing fear: Communication about agricultural biotechnology. *AgBioForum* 1 (1): 17-21.

Einsiedel, E.F., and J. Medlock. 2005. A public consultation on plant molecular farming. *AgBioForum* 8 (1): 26-32.

Ekman, P., and R.J. Davidson, eds. 1994. *The nature of emotion: Fundamental questions*. New York: Oxford University Press.

Fischhoff, B., and I. Fischhoff. 2001. Public's opinions about biotechnologies. *AgBioForum* 4 (3 and 4): 155-62.

Fox, M.W. 1992. *Superpigs and wondercorn: The brave new world of biotechnology . . . and where it all may lead*. New York: Lyons and Burford.

Gaskell, G., N. Allum, W. Wagner, T.H. Nelson, E. Jelsoe, M. Kohring, and M.W. Bauer. 2001. In the public eye: Representations of biotechnology in Europe. In *Biotechnology 1996-2000: The years of controversy,* ed. G. Gaskell and M.W. Bauer, 53-79. London: Science Museum.

Gaskell, G., P. Thompson, and N. Allum. 2002. Worlds Apart? Public opinion in Europe and the USA. In *Biotechnology: The making of a global controversy,* ed. M.W. Bauer and G. Gaskell, 351-75. Cambridge: Cambridge University Press.

Hallman, W.L., W.C. Hebden, C.L. Cuite, H.L. Aquino, and J.T. Lang. 2004. *Americans and genetically modified food: Knowledge, opinion and interest in 2004.* Publication no. RR-1104-007. New Brunswick, NJ: Food Policy Institute, Cook College, Rutgers University.

Hoban, T.J. 1997. Consumer acceptance of biotechnology: An international perspective. *Nature Biotechnology* 15 (March): 232-34.

Kahneman, D., P. Slovic, and A. Tversky. 1982. *Judgment under uncertainty: Heuristics and biases.* New York: Cambridge University Press.

Kershen, D.L. 2004. Legal liability issues in agricultural biotechnology. *Crop Science* 44: 456-63.

Kingdon, J.W. 1984. *Agendas, alternatives, and public policies.* New York: HarperCollins.

Kirk, D.D., and K. McIntosh. 2005. Social acceptance of plant-made vaccines: Indications from a public survey. *AgBioForum* 8 (4): 228-34.

Krimsky, S., and R. Wrubel. 1996. *Agricultural biotechnology and the environment: Science, policy and social issues.* Urbana, IL: University of Illinois Press.

Lerner, J.S., R.M. Gonzalez, D.A. Small, and B. Fischhoff. 2003. Effects of fear and anger on perceived risks of terrorism. *Psychological Science* 14 (2): 144-50.

Lewis, M., and J.M. Haviland-Jones. 2004. *Handbook of emotions,* 2nd ed. New York: Guilford Press.

Marcus, G.E., W.R. Neuman, and M. MacKuen. 2000. *Affective intelligence and political judgment.* Chicago: University of Chicago Press.

Midden, C., D. Boy, E. Einsiedel, B. Fjaestad, M. Liakopoulos, J.D. Miller, S. Ohman, and W. Wagner. 2002. The structure of public perceptions. In *Biotechnology: The making of a global controversy,* ed. M.W. Bauer and G. Gaskell, 203-23. Cambridge, UK: Cambridge University Press.

NRC (National Research Council). 2002. *Environmental effects of transgenic plants: The scope and adequacy of regulation.* Washington, DC: National Academies Press.

–. 2004. Biological confinement of genetically engineered organisms. Washington, DC: National Academies Press.

NSF (National Science Foundation). 2004. *Science and engineering indicators 2004.* Arlington, VA (NSB 04-01), May.

OSTP (Office of Science and Technology Policy). 2002. *Proposed federal actions to update field test requirements for biotechnology derived plants and to establish early food safety assessments for new proteins produced by such plants: Notice. Federal Register* 67 (149): 50577-80.

Pew Initiative on Food and Biotechnology. 2003. *Public sentiment about genetically modified food.* Report by Mellman Group, Inc./Public Opinion Strategies for the Pew Initiative. Washington, DC: Pew Initiative on Food and Biotechnology. September. http://pewagbiotech.org/research/2004update/overview.pdf.

–. 2004. Factsheet: Genetically modified crops in the United States. http://www.pewtrusts.org/news_room_detail.aspx?id=17950.

Rosenwald, M.S. 2005. Syngenta says it sold wrong biotech corn. *Washington Post.* March 23: E01. www.washingtonpost.com/ac/wp-dyn/A58449-2005Mar22.

Schattschneider, E.E. 1975. *The semisovereign people: A realist's view of democracy in America.* New York: Harcourt Brace Jovanovich College Publishers.

Schubert, J.N., P.A. Stewart, and M.A. Curran. 2002. A defining presidential moment: 9/11 and the rally effect. *Political Psychology* 23 (3): 559-83.

Slovic, P. 1992. Perception of risk: Reflections on the psychometric paradigm. In *Social Theories of Risk,* ed. S. Krimsky and D. Golding, 117-52. Westport, CT: Praeger Press.

Stewart, P.A., and A.J. Knight. 2005. Trends affecting the next generation of US agricultural biotechnology: Politics, policy and plant-made pharmaceuticals. *Technology Forecasting and Social Change* 72 (5): 521-34.

Stewart, P.A., and W.P. McLean. 2005a. Fear and hope over the third generation of agricultural biotechnology. *AgBioForum* 7 (3): 133-41.

–. 2005b. Public opinion toward the first, second and third generations of plant biotechnology. In *Vitro Cellular and Developmental Biology—Plant* 41 (6): 718-24.

Stewart, P.A., and A.A. Sorensen. 2002. Federal uncertainty or inconsistency? Releasing the new agricultural-environmental biotechnology into the fields. *Politics and the Life Sciences* 19 (1): 77-88.

Taylor, M.R., J.S. Tick, and D.M. Sherman. 2004. *Tending the fields: State and federal roles in the oversight of genetically modified crops*. Washington, DC: Pew Initiative on Food and Biotechnology.

USDA-APHIS (US Department of Agriculture–Animal and Plant Health Inspection Service). 2003a. Field testing of plants engineered to produce pharmaceutical and industrial compounds. *Federal Register* 68, no. 46: 11337-40.

–. 2003b. *United States Department of Agriculture pre-briefing for reporters on USDA's federal register notice on field testing of pharmaceutical-producing plants*. March 6. http://www.usda.gov/news/releases/2003/03/084.htm.

–. 2003c. Biotechnology regulatory services: Compliance and enforcement. http://www.aphis.usda.gov/brs/compliance.

USDA-OIG (US Department of Agriculture–Office of Inspector General). 2005. *Audit report: Animal and Plant Health Inspection Service controls over issuance of genetically engineered organism release permits*. Audit 50601-8-Te December.

Wright, S. 1994. *Molecular politics: Developing American and British regulatory policy for genetic engineering*. Chicago: University of Chicago Press.

8
Forestalling Liabilities? Stakeholder Participation and Regulatory Development

Stuart Smyth

The adoption of genetically modified (GM) crops continues to expand each year. James (2006) shows that the global adoption of GM crops grew by 20 percent between 2003 and 2004, 11 percent between 2004 and 2005, and by 13 percent between 2005 and 2006. James estimates that global production of GM crops totalled 250 million acres in 2006 and that production occurred in twenty-two countries. This production in 2006 included five countries within the European Union, namely the Czech Republic, France, Germany, Portugal, and Spain.

With the first decade of GM crop production just completed, studies are being released showing substantial positive impacts. For example, the commercial production of GM crops globally has generated a net benefit of US$19 billion over the nine years between 1996 and 2004. If the benefits of second crop production in Argentina are factored in, the level of benefit rises to US$27 billion (Brookes and Barfoot 2005). In addition to the economic benefits derived from GM crops, there has been a substantial reduction in the environmental footprint created by intensive production agriculture that has resulted from the use of GM crops. Brookes and Barfoot (2005) have estimated that since 1996 the cumulative net reduction in the environmental footprint is 14 percent, due to fewer pesticide applications. They claim that during this period there has been a reduction of 6 percent in the total volume of active pesticide ingredient applied to agriculture crops. The demand for, and benefit of, GM crops continues to increase.

At the same time, the rejection of this technology has also progressed, to the point that Denmark and Germany have passed legislation that places the entire financial liability of GM adoption on the producers that adopt the technology (Smyth and Kershen 2006). In late 2005, the Danish government instituted a compensation scheme whereby producers wanting to grow GM crops will have to pay a tax of 100 Danish kroner per hectare, which will be used to compensate any potentially disadvantaged organic or conventional crop producer.[1] German law imposes civil liability on producers of GM crops for the death, injury, impairment of health, or property damage of other persons resulting from the properties of a genetically engineered organism. In November 2005, citizens of Switzerland voted 56 percent in favour of establishing a five-year moratorium on the use of GM crops in that country. While it is difficult to determine if the opposition to biotechnology is increasing, it is possible to argue that opposition is become more entrenched.

There is substantial literature on the relationship between broader participation in, and transparency of, policy development and consumer support of the ultimate policy outcome (Barling et al. 1999; Einsiedel 2005; Knoppers and Mathios 1998; Rowe and Frewer 2000). Generally, the greater the level of involvement and openness, the greater the level of policy support, whether the policy favours or opposes GM crop development. Such is the case in Denmark, where citizens have been very engaged in the policy process regarding the production of GM crops in Denmark. The aforementioned Danish regulations are largely derived from the Danish public's aversion to GM crop technologies. One area that remains to be explored is the relationship between greater inclusivity and commercialization liabilities.

Commercialization liabilities can be defined as scientific or socio-economic arguments for seeking financial compensation for actions resulting from the commercial production of a new technology. This would be instances where the science behind the regulations was faulty, harm was suffered, and lawsuits occurred. An ongoing court case in Saskatchewan exemplifies the socio-economic claims for liability, where the province's organic producers are attempting to seek financial compensation for the alleged loss of organic canola markets.

In this chapter I examine the participation of stakeholders in the development of regulations for the initial GM crops approved for commercial production in Canada, as well as the process used to develop regulations for an emerging technology: plant molecular farming (PMF). In spite of regulator efforts to be inclusive and to consult with industry, liability challenges remain. What has created particular challenges for regulators and industry alike is the emerging issue of socio-economic liabilities.

Inclusiveness and Stakeholder Participation

Policy development theorists have suggested that the greater the inclusiveness and transparency of the policy development process, the more likely the policy process results in robust social outcomes. Modern society is a society of regulation and one that, for the most part, views increased regulations as a positive (Phillips, Smyth, and Kerr 2006). What remains to be explored is the relationship between inclusiveness and commercialization liabilities. The involvement of diverse groups with different interests is an important step in potentially forestalling commercialization liabilities.

The development of regulatory frameworks has increasingly moved toward greater inclusivity. Does greater consultation and engagement result in a more robust regulatory framework in terms of anticipating potential problems, including liability challenges?

The report of the External Advisory Committee on Smart Regulation (Government of Canada 2004) recognized the importance of facilitating greater inclusion in the development of regulatory policy. The report maintains that increased openness, inclusivity, and public consultations are commitments of the government: "A regulatory culture that emphasizes and encourages openness, transparency and inclusiveness is a prerequisite for building public trust in Canadian regulation and the integrity of the process" (136).

Increasingly, inclusion and participation have become common standards for public policy and are being incorporated into international agreements (e.g., the Convention on Biological Diversity). However, inclusion is becoming difficult to achieve given the dense and complex web of actors and relationships involved in assessing risks and managing liabilities. Even when governments commit to maintaining full inclusion and participation, the sheer volume and complexity of the data, methods, and procedures for evaluating new technologies are making it very difficult to figure out the sources and uses of evidence. Compounding this is the frequent cross-referencing of evidence between domains and the all-too-frequent imposition of secrecy provisions triggered by "confidential business information."

Given the complexity of regulatory development for emerging technologies and the broader knowledge bases required, policy development processes are increasingly more integrative of business, academic, and civil society organizations. While this process arguably produces regulations that can be more effective, what has not been done is to examine whether these new processes are able to reduce or minimize the occurrence of marketplace liabilities once commercialization has taken place.

Developing Regulations for First Generation GM Crops

The development of Canadian regulations for the initial GM crops was based on a science evidence base and all subsequent regulatory changes have continued to be based on this. The regulations are based on the end product that is established, not on the process used to create the product. To this end, Canada developed regulations for plants with novel traits (PNTs). Plants that are classified as PNTs are plants that have been modified via genetic engineering or mutagenesis, as well as plants that do not have a history of production and safe consumption in Canada.

Table 8.1

Legislation governing biotechnology

Agency	Product	Act
CFIA	Plants with novel traits Novel fertilizers and supplements Novel livestock feeds Veterinary biologics	Seeds Act Fertilizers Act Feeds Act Health of Animals Act
Environment Canada	All animate products of biotechnology for uses not covered under other federal legislation	Canadian Environmental Protection Act, 1999
Health Canada	Novel foods Pest control products	Food and Drug Act Pest Control Products Act

Source: CFIA 2005a

The regulation of products created via biotechnology is the responsibility of the Canadian Food Inspection Agency (CFIA), Environment Canada, and Health Canada (see Table 8.1). Using the novel foods guidelines, the CFIA is responsible for plants, animal feeds, fertilizers, and veterinary biologics. The Office of Food Biotechnology has been established within the CFIA to coordinate the safety evaluation of novel foods. Environment Canada acts as a regulatory safety net for products of biotechnology, with a regulatory mandate for all animate products of biotechnology for uses not covered under other federal legislation. Environment Canada regulates biotechnology within the scope of the 1999 Canadian Environmental Protection Act. Through the Food and Drugs Act, Health Canada oversees the regulation of foods, drugs, cosmetics, medical devises, and pest-control products. All safety assessments are conducted based on scientific principles developed through expert international consultations with the World Health Organization, the UN Food and Agriculture Organization, and the Organisation for Economic Co-operation and Development (Harrison 2001).

All novel trait products, before receiving registration approval, are thoroughly tested by CFIA, Environment Canada, and Health Canada officials using scientific principles. Officials from all departments work together on a new variety application. Officials do not redo the scientific experiments and research information that is submitted by the applying company; rather, they analyze all of the data that are submitted and may redo portions of the experimentation to corroborate results. Frequently, government officials will ask the submitting company to provide them with additional information on specific segments of the application, which may result in the company conducting further scientific experiments. Upon review of the information, the variety is accepted if all conditions are met and rejected if any condition is not deemed to be acceptable.

The Scientific and Governance Approach to the Initial Genetically Modified Crop Regulations

By 1988, transformed plants with new transgenic traits were available and ready for testing. All new varieties of grain and oilseeds generally flowed through the same system in Canada, with higher levels of oversight on those that involve novel traits. Private and public breeders were responsible for managing their research programs as long as the materials remained in isolated conditions (e.g., in laboratories or under glass). Once the breeder had developed a cultivar that was stable and unique and was ready to have it examined for registration, the formal system took over.

The first step was to make an application to the regulator–at that time, Agriculture Canada, now the CFIA–to undertake confined field trials to test the varieties in the environment. The agency examined the traits involved and established guidelines for the trials, including isolation distances, weed control provisions, and auditing. The trials were designed to provide the evidence to evaluate the environmental risks of the new cultivar and to assess its agronomic merit (e.g., yield, disease resistance, time to maturity, quality, and other traits). This involved developing with the academic community and industry a regulatory directive on the biology of the species. The regulators also needed to see evidence on the characterization of the transformation system, the nature of the carrier DNA, genetic material delivered to the plant, the components of the vector, and a summary of all genetic components. In addition, the regulator required an array of data to assess the inheritance and stability of the genetic modification (e.g., Mendelian segregation) and a description of the novel traits (e.g., Southern analysis and qualitative ELISA analysis of the gene expression levels).[2]

Once confined field trials were authorized, they were undertaken following a strict set of guidelines and standards, which, though national in application, were drawn from international evidence of the appropriate risk management procedures and the latest international biosafety evidence. Although the regulators were responsible for auditing and enforcing the rules on trials, these trials were usually managed directly by the research firm or by a contractor (in Canada, the various research farms operated by Agriculture Canada have managed many of the trials under contract with the companies). Table 8.2 demonstrates the rapid growth in field trials that took place leading up to the first commercialization of GM canola in

Table 8.2

Confined field trials conducted in Canada, 1988–94

		Trial locations (%)	
Year	Total trials	Prairies	Saskatchewan
1988	14	93	43
1989	44	80	66
1990	76	74	64
1991	174	89	61
1992	298	85	44
1993	489	82	44
1994	774	87	47

Source: CFIA 2006

1995. The table also shows that much of the early research was done in the Prairies, especially Saskatchewan.

By 1992, the companies conducting field trials had gathered enough data to demonstrate intergenerational stability, agronomic efficacy, and commercial promise and began to develop their regulatory package of evidence to present to the regulators to assess the safety of the products. In Canada, this required extensive data on the toxicity of the novel gene products (e.g., a series of toxicity studies with humans, animals, and non-target species). The product proponents also had to provide scientific studies on the nutritional aspects of the novel trait and plant for both humans and livestock and comparisons of the amino acid sequences of the novel trait to known allergen proteins. Finally, the proponents were required to provide a package of studies on the environmental impact of the novel traits on soil, weeds, wild relatives, and non-target organisms. McHughen (2000) published a histogram of the volume of data required to satisfy regulators of the health and safety of transgenic crops (in his case, a transgenic flax variety)–the pile of studies and reports exceeded three feet for the transgenic product, versus an average of about thirty pages for a conventionally bred variety. Researchers at the University of Saskatchewan, where the transgenic flax was being developed, were frustrated by the lengthy process required to obtain variety approval.

The results of the field trials, and food, feed, and environmental reviews were then examined by the appropriate regulators. In Canada, Health Canada undertook the food-safety review, while the environmental and animal health reviews were conducted by forerunner agencies of the CFIA. In each case, they had enabling standards embedded in legislation or regulation that needed to be made specific for each product or technology.

That process involved extensive negotiation between the regulator and the product proponent, supplemented with reference by the regulator to experts in other national regulatory systems and to those outside the regulatory system.

Finally, the initial GM crop varieties (three new trait canola varieties) were assessed by a committee of researchers operating under the authority of the Seeds Act–they analyzed the candidate varieties against a "check" variety–and then the committee authorized them for sale to farmers. Most other countries do not have this regulatory step. At that point, a blended public-private quality-control system took over. Most new varieties were multiplied either in contra-season locations (e.g., Chile, Australia, southern US states) or by registered seed growers in Canada. The Seeds Act identifies the Canadian Seed Growers' Association (CSGA), the umbrella organization of provincial growers associations, as the official seed pedigree agency in Canada responsible for certifying all new varieties and ensuring they meet the standards set in the act. They establish the rules for producing foundation, registered, and certified seeds and then undertake the audits and conformity measures of growers that are necessary to ensure the standards are achieved.

The significance of the above processes lies in the traditional governance system being based on an extensive horizontally based public-private regulatory system. As discussed, the Seeds Act is the first point of quality assurance, as new varieties must on average at least equal the quality of previous varieties. For the initial GM canola varieties, this was administered by the Western Canada Canola/Rapeseed Recommending Committee, a committee of more than thirty public and private breeders that evaluates new varieties against the check varieties and recommends varieties for release. This standard has been backstopped by the Canola Council of Canada trademark on canola, which specifies that products must have at most 2 percent erucic acid and 30 micromoles of glucosinolates per 100 grams of dried meal. Furthermore, the new variety approval system periodically raises the bar for new varieties by choosing a new criterion as the base for standards, e.g., setting oil and meal properties, yields, and disease resistance. Once the varieties are approved, the Canadian Seed Trade Association manages the seed multiplication system, specifying the tolerances for substandard materials, and the retail seed business, by overseeing the sale of seeds by registered name.

The important observation from the development of regulations for first generation GM crops is that the regulators were openly accepting of industry stakeholder involvement and willing to allow the CSGA to be fully in charge of overseeing the commercialization of this new crop technology. At the time of commercialization, in the mid-1990s, the regulators operated from the perspective that once the scientific risks were satisfactorily addressed, the technology was allowed to proceed unimpeded by regulatory interference. As will be shown later in the chapter, this has not been the case with the development of PMF crop technologies.

The Process of Developing PNT Regulations in Canada

Following the first field trials that took place in 1988, the initial workshop designed to address the regulatory framework that would be required for the successful commercialization of these new crop technologies was held. The Canadian Agricultural Research Council (now known as the Canadian Agri-Food Research Council) organized a workshop titled Regulation of Agricultural Products of Biotechnology, for December 4-7, 1988, in Ottawa. There were 108 attendees: 65 from the various government agencies and research organizations, 27 from numerous private industry firms, 14 from Canadian universities, and 2 from the US Department of Agriculture.

The workshop was divided into four major sessions. The first focused on the need for regulating new technologies and the existing state of regulations in the United States, the European Union, and Canada. The second session focused on the regulatory situation in Canada and discussed the Seeds Act, the Plant Quarantine Act, the Animal Disease and Protection Act, the Feeds Act, the Fertilizers Act, and the Pest Control Products Act. The third focused on the science behind GM plants, animals, and microbes, while the fourth was a multi-stakeholder perspective on issues and concerns about the regulation of biotechnology.

This workshop produced the following key recommendations, which provided the basis for the development of the PNT regulatory framework:

- Those plants that possess characteristics or traits sufficiently different from the same or similar species should require an assessment of risk.
- The product, not the process, should be regulated.
- The categories of novel herbicide tolerance, novel pesticidal properties, novel stress tolerances, and novel compositional changes were raised as categories of concern (CARC 1988).

Over the next three years, the director of the Animal and Plant Health Directorate within the Food Production and Inspection Branch of Agriculture Canada would convene periodic ad hoc meetings of varying representation to discuss pertinent issues. It was not until 1992 that a formalized structure was put in place to deal with the regulatory changes that would be required. It was decided in April 1992 that a standing advisory committee would be established with these objectives:

- provide information and guidance on the regulation of plant biotechnology

- assist in the development of a consistent regulatory approach
- assess and evaluate regulatory requirements for field testing and commercialization of genetically modified plant material (Agriculture Canada 1992, 1).

The Plant Biotechnology Advisory Committee would have formalized representation as well, and the membership consisted of representatives from eleven agriculture-related societies and associations.[3]

In 1992, the Food Production and Inspection Branch contracted with Dr. Wally Beversdorf (chair of the Department of Crop Science at the University of Guelph) to develop draft protocol and assessment criteria for unconfined release of PNTs. This initiative, when taken in consideration with a series of workshops held across Canada between January and March 1993, produced a draft set of regulations. A workshop was held November 8-10, 1993, in Ottawa to discuss the draft regulations. The draft regulations were shared with attendees before the workshop and were titled *Assessment Criteria for Determining Environmental Safety of Genetically Modified Plants.*

The workshop consisted of numerous presentations from a variety of key stakeholders who had been invited to participate. Not only was there a diverse representation of speakers, there was a very broad spectrum of attendees participating in the workshop discussions. The key objectives of the workshop were identified as building consensus on the approach to regulate PNTs; maintaining consistency with existing regulations; sharing of information; and developing working relationships (CFIA, undated).

The principles of the federal regulatory framework were identified as building on existing legislation and institutions; upholding health and environmental safety standards; harmonizing with national priorities and standards; using risk-based assessments and methodologies; assessing products, not processes; and developing a favourable climate for investment, development, and innovation by adopting sustainable products and processes.

Much attention was given to the concept of substantial equivalence for products derived through biotechnology, especially the difference between "familiarity" and "substantial equivalence." Familiarity was described as an extensive knowledge of factors relating to the production of a particular crop species that allows for decisions pertaining to safety to be made, whereas substantial equivalence would apply to those new crop types whose safety could not be identified from a standard risk assessment. In essence, it was proposed that products found to be not substantially equivalent would be subject to an additional level of regulation.

Additional discussions stressed that time should not be a factor in approving these new technologies; rather, safety should be the chief concern, and safety should be proven regardless of the time taken to do so. International acceptance of the products was raised as an issue of importance for these new crop technologies. Participants acknowledged that there are risks, but the focus must be given to identifiable, science-based risk, not hypothetical, socio-economic risks. Finally, the importance for regulatory harmonization within North America was discussed, to ensure that the ultimate regulatory framework would not hinder the competitiveness of this emerging industry in Canada.

In March 1994, a follow-up workshop was convened by the Plant Products Division (PPD) of Agriculture Canada with the objective of reviewing the regulations that had been redrafted following the November 1993 workshop. The PPD wanted an expert review of specific guidelines that had been developed to provide for the unconfined release of new canola and flax varieties prior to the regulations being released. The PPD called the members of the Plant Biotechnology Advisory Committee (see footnote 3 for the list of members) together to provide their insights. Although this workshop focused on unconfined release for canola and flax, the group also held initial discussions on unconfined release of corn, soybeans, potatoes, and *rapa* canola. The feedback from the committee members was incorporated into the document released for public comment, and the biotechnology industry believed it had cleared the final regulatory hurdle.

In June 1994, those involved in the development of the regulatory framework were surprised and alarmed when the Feed Section of the Plant Products Division informed participants of the Plant Biotechnology Advisory Committee that they would initiate the development of their own regulatory guidelines for the use of GM plant material in livestock feed. The Feed Section sought participation by experts to form the Advisory Committee on Transgenic Plants as Livestock Feed. Membership of this committee consisted of sixteen experts with varied expertise in biotechnology.[4] A workshop was held in Ottawa on September 21-22, 1994, and draft guidelines were developed and sent out for public comment on November 22, 1994. Twenty-five comments were received and incorporated into a revised draft of the regulations that was sent back to the members for comment. Comments were due back from committee members by March 15. It is interesting to note that the first decision document approving unconfined release of a PNT crop (AgrEvo's ammonium-tolerant canola) was approved on March 10, 1995, five days before the end of the comment period and well in advance of the final approval of regulations developed by the Feed Section. The second was given to Monsanto's Roundup herbicide-tolerant canola on March 24, 1995. Figure 8.1 summarizes the involvement of the various regulatory actors over the process.

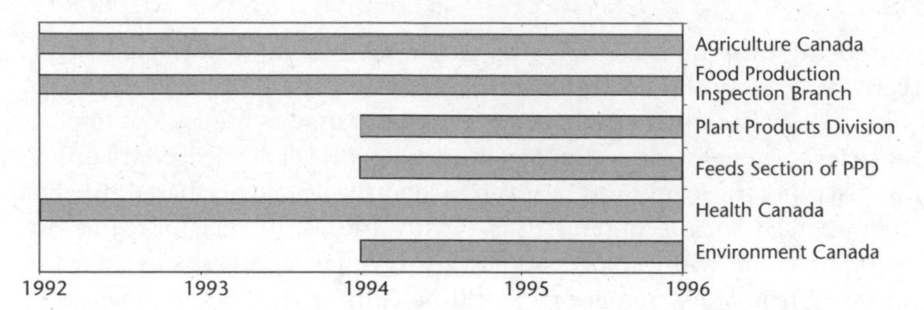

Figure 8.1 Regulatory actors involved in developing regulations on plants with novel traits

The Process of Developing Plant Molecular Farming Regulations in Canada

The CFIA has defined plant molecular farming (PMF) as the use of PNTs for the production of a pharmaceutical or industrial adventitious trait rather than for food or feed use. In 2003, the CFIA stated that "since PNTs for molecular farming may present greater potential for environmental or human health risks, the Government of Canada may put even more stringent restrictions on the use of these novel plants than for other PNTs" (CFIA 2003). Officials within the CFIA have acknowledged that there are public concerns about the potential risks associated with PMF, yet the CFIA clearly maintains that market issues are not a consideration when undertaking a product approval.

The regulation of PMF has been a challenge for the regulatory community in Canada from the onset of this innovative technology. Criticisms of the regulatory process have come from both industry and the NGO (nongovernmental organization) community. Three events have been held to address the development of a PMF regulatory framework: the first was the CFIA Multi-Stakeholder Consultation on Plant Molecular Farming held October 31-November 2, 2001; the second was the Technical Workshop on the Segregation and Handling of Potential Commercial Plant Molecular Farming Products and By-Products held March 2-4, 2004; and the third was the Technical Workshop on Developing a Regulatory Framework for the Environmental Release of Plants with Novel Traits Intended for Plant Molecular Farming held March 2-3, 2005.

The first multi-stakeholder consultation had two objectives. The first was to review and develop specific guidelines for field testing and precommercial release of PMF; the second was to discuss the procedures for Health Canada's regulatory involvement (CFIA 2001).

What was striking about this consultation was that a wide variety of stakeholders were invited to participate in the workshop: seven public interest groups; thirteen from agriculture and agribusiness; thirteen from academia; nine from the biotechnology industry; and provincial and federal government representatives. Having sent seventy-five invitations to forty-four firms, organizations, and government representatives, there were fifteen participants in the consultation process.

The workshop focused on three main themes for presentation and discussion: human health, animal health and feed, and environmental questions. Numerous risks were identified by the participants, including exposure risks for those involved in handling the plants, comingling with like food crops and possible consumption by humans, and emotional reaction of the public to the concept of utilizing human genes in plants. To mitigate these risks, it was generally agreed that the CFIA should use a case-by-case assessment strategy and that there was a need for good agricultural practices and long-term monitoring. Some participants found it acceptable to use food and feed crops, while others felt that food crops should be avoided. It was suggested that where plants about which limited physiological knowledge exist, use of these plants should not be allowed. Many were of the opinion that the current segregation systems were not rigorous enough to handle the bulk movement of PMF crops and that these systems needed to be strengthened by the use of licensing requirements and independent third-party auditing. Response on risk assessment varied widely, as some were of the opinion that full risk assessments were needed for any and all applications and conditions, while others believed that the circumstances of the technology application should dictate whether a full risk assessment was required.

Discussions were also held on outcrossing and gene flow, and mitigation strategies such as isolation distances, border rows, and genetic use restriction technologies were suggested. Entry into the food-supply chain was noted as a concern, and it was suggested that persistence in soil and danger to non-target organisms could be controlled by the use of tissue-specific or post-harvest inducible promoters. The majority believed that PMF should be permitted outside containment, only if toxicological and persistence studies showed that risks were within defined tolerance levels. There was a suggestion that the CFIA develop environmental monitoring protocols and procedures. The majority also agreed that the post-harvest land use restrictions in place for field trials should be applied to any commercial production.

The result of this multi-stakeholder workshop was that the CFIA released an interim amendment to the previously issued Regulatory Directive 2000-07. This amendment was targeted specifically at confined field trials of PNTs for PMF. *One of the key recommendations was that companies engaged in this form of research should avoid using major food or feed crops as a platform for their technology.* Other key recommendations were that crops that are pollinated by bees for commercial honey production should be avoided, while the use of fibre crops was encouraged. Also recommended was that plant platforms be amenable to confinement strategies and the use of genetic mechanisms to minimize environmental exposure. Isolation distances were increased, and a CFIA official had to witness the disposal and destruction of residual plant material. It was evident that the stakeholder participants were increasingly conscious of the potential risks and their ramifications in the public arena.

The amendment to Regulatory Directive 2000-07 achieved the first objective of the workshop; the second objective, procedures for Health Canada's regulatory involvement, was also met, as the discussions were a factor in Health Canada's revision of the Novel Food Guidelines that took place in 2002.

The second event coordinated by the CFIA occurred March 2-4, 2004, in Ottawa. Participants consisted of thirty representatives from various Canadian government departments and agencies, seventeen from the agriculture and biotechnology industries, three from the US Department of Agriculture, and one from the NGO community, for a total of fifty-one participants.

The format and structure of this workshop differed considerably from the previous event. For this workshop, a consultant was hired to produce a background paper on the potential segregation of PMF products. The report was reviewed by the participants, and discussions were held on the report. The report noted that the grain-handling and transportation system was not equipped for, nor could the system guarantee the segregation of, PMF material. It was suggested that the design of each identity preservation system, whether for PMF crops or not, must be collectively developed by stakeholders (including growers, processors, and consumers) and not by the PMF industry or regulators. Additional discussions focused on regulatory oversight and highlighted the importance of developing validated methods to help establish levels of risk and determine a regulatory trigger for PMF.

The result of this workshop was that it allowed the CFIA to proceed with two initiatives. The first was that the CFIA and Health Canada would work toward developing a memorandum of understanding that would outline "each organization's respective responsibilities in the assessment and authorization of PNTs intended for PMF" (CFIA 2004, 6). The second initiative was to work in collaboration with the US Department of Agriculture's Animal and Plant Health Inspection Service (APHIS) to develop a code of best practices for PMF.

The third workshop, held March 2-3, 2005, involved fifty-seven participants: twenty-nine from Canadian government departments and agencies, twenty-one from the agriculture and biotechnology industry, four from academia, and three from American government agencies (CFIA 2005b).

The workshop was designed to solicit feedback from participants on draft guidelines for assessing environmental safety and to address some of the technical biosafety issues that need to be resolved prior to a regulatory framework being completed. One of the key regulatory requirements seen by industry as holding back the advancement of commercial growth in the industry is the CFIA regulation that limits the maximum production to 125 acres spread over fifty growing sites across Canada. The PMF industry views the development of a comprehensive regulatory framework for large-scale commercial production as one way for Canada to position itself as a world leader in the development of PMF technologies (BIOTECanada 2004).

The participant feedback on the preliminary draft guidelines was mixed. Some participants expressed the view that the CFIA was being reasonable in its approach to drafting the guidelines, while others failed to see how the proposed guidelines could be flexible enough to accommodate large-scale production. It was suggested that one way to increase the flexibility of the proposed regulations would be to have environmental factors assessed using the CFIA's existing criteria for environmental safety assessments (i.e., weediness, gene flow, plant pest potential, impact on non-target organisms, and impact on biodiversity). Another suggestion was that isolation distances for the production of PMF crops could be undertaken using the existing distances established by the Canadian Seed Growers' Association (CSGA). Perspectives were equally split on this: some felt that the CSGA distances were adequate, some maintained that they were just a starting point for the development of more rigorous distances, while yet others argued that CSGA distances should not be used at all in the development of distances for PMF crops. Isolation distances are especially important to the production of PMF crops, and there needs to be a mitigation of both outflow and inflow of materials. The economic value of a PMF crop would be negatively impacted should it become contaminated with a neighbouring producer's crop. This is an important issue, as the production of a PMF crop is based on delivering the highest level of purity possible, and outside contamination would reduce the premiums paid to the producer.

Using the feedback from the workshop, the CFIA revised the draft guidelines. Following this revision, a web-based consultation was carried out in early 2006 to garner input from a wider audience.[5]

This assessment of the regulatory development process highlights an interesting governance challenge. There is clearly a larger number of federal regulatory departments and divisions involved with the development of emerging technologies, such as plant molecular farming. Eight branches from three departments are involved in the regulatory development process, suggesting greater bureaucratization and increasing knowledge-base requirements. Figure 8.2 shows the involvement of the various regulatory actors.

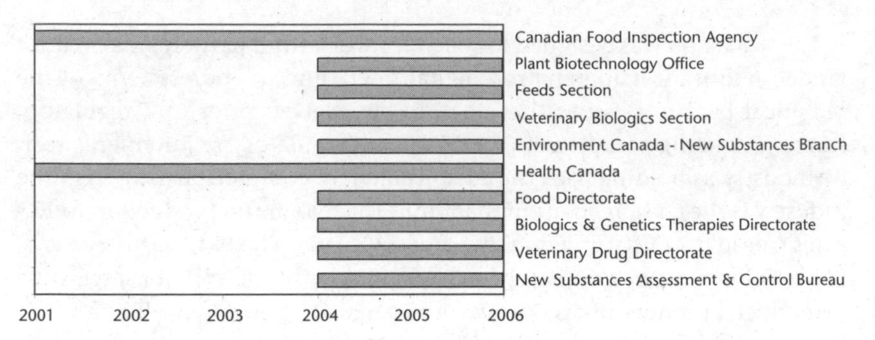

Figure 8.2 Regulatory actors involved in developing regulations on plant molecular farming

Another challenge is the regulatory expansion into market issues. Despite CFIA claims that market factors are not part of the regulatory process, handling and segregation are clearly market issues, and the workshop that was held on this topic raises questions about the expanding mandate of the agency–or at least its broader scope of interest. The regulatory authority of the various government agencies ceases at the point of variety approval, and it is up to the privately run commercial grain-handling system to accommodate new crop varieties. This extension of regulatory mandate beyond science-based risk assessment for variety approval to consider market issues is another notable change in the regulatory process.

A final observation of the difference between the two consultation processes is that the degree of inclusion expanded for the PMF process. Regulators within the CFIA made greater efforts to invite a wider and more diverse group of participants to the workshops. This would indicate that the CFIA is taking steps to improve on the PNT process and has identified inclusiveness through broader stakeholder participation as a major component of this process. This broadened input into regulatory governance is notable because it was not a significant feature of the earlier policy development process on plants with novel traits.

Observations about Liability Management

Given the high level of participation in the regulatory development process for PNT crops, one would expect that acceptance would be high for these crops and that the products would enter the market with little to no fuss. This has not been the case, and liability challenges have arisen.

The liability management issue can be approached from two divergent perspectives: scientific and socio-economic. From the scientific perspective, the management of commercial liabilities has been handled virtually without incident. The development of the PNT regulations was sufficiently thorough, so that the only issue that has developed over more than a decade of commercial production is a few cases of cross-pollination between GM crops and non-GM crops. The control of unwanted GM volunteers has been easily and cost-effectively managed (Smyth, Khachatourians, and Phillips 2002), and to this point there has been but a single court case regarding a lawsuit against a seed development firm in Canada over the undesired presence of GM canola. Louise Schmeiser filed claim with the Provincial Court of Saskatchewan (Humboldt) seeking Cdn$140 in compensation for the removal of volunteer canola from an allegedly organic garden. Following trial proceedings, the court ruled that Schmeiser had failed to prove her claim and denied her compensation (*Schmeiser v. Monsanto*).[6]

There have been problems from an intellectual property perspective with some of the GM crop varieties in that some producers adopted the stance that the research used to develop these new varieties was essentially worthless and not deserving of a return on investment, and these producers decided to infringe on the firm's intellectual property. The best documented case of this is the Supreme Court of Canada decision in favour of Monsanto Canada (in the case of *Monsanto v. Schmeiser* [2004]), whereby the court found that, under Canadian law, firms are legally entitled to hold intellectual property on the process to create a new plant variety and to enforce the protection of that intellectual property.[7] The issue of patent infringement is not unique to GM plants but is common in all forms of manufacturing industries.

Scientifically, the knowledge about the potential for cross-pollination between canola varieties was well known and ultimately expected and, from a financial liability perspective, has not created difficulties for the biotechnology industry. However, difficulties have developed on the socio-economic side.

The most significant socio-economic liability that arose from the commercial release of GM crops was the moratorium that was enacted in the European Union from 1998 to 2004. This disrupted international trade markets and ultimately led to Canada and the United States filing a case of trade violation to the World Trade Organization. In early 2006, this case was ruled in favour of North America, and the claims of trade violations were validated. Although the moratorium itself was lifted in 2004, it is a slow process for the development of biotechnology in Europe. Europe has started to give regulatory approval to new traits but has yet to come to any consensus on the production of GM crops. In fact, producing GM crops has always been contentious within the European Union.

Socio-economic liability claims have also occurred within Canada. Saskatchewan organic farmers brought a class action on behalf of all organic grain farmers in the province against Bayer CropScience, Inc., and Monsanto Canada, Inc., for damages arising from the commercialization of transgenic canola.[8] Although the plaintiffs at first focused on loss of organic certification, they refocused their claims to concentrate on loss of the European market for organic canola and the loss to organic farmers of the practical option of choosing to grow organic canola.

In a lengthy opinion, Judge G.A. Smith discussed and ruled on the main liability action put forward by the organic growers: negligence and nuisance. The judge denied the negligence claim on three grounds. First, the organic growers did not adequately demonstrate that markets rejected organic products or that farmers declined to grow organic grains because of the adventitious presence of transgenic material. Second and more serious, the organic growers had not proven a causal relationship between GM and organic canola, especially physical damage to property. Third, even if the first two reasons could be overcome, pure economic loss was insufficient for recovery under a negligence claim.

Judge Smith maintained that the companies would be liable only if an escape of a substance came from property the companies owned or controlled. The judge ruled that a commercial release of the approved transgenic crop could not reasonably be considered an escape.

An important consideration in the decision was the evidence presented that some organic producers in Saskatchewan were still able to produce and export canola that met all the international standards for organic canola, especially those pertaining to comingled GM content. Based on these producers' ability to produce and export organic canola, clearly the ability to produce, and the export market for, organic canola had not been destroyed by the commercialization of GM canola, as was claimed. The growers have the right to appeal, but at the time of writing, this case had not yet proceeded to the higher court.

Socio-economic liabilities resulting from the initial release of GM crops appear to be mostly along the lines of potentially foregone economic rents. Some producer groups opposed to the use of GM crops have tried (unsuccessfully) to extract financial resources from large multinational seed development firms in an arguably somewhat frivolous manner.

The relationship between broader stakeholder inclusion and commercialization liability would appear to be very important for scientific aspects of liability claims. The PNT regulations have not received serious challenges and were developed within a framework that has substantial scientific merit. Inclusion and the participation of experts are crucial to the long-term success of a regulatory framework.

The relationship between regulatory inclusiveness and socio-economic-based liabilities is murkier. What is unique to this situation is the potential to sue the commercializers of the new, innovative technology for market loss. As has been witnessed by the court cases to date, socio-economic damages are difficult to substantiate and even more difficult to prove in a court of law. Although it will be possible to determine what forms of opposition may exist within a society to a new, innovative technology, it will not be possible to predict, with any reliability, the potential for socio-economic liability lawsuits.

The limitations of science-based regulation become clear when socio-economic questions arise. Stakeholder involvement may help to anticipate these socio-economic challenges and assist in developing a regulatory framework that can inspire public confidence.

Lessons in Liability Management for Emerging Technologies

The management of commercialization liabilities is most appropriately done through the development of comprehensive regulations. The development of PNT status for the initial commercialization of GM crops has helped to ensure that liabilities have been minimized. The process that is ongoing for the development of PMF regulations is even more rigorous than those developed previously, and the federal regulatory bodies are clearly cognizant of anticipating possible commercialization liabilities. Regulators have tried to do this through broader stakeholder inclusion and participation.

What we can conclude from the stakeholder involvement in the development of PNT regulations is that stakeholders providing input and advice on science-based risks is a crucial process. The stakeholders were able to review the proposed regulations and work with the regulators to ensure that the final outcome created a framework that ensured that no science-based liabilities developed once the technologies were commercialized. The challenge that faces the industry stakeholders and regulators in developing a PMF regulator framework is the looming issue of socio-economic liabilities. The process that regulators are using attempts to address not just science-based issues but also potential market issues in order to forestall the occurrence of socio-economic liabilities.

Considerable experience has been gained from earlier regulatory development, which is now being applied to the PMF process. Federal regulators and industry firms are recognizing that social issues need to be identified within the process, and although this is not without its own challenges, it is important that they have even been recognized. Efforts to address marketplace liabilities have been added to the regulatory development process and, when coupled with recent court opinions, make such consideration unavoidable.

There are three important observations about liability management for emerging technologies. First, the issue of containment will be crucial, especially for aspects of the PMF industry. If a firm were producing a toxic protein within a plant, any escape of this plant would have a devastating impact on this emerging industry. Given the low volume required for much of the PMF industry, it is expected that this demand could be met by greenhouse production. Where this is not possible, a closed-loop production system will need to be developed and monitored.

Second, it will be of paramount importance for both emerging industries and those responsible for regulation to ensure that food safety is maintained. Although Canadian consumers have experienced food-safety challenges in the past several years from mad-cow disease and avian influenza, these challenges can be viewed as coming from outside the typical production system. Bovine spongiform encephalopathy and the avian influenza strains have entered the Canadian agriculture system from outside our national borders. Therefore, consumers still trust, for the most part, the domestic production systems. Should a food-safety challenge arise because of an emerging technology from within the Canadian production system, the consumer response could be expected to be vastly different.

Third, regulations are a key component for minimizing commercialization liabilities, and a key feature of this is for the regulatory bodies to publicly promote and increase awareness of Canadian regulations to enhance public confidence. The Canadian public is indicating through surveys that it wants to see a strong regulatory presence for emerging technologies such as PMF.

The transparency of regulatory development has improved immensely over the past fifteen years, as has the participation of stakeholder groups in regulatory development. Such shifts can be an important base for avoiding the greatest liability that emerging technologies could face–the erosion of consumer trust.

Notes

1 At the time of writing (June 2006), 100 Danish kroner was the equivalent of Cdn$19.
2 See http://www.inspection.gc.ca/english/plaveg/bio/subs/subexe.shtm for a detailed list of what this involves.
3 Membership consisted of the Canadian Seed Growers' Association, Canadian Seed Trade Association, Crop Protection Institute, Genetics Society of Canada, Canadian Society of Agronomy, Confederation of Canadian Faculties of Agriculture and Veterinary Medicine, Plant Biotechnology Institute, Expert Committee on Weeds, Canadian Society on Botany, Canadian Phytopathological Society, and National Seed Potato Bureau.
4 Environment Canada was sporadically engaged in the regulatory process at this time.
5 This online consultation was held in May 2006.
6 *Schmeiser v. Monsanto Canada*, No. 18/04, para. 51 (Sask. Provincial Ct. June 15, 2005). It would appear to be incredibly ironic that Schmeiser would seek compensation from Monsanto for removing unwanted volunteer canola when her husband was found guilty of illegally growing more than a thousand acres of Monsanto's canola, the very canola Schmeiser sought removal of.
7 *Monsanto v. Schmeiser* (2004), 1 S.C.R. 902, 2004 SCC 34 (May 21, 2004).
8 *Hoffman v. Monsanto Canada Inc.* (2005), S.J. No. 304, 2005. SKQB 225, Q.B.G. No. 67 of 2002 J.C.S.

References

Agriculture Canada. 1992. Letter of invitation to participate. Mimeograph.
Barling, D., H. de Vriend, J.A. Cornelese, B. Ekstrand, E.F.F. Hecker, J. Howlett, and J.H.

Jensen. 1999. The social aspects of food biotechnology: A European view. *Environmental Toxicology and Pharmacology* 7 (2): 85-93.

BIOTECanada. 2004. BIOTECanada Molecular Farming Committee white paper. Submitted to Industry Canada on October 28, 2004.

Brookes, G., and P. Barfoot. 2005. GM crops: The global socio-economic and environmental impact—The first nine years, 1996-2004. *AgBioForum* 8 (2 and 3): 187-96.

CARC (Canadian Agricultural Research Council). 1988. *Proceedings of the Workshop on Regulation of Agricultural Products of Biotechnology.* Ottawa: Queen's Printer.

CFIA (Canadian Food Inspection Agency). 2001. *CFIA Multi-stakeholder Consultation on Plant Molecular Farming.* Report of proceedings. http://www.inspection.gc.ca/english/plaveg/bio/mf/mf_cnsle.shtml.

–. 2003. *Plant molecular farming.* http://www.inspection.gc.ca.

–. 2004. *Technical Workshop on the Segregation and Handling of Potential Commercial Plant Molecular Farming Products and By-Products.* http://www.inspection.gc.ca/english/plaveg/bio/mf/worate/woratee.shtml.

–. 2005a. *Regulation of agricultural biotechnology in Canada.* Ottawa: Queen's Printer.

–. 2005b. *Technical Workshop on Developing a Regulatory Framework for the Environmental Release of Plants with Novel Traits Intended for Plant Molecular Farming.* http://www.inspection.gc.ca/english/plaveg/bio/mf/fracad/techenviroe.pdf.

–. 2006. *Confined field trials 1988-2005.* http://www.inspection.gc.ca/english/plaveg/bio/confine.shtml.

–. Undated. *Workshop on Regulating Agricultural Products of Biotechnology.* http://www.inspection.gc.ca.

Einsiedel, E. 2005. Public perceptions of transgenic animals. *Revue scientifique et technique* 24 (1): 149-57.

Government of Canada. 2004. *Smart regulation: A regulation strategy for Canada.* Report of the External Advisory Committee on Smart Regulation. http://www.brad.ac.uk/irq/documents/archive/Canada_Smarter_Regulation_Report.pdf.

Harrison, B. 2001. Regulating novel foods in Canada. Presentation to the conference "The Convergence of Global Regulatory Affairs: Its Potential Impact on International Trade and Public Perception," Saskatoon.

James, C. 2006. *Global status of commercialized biotech/GM crops: 2006.* International Service for the Acquisition of Agri-biotech Applications briefs no. 35-2006. International Service for the Acquisition of Agri-Biotech Applications. http://www.isaaa.org/RESOURCES/PUBLICATIONS/BRIEFS/35/EXECUTIVESUMMARY/default.html.

Knoppers, B., and A. Mathios. 1998. *Biotechnology and the consumer.* Dordrecht, The Netherlands: Kluwer Academic Publishers.

McHughen, A. 2000. *Pandora's picnic basket: The potential and hazards of genetically modified foods.* Oxford: Oxford University Press.

Phillips, P.W.B., S. Smyth, and W.A. Kerr, eds. 2006. *Governing risk in the 21st century: Lessons from the world of biotechnology.* Hauppauge, NY: Nova Science Publishers.

Rowe, G., and L. Frewer. 2000. Public participation methods: A framework for evaluation. *Science, Technology and Human Values* 25 (1): 3-29.

Smyth, S., and D.L. Kershen. 2006. Legal liability regimes from comparative and international perspectives. *Global Jurist Advances* 6 (2): 1-94.

Smyth, S., G.G. Khachatourians, and P.W.B. Phillips. 2002. Liabilities and economics of transgenic crops. *Nature Biotechnology* 20 (6): 537-41.

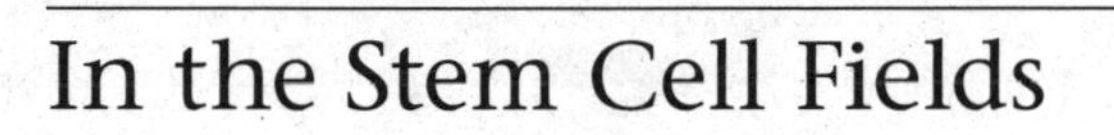

In the Stem Cell Fields

9 When Human Dignity Is Not Enough: Embryonic Stem Cell Research and Human Cloning in Canada

Tania Bubela and Timothy Caulfield

Infringement of human dignity is often posited as a primary justification for curtailing, regulating, or prohibiting areas of scientific inquiry, particularly in new and emerging technologies (Caulfield and Brownsword 2006; Caulfield and Chapman 2005; Caulfield 2003). This use of human dignity as a form of general condemnation of controversial biotechnologies such as human embryonic stem cell (hESC) research, human cloning, and human gene patents marks a significant departure from its use as the grounding principle of international human rights.[1]

In this chapter we explore the formulations of human dignity in the Canadian debates on stem cell research and their role in shaping Canadian legislation (*The Assisted Human Reproduction Act [AHRA]*) and the statutory prohibition of reproductive and therapeutic cloning, otherwise known as somatic cell nuclear transfer (SCNT). Canadian law permits hESC research using surplus embryos from in vitro fertilization (IVF) but prohibits the creation of embryos for research purposes and research on embryos beyond fourteen days. All research protocols must pass through the Assisted Human Reproduction Agency, also known as Assisted Human Reproduction Canada, which was established in 2006. That agency is partly modelled on the United Kingdom's Human Fertilisation and Embryology Authority. In addition, guidelines put out by the Canadian Institutes of Health Research apply to any hESC research funded by the CIHR. Those guidelines also limit hESC research to the use of surplus IVF embryos.

This *AHRA* is among the most restrictive laws worldwide, with the greatest penalties, contrary to assertions of policy makers in Canada. In 2004, at the Senate Standing Committee on Social Affairs, Science and Technology in Ottawa, Senator Morin incorrectly stated that "the immense majority of countries who have passed legislation recently do ban both reproductive and therapeutic cloning." Indeed, regulatory moves are united only with regard to the banning of reproductive cloning. A recent survey of laws found that although SCNT is not permitted in seventeen of thirty countries, it could be permitted in up to thirteen countries. Indeed, a parliamentary committee in Australia has recently recommended that SCNT be made a regulated activity, and Germany, formerly strongly opposed to hESC, has also recommended allowing SCNT (Wroe 2005). SCNT is legislatively permitted in Belgium and the United Kingdom and is unregulated outside of federal funding in most states in the United States. In this chapter, which builds on previously published work (Caulfield and Bubela 2007), we explore the justifications for the Canadian law, particularly as those justifications relate to human dignity as a rationale for the ban on SCNT.

How Dignity Is Used in Science Policy Debates

The traditional concept of human dignity acknowledges the intrinsic worth of all humans; it is an engine of individual empowerment, reinforcing individual autonomy and the right to self-determination (Brownsword 2003; Caulfield and Brownsword 2006).[2] It gained strength after the Second World War and, as a result of Nazi atrocities, was used to "contain the power of states over persons"(Brownsword 2003). Some commentators would even argue that human dignity in the context of individual autonomy is the only appropriate normative use of the idea of dignity (Macklin 2003).

The human rights concept of human dignity remains an important component of current bioethical debate, as it emphasizes individual autonomy. This view is often seen in contrast to a utilitarian standpoint, which emphasizes the public good. A rights-based framework exists in traditional research ethics and some science policy. For example, discourse on the regulation of human genetic databanks balances individual rights to autonomy and choice, articulated through consent processes, against potential public health benefits. A 2003 World Health Organization report suggests that genetic databases create the need to balance "human dignity and human rights as against public health, scientific progress and commercial interests in a free market" (WHO 2003). And, of course, the concept of human dignity permeates research ethics policy (e.g., the Helsinki Declaration).

More recently, however, in the context of biotechnology, human dignity has taken on a new meaning, one of general condemnation of a given technological development (Beyleveld and Brownsword 2001; Brownsword 2003). Employed by what has been termed the "dignitarian alliance," a group informed by a wide variety of religious, philosophical, and political traditions,[3] this new formulation demands regulatory restraint whenever biotechnology is seen as compromising human dignity.

This conception of human dignity as constraining force now serves as an all-purpose justification for opposing technological development and often amounts to little more than a general sense of social unease, a politically palatable articulation of the "yuck factor" (Caulfield and Chapman 2005). Dworkin (2000, 443) has also noted that reliance on basic principles such as human dignity, in part, reflects an amorphous social angst, an expression of "some deeper, less articulate ground for that revulsion, even if they have not or perhaps cannot fully articulate that ground, but can express it only in heated and logically inappropriate language, like [a] bizarre reference to 'fundamental human rights.'"

Such use generally lacks rigour when employed in science policy because it is rarely explained how to judge whether human dignity has been infringed or degraded. For example, a 2002 report by the President's Council on Bioethics is titled *Human Cloning and Human Dignity: An Ethical Inquiry,* yet it fails to conceptualize human dignity or address the specific ways in which human cloning may impinge on human dignity.

Costa Rica's past proposal to the UN for an international treaty banning cloning stands as another good example. The Costa Rican draft convention sought to "ensure respect for the dignity and basic rights of the human being" in the face of the "threat posed by experiments in the cloning of human beings."[4] Likewise, in the area of stem cell research, opponents refer to the dignity implications as a rationale for limiting research on human embryos. In Europe, it is an underpinning of the *ordre public* (public order and morality) restriction of patent law, which has been used to deny patents on cloning technologies and hESCs. The United Nations Educational, Scientific and Cultural Organisation's Universal Declaration on the Human Genome and Human Rights recommends a ban on "practices which are contrary to human dignity, such as reproductive cloning."[5] Both Japan's 2001 stem cell research guidelines and Canada's recent legislation covering research involving human reproductive material claim the protection of human dignity as a primary objective of the regulatory regimes. Canada's primary research ethics document, the *Tri-Council Policy Statement,* declares that the "cardinal principle of modern research ethics, as discussed above, is respect for human dignity."[6]

The new dignitarianism employed by the "dignitarian alliance" represents a range of philosophical and religious views (Caulfield and Brownsword 2006). Brownsword (2004a, 19) explains that the appeal of dignitarianism is its support of "conservatism, constancy, and stability," as neither utilitarian nor human rights perspectives can. Instead, rights based articulations of dignity have "acted as such a dynamic and progressive force for change, it might seem incongruous to enlist this same idea in defence of the status quo. Yet, as the pace of biotechnology accelerates, we should not underrate the felt need to find a way of registering our concern that we should at least have the opportunity to hang onto those parts of the human condition that are familiar and reassuringly 'human.'

Rather obviously, the notion of 'human dignity' fits this particular bill" (Brownsword 2004a).

The Problem with "Human Dignity"

The problem with the concept of "human dignity" lies in a lack of shared values that inform the various dignitarian positions. This does not mean that these are not valid positions worthy of public debate, but in a pluralistic society it is questionable whether they should be used as a foundation for regulatory policy. As stated by Beyleveld and Brownsword (2001, 680), "From any perspective that values rational debate about [new technologies], it is an abuse of the concept of human dignity to operate it as a veto on any practice that is intuitively disliked."

The policy-making role of human dignity becomes more questionable when it is used as a form of general condemnation for two principal reasons: one concerning regulatory effectiveness and the other regulatory legitimacy (Brownsword 2004b; Caulfield and Brownsword 2006). First, it is a precondition of effective regulation that the rationale and purpose of the regulation are clearly stated. Yet, so long as human dignity is a contested concept, open to different interpretations, regulatory references to "respect human dignity" cannot give clear guidance to regulators and the scientific community whose activities are curtailed. In a culture where researchers and funding bodies are anxious to achieve regulatory compliance, such uncertainty can operate as a chilling factor.

Second, most modern societies are pluralistic and, accordingly, consensus is difficult to obtain, whether about human dignity or other complex social and ethical issues introduced by scientific innovations. There is not even agreement about the foundation of human dignity–whether it is faith-based or secular–let alone what human dignity entails. If regulators declare a position that reflects the views of only the dignitarian alliance, for example by prohibiting SCNT, that will seem partial and undemocratic, particularly if more compelling justifications are absent. As a result, the use of dignity will not necessarily represent a broadly accepted social value, but, instead, it may express a particular worldview that may not reflect majority opinion.

Concerns about the uses of dignity are magnified when dignity becomes the justification for a state's use of its coercive criminal-law powers to prohibit particular avenues of scientific inquiry. This is the most extreme collision between the autonomy of individuals engaged in scientific inquiry, either as researchers or participants, and the coercive power of the state based on a constraining vision of human dignity. One can argue that criminal law should be used only where there is a degree of social consensus, as stated in the Supreme Court of Canada decision *R. v. Labaye* [2005] 3 S.C.R. 728. However, because human dignity is viewed as a foundational concept, its use may imply a degree of social consensus that simply does not exist (Caulfield and Chapman 2005). If something is said to infringe human dignity, one would expect a degree of agreement that this is so.

Finally, if public debate about biotechnology is to be framed in terms of issues relating to human dignity, the public will become fully engaged with the key questions only if the advocates of respect for human dignity are absolutely clear in declaring their meaning.

Dignity Shaping Science Policy in Canada

In Canada, the recent enactment of the Assisted Human Reproduction Act has criminalized a range of research endeavours, including human cloning, SCNT, and germline alteration. One of the primary declared principles under which the prohibitions were enacted is a need to protect and promote human dignity and rights, along with other more usual justifications for regulatory restrictions, such as health and safety.[7] This position in Canada is particularly interesting because, as a pluralistic society, it does not have a dominant religious tradition or historical precedent (such as fears of eugenics in Germany and Austria) that can clearly explain the bans. As such, it stands as an interesting case study of the policy-making process and the debate about the extent to which a society should use the law to enforce its moral judgments, a question framed by Lord Patrick Devlin (1965) in *The Enforcement of Morals*.

In the Canadian context, the question can be asked, whose dignity? Not all participants in the Canadian, and indeed in the international, stem cell research debate agree on the moral status of the embryo and, as such, they cannot reach agreement on the degree to which research involving the destruction of embryos challenges human dignity. Nevertheless, the moral status of the embryo, at the political level, became the primary driver behind the Canadian policy debate. This justification leads to the conclusion that Canadian policy on SCNT has been shaped by one constituency's view of the moral status of the embryo. Here, we explore Canada's decision to criminalize human cloning, both therapeutic and reproductive, and examine the role that "dignity" played in shaping that policy.

SCNT is a research method that involves the creation of an "embryo" (or, as some would prefer, an "embryo-like entity") by fusing an enucleated egg and genetic material from another individual. This is the technique that has been used to create a number of animal clones, including Dolly the sheep (Wilmut et al. 1997) and, most recently, a dog (BBC News 2005). It is hoped that the technique could be used to create stem cell lines that have the same genetics as the individual in need of therapy or transplantation, thus avoiding immune rejection problems (i.e., it is hoped that SCNT will allow for the creation of genetically matched replacement tissue). It is also a technique that may help in the development of xenotransplantation techniques and the creation of cell lines that could act as models for the study of particular diseases.

In reality, SCNT remains a relatively marginal scientific activity. A handful of research teams have a significant profile as a result of their work on SCNT (Medical News Today 2005). Some of this work, such as that done by Korean Woo-Suk Hwang, has generated considerable scientific and ethical controversy (BBC News 2005; Magnus and Cho 2005). In addition, its potential scientific and therapeutic value continues to be debated. It has been noted, in fact, that there is a degree of inappropriate hype surrounding all aspects of this research area (Theise 2003). Nevertheless, because of its divisive nature, and because some countries have chosen to take legislative action, SCNT provides an ideal opportunity to explore the challenges associated with making science policy in a morally contested area.

Legislative Background

Canada's law has a long history. In 1993, the Royal Commission on New Reproductive Technologies issued a report titled *Proceed with Care*.[8] This report called for the regulation and oversight of reproductive technologies throughout Canada. The focus, like the United Kingdom's 1984 Warnock Report before it, prepared for the Department of Health and Social Security, was on emerging reproductive technologies, such as prenatal diagnosis, and the regulatory emphasis was on, understandably, the safety of women and children.

As a result of this report, a series of federal bills were put forward, starting with Bill C-47 in 1996 (Caulfield 2001; Campbell 2002; Bernier and Gregoire 2005). This law died as a result of the calling of a federal election in 1997.[9] After a legislative hiatus, the issue resurfaced in May 2001, when the government released the draft *Proposals for Legislation Governing Assisted Human Reproduction*.[10] The unusual step of releasing draft legislation was meant to stimulate national debate and serve as a framework for discussion and analysis by the House of Commons Standing Committee on Health. The committee heard submissions from interested parties ranging from academics to professional associations. Its report, *Assisted Human Reproduction: Building Families,* was submitted in December 2001 and recommended that legislation be introduced on a priority basis.[11] On May 9, 2002, the government introduced Bill C-56, An Act respecting assisted human reproduction, in the House of Commons (Caulfield 2002).[12] This proposed bill died on an Order Paper. After the parliamentary summer break, this bill was reintroduced as Bill C-13 on October 9, 2002, received second reading on October 9, 2002, was sent back to committee, received third reading on October 28, 2003, but died on the Order Paper when an election was called.[13] Finally, Bill C-6, a virtual replica of Bill C-13, was introduced and passed first, second, and third reading in the House of Commons on the same day, February 11, 2004.[14] It also rapidly passed in the Senate and received Royal Assent on March 29, 2004. Bill C-6 became the Assisted Human Reproduction Act.[15]

The main debates in Parliament, therefore, occurred after the introduction of Bill C-56. Although the proposed legislation dealt with all aspects of assisted reproduction, from commercial surrogacy to regulation of IVF clinics and storage of human reproductive material, human reproductive cloning and embryonic stem cell research dominated the political debates. Indeed, the press frequently referred to the legislative proposal as the "cloning and stem cell bill" (Greenaway 2002, A2). All votes in Parliament were free votes, meaning that MPs were free to vote according to their conscience and not according to party dictates.

The act creates two broad categories of activities: those banned by a statutory prohibition and those controlled and that may be carried out only in accordance with the legislation and the regulations.[16] SCNT and reproductive cloning fall into the former category, as do creating embryos for research purposes, embryo research beyond fourteen days, sex selection, germline alteration, creating chimeras or hybrids, commercial surrogacy, and the selling and buying of sperm and ova.[17] The maximum punishment for engaging in prohibited activities is a fine of Cdn$500,000 and ten years' imprisonment.[18]

The act also sets up a regulatory scheme, including the creation of the Assisted Human Reproduction Agency to oversee the licensing of regulated activities. The agency's objectives are "(a) to protect and promote the health and safety, and the human dignity and human rights, of Canadians, and (b) to foster the application of ethical principles" (s. 22). Regulated activities include embryonic research, non-commercial surrogacy and the donation of sperm and ova, and the storage and creation of human reproductive material. The act, however, takes the unusual step of requiring the minister of health to lay proposed regulations before both houses of Parliament, to be reviewed by the appropriate committee of each house. The agency has not yet been created, so no regulations are currently available for review (Nelson 2006). The final noteworthy feature of the act is that it is to be reviewed by Parliament within three years of the creation of the Assisted Human Reproduction Agency. This is due to occur next year, but there will likely be little political will to reopen the debate in the current Conservative minority Parliament.

Approach—Tracing Rationales

Our approach was designed to trace the evolution of rationales for the ban on SCNT from the bureaucracy (Health Canada), which sponsored the legislation, through consideration by the Standing Committee on Health, to the parliamentary debates and the legislation itself. We focused primarily on debates by elected officials, particularly the parliamentary debates.

We reviewed the justifications found in publicly available documents from Health Canada, and in news releases announcing the review of the draft legislation.[19] Although other internal policy documents may have been available, our selection represents the public rationale of policy development in Health Canada. Likewise, our analysis of submissions by interested parties to the committee and the committee report reflect publicly stated positions.

We reviewed transcripts of oral submissions to the committee, starting with the introduction by the minister of health on May 3, 2001, of the legislation to be reviewed, and ending with the release of the report in December 2002.[20] We searched for the word *dignity* in all submissions and catalogued its use. We then reviewed all references to SCNT in the report and documented the committee's justification for the ban.

The parliamentary analysis, the largest component of our study, involved the creation of a full-text database of all references to stem cells in the parliamentary Hansard. We assigned codes to all of the text using the qualitative analysis software ATLAS.ti. This enabled us to glean an overall sense of the main topics and the tone of the debates. Our analysis focused on the main debates on the legislation from October 2002 to June 2004. Here we present an analysis of the use of the codes that dealt with the two conflicting notions of dignity: dignity in the context of human rights or dignity as empowerment, and dignity as a constraint, used by the dignitarian alliance.

Rationales for the Ban on SCNT

What are the most common arguments against SCNT, and how do they play out in the context of the Canadian law and the associated parliamentary debate?

The Bureaucracy

Health Canada's rationales for the ban are relatively straightforward: commodification concerns, social consensus, and the protection of the health and safety of Canadians. There are no explicit references to the moral status of the embryo as a justification for a regulatory response. Indeed, even the stated commodification concerns are relatively non-specific (though one can presume they include concerns relating to the moral status of the embryo). To be fair, these issues may have been discussed during Health Canada meetings and, as such, may have informed the writing of the documentation.

Some of the position of the bureaucracy was further elucidated through testimony of senior Health Canada bureaucrats to the Standing Committee on Health. Ian Shugart, assistant deputy minister of Health Canada, supported the ban, noting the "special status" of human embryos.[21] Others testified that there is no current need for SCNT and no international consensus on its use. They noted health and safety concerns but related these issues to the need for more research on point. The Health Canada Advisory Committee on the Interim Moratorium on Reproductive Technologies indicated that there had been serious deliberation of whether to include the prohibited activities in the list of regulated activities instead. However, Madeline Boscoe, executive coordinator of the Canadian Women's Health Network, in describing the policy process that led to the line in the sand between regulated and prohibited activities, said: "Often when we're trying to develop policy in what looks like chaos, these first statements, these first ideas, become a reference point for those who come after us to say, that's what they were thinking about; they were talking about human dignity there."[22]

Further, Rhonda Ferderber, director of the Special Projects Division of Health Canada's Policy, Planning and Priorities Directorate, explained: "You were asking a question around what kinds of activities or criteria we engaged in when we specified what would be prohibited versus this set of controlled activities. Very simply, all the prohibitions are grounded in ethical principle. That was a key criterion in helping us decide what would be on the list of prohibited activities. These are the principles of human dignity and integrity of the human genome, and issues in opposition to the commercialization of reproduction."[23]

Submissions to the Standing Committee

Only fourteen out of seventy-one submissions directly addressed SCNT, though a total of thirty-seven witnesses indicated their position in questioning by committee members. The minister of health, Allan Rock, set the stage in introducing the draft legislation for consideration by the committee, indicating that "as to what we believe should be prohibited, there's broad consensus on practices that are found on this list. Every one of the prohibitions we propose, such as human cloning, is on that list of prohibitions because it's inconsistent with human dignity." The minister made it clear that science could not be left "to its own devices" and "pursued to whatever ends it may reach. If science alone is to predominate then I think what we're leaving out are human values that are shared by all Canadians." That common awareness and acceptance includes a recognition that "we have to go beyond ourselves, beyond just a technical laboratory analysis, and recognize that we're dealing with a dignity of human life and that we should govern the procedures accordingly."[24]

Surprisingly, there were few champions for SCNT in the scientific community. Only twelve witnesses supported a regulatory environment for SCNT. Researchers and scientific experts focused their submissions on the need for embryonic stem cell research per se and seemed concerned that such research could be severely limited or even banned. Researchers seemed willing to concede the issue of SCNT if that ensured that the moratorium on embryonic stem cell research would be lifted and the legal and regulatory uncertainty in that field of research could be settled.[25]

The president of the Canadian Institutes of Health Research, Alan Bernstein, stated that there was a risk that SCNT would not work because of accumulated errors.[26] He added that the therapeutic benefits of SCNT remain theoretical and Canada's position represents the middle ground on the moral status of the embryo, providing "a balance that is respectful of, on the one hand, human dignity, the value of the individual, and the perspective of those who are against the use of human embryos in any kind of research, and those with serious life-threatening diseases." It was not explained, however, why a criminal ban was the appropriate legal instrument to address the issues associated with SCNT.

The fundamental position of the Canadian Institutes of Health Research and two other Canadian granting councils was explained as follows: "If you look at our tri-council policy statement, you will realize that one of our paramount principles is human dignity. If you look at chapter 9, you will realize that we are against it. It's not ethically acceptable to create an embryo only for research. But we also realize that the use of embryos in this type of research might be very beneficial for the community, for people who are suffering from those diseases, so we are willing to use embryos that are not in a family plan of those who gave this material."[27]

Professor Bartha Maria Knoppers of the Faculty of Law, Université de Montréal, most clearly articulated the human dignity as empowerment argument. She testified that SCNT should not be criminalized but, rather, placed on a moratorium list to be reviewed under the regulations. In discussing the preamble, Professor Knoppers stated: "At the same time, I am a little disappointed that we use–and it is the fashion, it happens everywhere in the world, not just in Canada–the term human dignity everywhere in the text . . . We are therefore putting human dignity on the same level as basic rights even though we know very well that human dignity is the source of those fundamental rights. It is human dignity that is inherent to a person, and it is because of that dignity that we have defined basic rights. That is what allows us to implement basic rights."[28]

Only a few witnesses discussed the dignity of those who could benefit from potential therapies and treatments, though none directly in the context of SCNT. The only faith-based representative cautiously in favour of SCNT because of its potential to alleviate suffering was Rabbi Irwin A. Zeplowitz. Human dignity inheres in humans "because they are made in God's image." In his view, all human tissue ought to be treated with proper dignity, meaning that "we ought to ask individuals for their consent in using any body part, by-product, or cellular material, and second, that strict guidelines should be in place to control the commodification of such material. A strong case can be made from Jewish teachings for prudent concern regarding the commercialization of human reproductive technologies."[29]

Judaism views all scientific research and medical advancement in the area of human reproduction in the larger context of justice for the individual and community, human dignity, equitable distribution of resources, and compassion for those who want healthy children. Canadian law should allow for the advancement of science but tread cautiously so that in seeking to heal we do not lose sight of the value of all the variety of God's children.

Opponents made five main arguments: (1) the embryo has moral status as "human life" and therefore should not be instrumentalized or commercialized, (2) the moral harm outweighs any benefits from the research; (3) research alternatives using adult stem cells should prevail because these are more stable, do not pose the same rejection problems, and are less morally contentious; (4) the respect for human dignity should trump any research benefit associated with SCNT (these arguments implicitly presumed a moral status); and (5) the slippery slope argument that SCNT will lead to reproductive cloning.

The dignitarian position was primarily expressed by representatives of Christian groups, a total of nine witnesses, the single largest group that spoke to the issue of dignity. The common viewpoint among these was that life begins at conception, embryos should have full moral status, and therefore any research involving the destruction or creation of embryos for research purposes violated the dignity of human life, which includes embryos. This position was articulated by Bruce Clemenger, director of the Centre for Faith and Public Life, Evangelical Fellowship of Canada, who claimed a social consensus in Canada:

> We believe human beings are created in the image of God and have dignity and worth. This is our foundation for the principle of human dignity. There is no such thing as an unworthy life, because our worth is not determined by what we can do or accomplish, or even by the pleasure we experience, but rather by who we are in relation to God and to each other. It is not up to us to decide whether human beings have dignity or not. We therefore believe human life must be valued, respected, and protected through all its stages. This applies equally to people with disease or with disability. The affirmation of the dignity of human life is shared by Canadians and is reflected in a variety of ways in Canadian law.[30]

This viewpoint was shared by representatives of the Catholic Church. Archbishop Terrence Prendergast stated: "Now our major concerns. First, the major concern we have about this draft legislation is that while it deals with the beginning of human life, it does not define the human embryo as a human being or protect it with full moral status. The Catholic Church believes human life is God's most precious gift to us; each human being is created in the image of God; and each human being has inherent worth and dignity. Therefore, each human person must be respected and protected as a person, however small and fragile."[31]

The Report of the Standing Committee on Health

The only explicit justification for the statutory prohibition of SCNT in the report comes as an afterthought to a strong and unequivocal statement about reproductive cloning: "The Committee feels strongly that the potential adverse effects, whether physical, psychological or social, for the resulting children are sufficient reason to prohibit reproductive cloning. *In addition, 'therapeutic cloning' should be banned as it is unsafe and commodifies the embryo"* (emphasis added).[32] The report does not cite any evidence for why safety is a special issue in the context of SCNT, nor does it discuss the origins of this concern. It confirms that commodification is an issue in the context of SCNT, not because of concern for women's health or reproductive rights, but because of the moral status of the embryo. The report states, for example, that "the Committee . . . concurs that there must be a measure of respect and protection for the embryo that is based on its potential for personhood,"[33] and that the committee respects "the deep desire communicated to the Committee by many Canadians that human embryos and other 'reproductive materials' be accorded the respect and dignity which is their due."[34]

In justifying the use of statutory prohibitions, the report relies on social consensus and safety, without providing evidence for either, stating "an outright statutory ban signals more clearly that certain activities are either unsafe or socially unacceptable."[35]

Parliament

Our qualitative analysis of the parliamentary debates was designed to give us an overall sense of the tone of the debates on stem cell research, highlight the main features of the debate, and direct us to salient and representative quotes on the issues. With that in mind, the magnitude of the study–which included the coding of all parliamentary references to stem cell research from 1994 to 2004 and the characterization of 3,848 quotes–allows us to draw quantitative observations about the main features of the debates. We present a full analysis of the parliamentary debates in Caulfield and Bubela (2007). Here, we discuss parliamentary references to human dignity in the context of embryonic stem cell research, human cloning, and SCNT.

Dignity and the Moral Status of the Embryo The majority of references to dignity in the parliamentary debates were associated with the moral status of the embryo—that the embryo has an inviolable dignity, making embryonic stem cell research or SCNT unacceptable. These arguments had numerous premises: that humans or human life is imbued with innate dignity, and that embryos are human life and therefore the destruction of human life violated dignity and so is ethically impermissible. To cite one example, an MP suggested: "Embryonic stem cell research inevitably would result in the death of the embryo. Life would not go on. For many Canadians this would violate the commitment to respect human dignity, to respect integrity, and to respect human life."[36]

Some of the most significant categories of quotes explicitly granted full moral status to the embryo, referring to it as "life" or a "human being" whose destruction could not be tolerated. Indeed, the full lexicon and rhetoric of the anti-abortion lobby was used to personalize and humanize the embryo, stating, "this is a bill dealing entirely with what I would consider the life of a baby."[37]

The Canadian Alliance–the official Opposition and a relatively conservative political party–formalized the dignitarian position, calling for the overarching consideration of the Assisted Human Reproduction Agency to be "respect for human life," which would "require respect and protection for the human embryo not simply because of its potential but because of the fact that it is human life."[38] The Opposition called for a moratorium on all hESC research not because it "commodifies the embryo" but because of the destruction of the embryo, "which is contrary to the ethical commitment to respect human individuality, dignity, integrity, and life."[39]

These arguments may be contrasted against the government's position that met three goals in this legislation: "protect the health and safety of Canadians; ensure the appropriate treatment of human reproductive materials; [and] protect the dignity and security of all persons, especially women and children" (Minister of Health 1996). Here the dignity concerns women involved in IVF treatment and research, and the children born of such treatments.

The majority of references to commodification concerned other aspects of the legislation, such as IVF, surrogacy, and reproductive material used for research but not embryonic stem research specifically. However, approximately 10 percent concerned the commodification of the embryo or "human life," which included the embryo and the use of excess IVF embryos for embryonic stem research. Surprisingly, almost no reference was made to commodification as it related to the health and safety of women (e.g., that women will be exploited or inappropriately coerced into providing eggs for SCNT).

Arguments against embryonic stem research and, implicitly, SCNT were also couched as arguments in favour of focusing research on adult stem cells as the means of protecting the dignity of human life. For example: "Research in adult stem cell and umbilical cords has indicated many things. There are a number of ways we can deal with this kind of research in a manner that does not manipulate human life and does not deal with the creation of life or the destruction of such. I would encourage members to do everything we can to go down that path rather than the path of creating embryonic cells to be used as research, or the cloning of human beings."[40] Another argument raised concern about a slippery slope to reproductive cloning. These arguments, however, were most often put forward by people who also advocate a ban based on the moral status of the embryo. The slippery slope argument was used as a secondary line of support.

Social Consensus on Dignity and the Moral Status of the Embryo The issue of social consensus emerged frequently. But claims of social consensus were rarely based on polling information or other evidence. In other words, claims of social consensus were simply asserted or implied. And, in general, they were used as arguments against embryonic stem-cell research. There were very few references to SCNT, and most MPs, rightly or not, treated SCNT and reproductive cloning as one technology—stating simply that the Canadian public supported a ban on "human cloning."

Other MPs used letters from constituents and communication with Canadians to shore up their own dignitarian position on the moral status of the embryo: "I rise to present a petition signed by residents of Huron-Bruce. The petition deals with the proposed assisted human reproduction act . . . The petitioners are calling on parliament to enact legislation that respects the dignity of human life by completely prohibiting the destruction of human embryos. They request that parliament give consideration to providing financial resources and support for research into adult stem cell potential." However, one MP, recognizing the reality of public support, stated that he could not support embryonic stem cell research because he believes "human life starts at conception and includes an embryo," even though "if polls were taken today, most Canadians would support embryonic stem-cell research."[41]

Analysis of the Use of Dignity in Canada

It is no surprise that the dignitarian position premised on the moral status of the embryo was a dominant theme in the legislative debate. SCNT requires both the creation of an embryo and its destruction. For those who hold strong views regarding the moral status of the embryo, these activities are clearly unacceptable, as the embryo is viewed not as a cluster of cells but as an entity with moral status worth protecting (Javitt, Suthers, and Hudson 2005). Indeed, as revealed by the UN debate on human cloning, this is the ethical position that informed most national policies advocating a ban on SCNT.[42] Many, but not all, of the countries favouring a comprehensive ban on all forms of cloning, including SCNT, have a religious tradition that undoubtedly informed the policy position (Pattinson and Caulfield 2004).

Unlike the legislation in many of these nations, however, the Canadian legislation did not start as a policy founded on the protection of the embryo. On the contrary, most of the early documents, such as those from Health Canada, seem careful to avoid any reference to the moral status of the embryo as a justification for regulatory response. Indeed, during the October 2001 hearings, the chair of the Standing Committee noted this situation: "I'm finding this piece of legislation overall to be very shortsighted in the sense that, I think in order to avoid all the potential arguments around abortion and what happens at the other end, they want to limit the discussion to what happens prior to the implantation of an embryo into a woman. I understand it, but I find it kind of irresponsible."[43]

However, if one read only the parliamentary debates, the submissions to the Standing Committee, and the Standing Committee report, one would be compelled to conclude that this was the primary purpose of the legislation.

Underlying the position that SCNT infringes human dignity is the often-stated claim that there is a social consensus supporting a ban, a rationale that exists at every stage of the political process, though stated in dignitarian terms of social consensus that the embryo has dignity and is deserving of full moral status in the parliamentary debates. It is, in fact, the only consistent theme in the entire debate. For example, Health Canada documentation states that there is a "broad consensus."[44] The report from the Standing Committee on Health suggests that the banned activities are clearly "socially unacceptable," and numerous MPs implied general social angst.[45] Despite such claims, all of the Canadian survey research available at the time demonstrates that there was, in fact, no social consensus.

Public opinion surveys and focus groups are, of course, inherently limited methodologies (Nisbet 2004). However, in the context of the Canadian political debate, social consensus was used as a justification *for* a ban. In such circumstances, one could argue, the onus lies with those parties using consensus as a justification to establish the validity of the rationale–particularly when the proposed regulatory tool is a criminal ban.[46] Such evidence was either absent or highly equivocal, thus making the one consistent rationale for a ban decidedly suspect. Recently, the Supreme Court of Canada has stated that criminal law should be applied only when individuals "have violated values which Canadian society as a whole has formally endorsed," and they cannot "be convicted and imprisoned for transgressing the rules and beliefs of particular individuals or groups."[47] It is especially difficult to gauge social consensus in a pluralistic society, which can "function only with a generous measure of tolerance for minority mores and practices."[48]

Conclusion

Clearly, parliamentary debates are creatures of political strategy, opportunity, and compromise. Those MPs who feel strongly about the moral status of the embryo focused on that aspect of the debate and maintained a strongly dignitarian position. For those who may have felt differently, there was little, politically, to be gained by openly disputing the moral status rationales or even by introducing alternative positions. As such, the issue of whether concern for the moral status of the embryo is an appropriate justification for a criminal ban was never debated.

In addition, despite more than a decade of debate, the true justification for the policy has remained unclear, a point noted by several authors (Devolder 2005; Young and Wasunna 1998; Caulfield, Knowles, and Meslin 2004; Pattinson 2002). Section 2 of the legislation provides principles meant to inform the administration of the act, including the promotion of health, safety, and dignity. But these principles are so broad and open to interpretation that they provide little true guidance (Pattinson 2002). Recently, Bernier and Gregoire (2005, 529) stated: "We believe that the Canadian government and the standing committee on health, in expressing a strong opposition to therapeutic cloning, are being too rigid without providing sufficient reasons to support their position . . . We believe the Canadian government's position on therapeutic cloning should not be driven by doubts and dogmatic fears." Also, Campbell (2002, 85) notes that "until a more comprehensive legislative justification is articulated, Parliament's activities in this area will be perpetually scrutinized and challenged, thereby revoking attention from the more important social and scientific issues sure to arise in the area of reproductive technologies."

If the law were legally challenged, Canadian courts would be required to divine the purpose from the documents and political debates studied in this chapter.[49] If the law is based on the moral status of the embryo, as much of our analysis indicates, we are left wondering whether, in a pluralistic society, it is appropriate to base a criminal ban on a "single view of embryonic life" (Childress 2001, 161; Brownsword 2003)–particularly when it is clear that other faiths hold different perspectives and there is no evidence that a majority of the Canadian public wants the ban. This is not to say that moral status concerns are not legitimate or worthy of consideration (e.g., Deckers 2005). However, not all social concerns should give rise to a criminal ban. As Charo (2004, 311) states: "Moral angst is one thing; federal criminalization of research or medical practice is another."

Interestingly, the United Kingdom, a pluralistic liberal democracy with a parliamentary tradition similar to Canada's, underwent the same debates in the House of Commons and the House of Lords. In those debates, conservative and Catholic MPs and Lords strongly advanced the same dignitarian position, premised on the full moral status of the embryo (Brownsword 2004a). How, then, was the regulatory outcome so different? The United Kingdom allows SCNT, and, in fact, recently launched the UK National Stem Cell Network.

Our analysis of the United Kingdom debates (Bubela and Caulfield, unpublished data) indicates that the greatest difference between the Canadian and the British debates was the willingness of British MPs to expend political capital in defending and explaining embryonic development and stem cell research (also noted by Parry 2003). The quality of debates in the United Kingdom was extremely high, and indeed the parliamentary debate itself, though not necessarily its outcome, was held up as a model of vital and intelligent debate in other contexts (Bubela and Caulfield, unpublished data). Exaggerated and erroneous claims by the dignitarian alliance were immediately contradicted and debated by the secretary of health and other MPs, with impressive scientific credentials. By contrast, the Canadian debates were characterized by silence from the ruling party and a woeful ignorance of science. For example, one MP described SCNT as follows:

> The key issue involved in cloning, once the possibility of cloning merely for the purposes of reproduction has been eliminated, involves mostly therapeutic considerations. Let us imagine someone with Parkinson's disease. If human cloning were possible, an embryo could be produced from an adult cell from a patient and someone's egg. A few months later, the embryo, which would be implanted in a woman's uterus, would develop into a foetus genetically identical to the patient. The foetus is aborted, the brain cells are extracted and grafted onto the patient's brain, which will not reject them because they are identical to its own cells.[50]

Along similar lines, another stated: "There is a group of people, the Raelians, running around . . . their vision is to perpetuate human life by creating a clone . . . If we took one of his cells, extracted the nucleus and put it into an ovum, one could stimulate it electrically and allow it to grow. The so-called therapeutic clone would be to take the immature model of Mr. Speaker and extract an organ . . . killing the clone in the process. That is so-called . . . therapeutic cloning."[51]

Even the scientific expert witnesses before the Standing Committee seemed more concerned with ending the moratorium that was then in place for hESC than for fighting the further battle of garnering support for SCNT. No voice spoke out strongly in favour of SCNT in Canada (Bubela, DeBow, and Caulfield 2005). As a result, the scientific community has been presented with restrictive legislation based on an ambiguous premise; the scientific community has been forced into a pragmatic compromise to "accept more restrictions than are strictly required by their position to protect a more important moral goal"–the utilitarian- and rights-based goal of pursuing the therapeutic potential of hESC research, while sacrificing the presently less important avenue of SCNT (Pattinson and Caulfield 2004).

Finally, policy makers need to understand that these policy debates will have relevance beyond the debates associated with SCNT. This policy dialogue may stand as a precedent for future science policy and for future controversies that implicate the moral status of the embryo, a point noted by Robyn Shapiro (2003, 398): "Even beyond the lives that may be affected by legislative resolution of the embryonic stem cell research debate, law makers should be keenly aware that their action will more generally help to shape the law governing research and the freedom of scientific inquiry in this country."

Notes

The general discussion in this chapter is largely based on several published papers: Caulfield and Brownsword 2006; Caulfield and Chapman 2005; Caulfield 2003. It follows on from the more detailed analysis of policy rationales for the Canadian ban on Somatic Cell Nuclear Transfer presented in Caulfield and Bubela 2007. We would like to thank Vinny Kurata, Shaun Pattinson, Lori Knowles, Darren Shickle, Suzanne Debow, Peter Borszcz, and Thomas Moran for their assistance and insight, and Genome Canada and the Stem Cell Network for funding support.

1 The preamble to the Universal Declaration of Human Rights, adopted by the United Nations General Assembly in 1948, states that "recognition of the inherent dignity and of the equal and inalienable rights of all members of the human family is the foundation of freedom, justice, and peace in the world" (United Nations General Assembly [1948] Universal declaration of human rights, General Assembly Resolution 217A [III], UN Doc A/810, New York: United Nations General Assembly Official Records).

2 The term *dignity* has been in use much longer than the Universal Declaration. In ancient Rome, the word *dignitas* was used to refer to social status and individual prominence, primarily of the upper classes. The concept has also played an important role for religious philosophers. For example, Christian commentators have long used *dignity* to express, among other things, the unique relationship humans have with God and the resultant special place humans have in the world order. Academic philosophers have given human dignity a range of meanings, including treating it as a virtue or as a description of personal conduct, as in dignified character (Caulfield and Brownsword 2006).

3 Roger Brownsword calls this alliance "dignitarian" "because its fundamental commitment is to the principle that human dignity should be compromised" and an "alliance" "because there is more than one pathway to this ethic—Kantian and communitarian, as well as religious. So, for example, if the dignitarian perspective was to be expressed in communitarian terms, it would be said that human dignity is a good that must not be compromised and that any action or practice that compromises the good is unethical irrespective of welfare-maximizing consequences (contrary to utilitarianism) and regardless of the autonomy rights or informed consent of the participants (contrary to human rights thinking)" (Brownsword 2004b).

4 United Nations General Assembly (April 2003) Annex 1 to the letter dated April 2, 2003 from the Permanent Representative of Costa Rica to the United Nations addressed to the Secretary-General: Draft international convention on the prohibition of all forms

of human cloning. UN Doc. A/58/73. http://daccessdds.un.org/doc/UNDOC/GEN/N03/330/84/PDF/N0333084.pdf?OpenElement.

5 United Nations Educational, Scientific and Cultural Organisation (1997 November) Universal Declaration on the Human Genome and Human Rights. http://portal.unesco.org/en/ev.php-URL_ID=13177&URL_DO=DO_TOPIC&URL_SECTION=201.html.

6 Medical Research Council of Canada, Natural Sciences and Engineering Research Council of Canada, Social Sciences and Humanities Research Council of Canada (2003 June) *Tri-council policy statement: Ethical conduct for research involving humans.* http://www.pre.ethics.gc.ca/english/pdf/TCPS%20June2003_E.pdf.

7 The purpose of the act is laid out in its Declaration of Principles. This in itself is unusual and carries greater legal force because it is enshrined in a statutory declaration as opposed to the usual preamble. The Declaration of Principles does not explicitly refer to the embryo but states broader principles such as:

1. Giving priority to the health and well-being of children born through the application of assisted human reproductive technologies
2. Taking appropriate measures "for the protection and promotion of human health, safety, dignity and rights"
3. Protecting "the health and well-being of women must be protected in the application of these technologies"
4. Promoting and applying free and informed consent
5. Ensuring that "persons who seek to undergo assisted reproduction procedures must not be discriminated against"
6. Prohibiting the "trade in the reproductive capabilities of women and men and the exploitation of children, women and men for commercial ends"
7. Preserving and protecting the "human individuality and diversity, and the integrity of the human genome."

Specifically, section 2(b) states: "The Parliament of Canada recognizes and declares that . . . (b) the benefits of assisted human reproductive technologies and related research for individuals, for families and for society in general can be most effectively secured by taking appropriate measures for the protection and promotion of human health, safety, dignity and rights in the use of these technologies and in related research."

8 Canada, Royal Commission on New Reproductive Technologies, *Proceed with care: Final report of the Royal Commission on New Reproductive Technologies*, vols. 1-2 (Ottawa: Canadian Government Publishing, 1993).

9 Bill C-47, An Act respecting human reproduction technologies and commercial transactions relating to human reproduction, 2nd sess., 35th Parliament, 1996.

10 See the Health Canada website, http://hc-sc.gc.ca/ahc-asc/alt_formats/cmcd-dcmc/pdf/media/releases-communiques/2001/legislation.pdf.

11 Canada, House of Commons, Standing Committee on Health, *Assisted human reproduction: Building families* (Ottawa: Public Works and Government Services Canada, 2001).

12 1st sess., 37th Parliament, 2002.

13 An Act respecting assisted human reproduction, 2nd sess., 37th Parliament, 2002.

14 An Act respecting assisted human reproduction and related research, 3rd sess., 37th Parliament, 2004.

15 S.C. 2004, c. 2.

16 Ibid. ss. 5-13.

17 Ibid. ss. 5-9.

18 Ibid. s. 60.

19 Health Canada, *Proposal for legislation governing assisted human reproduction: An overview* (2001), http://www.hc-sc.gc.ca/ahc-asc/alt_formats/cmcd-dcmc/pdf/media/releases-communiques/2001/repro_over.pdf; Health Canada, Assisted human reproduction: Frequently asked questions (2001), http://www.hc-sc.gc.ca/ahc-asc/media/nr-cp/2001/2001_44bk2_e.html. Health Canada, news release, "Rock launches review of draft legislation on assisted human reproduction to ban human cloning and regulate

related research" (May 3, 2001), http://www.hc-sc.gc.ca/ahc-asc/media/nr-cp/2001/2001_44_e.html.

20 For a complete index of the proceedings of the Standing Committee on Health, including transcripts of oral submissions, see http://www.parl.gc.ca/committee/CommitteeList.aspx?Lang=1&PARLSES=371&JNT=0&SELID=e21_&COM=218.

21 Ibid. at Ian Shugart, assistant deputy minister, Health Canada, Meeting 15 of the Standing Committee on Health, Assisted Human Reproduction on May 10, 2001.

22 Madeline Boscoe, executive coordinator, Canadian Women's Health Network, Meeting 17 of the Standing Committee on Health, Assisted Human Reproduction, May 3, 2001 (12:05). http://www.parl.gc.ca/committee/CommitteePublication.aspx?SourceId=55087 (accessed February 3, 2006).

23 Ibid. at Rhonda Ferderber, director, Special Projects Division, Policy, Planning and Priorities Directorate, Health Policy and Communications Branch, Health Canada.

24 See note 19 at Allan Rock, minister of health, Meeting 13 of the Standing Committee on Health, Assisted Human Reproduction, May 3, 2001.

25 For example, the submission by Janet Rossant of the Samuel Lunenfeld Research Institute, Mount Sinai Hospital, University of Toronto, emphasized that embryonic stem cell research was not dependent on cloning technology (ibid., Meeting 37 of the Standing Committee on Health, Assisted Human Reproduction, October 31, 2001); the submission by Michael Rudnicki, Canada Research Chair in Molecular Genetics, Ottawa Health Research Institute, rapidly conceded the point on questioning that regulations were more flexible and responsive to a rapidly evolving scientific field than criminal prohibitions enshrined in legislation (ibid., Meeting 25 of the Standing Committee on Health, Assisted Human Reproduction, September 27, 2001).

26 Ibid. at Alan Bernstein, president, Canadian Institutes of Health Research, Meeting 37 of the Standing Committee on Health, Assisted Human Reproduction, October 31, 2001.

27 Dr. Thérèse Leroux, director of ethics, Canadian Institutes of Health Research, Meeting 25 of the Standing Committee on Health, Assisted Human Reproduction, September 27, 2001 (12:10).

28 Prof. Bartha Maria Knoppers, Faculty of Law, Université de Montréal, Meeting 39 of the Standing Committee on Health, Assisted Human Reproduction, November 7, 2001 (16:15).

29 Rabbi Irwin Zeplowitz, Temple Anshe Sholom, Hamilton, Meeting 20 of the Standing Committee on Health, Assisted Human Reproduction, June 7, 2001 (11:38-13:22).

30 Bruce Clemenger, director, Centre for Faith and Public Life, Evangelical Fellowship of Canada, Meeting 20 of the Standing Committee on Health, Assisted Human Reproduction, June 7, 2001 (12:15).

31 Archbishop Terrence Prendergast, chair, Canadian Conference of Catholic Bishops, Meeting 44 of the Standing Committee on Health, Assisted Human Reproduction, November 26, 2001 (10:40).

32 Standing committee report; see note 11 at 10.

33 Ibid. at 5.

34 Ibid. at 1.

35 Ibid. at 9.

36 *House of Commons Debates* (February 11, 2003) at 3390 (David Anderson, MP).

37 *House of Commons Debates* (October 2, 2003) (John O'Reilly), 10:15.

38 Standing Committee report, see 11 at 78.

39 Ibid. at 79.

40 *House of Commons Debates* (February 11, 2003) at 3407 (Myron Thompson, MP).

41 *House of Commons Debates* (April 10, 2003) at 5344 (Paul Steckle, MP; Clifford Lincoln, MP).

42 See note 1.

43 House of Commons, Standing Committee on Health, *Evidence,* 37th Leg., Meeting 37, October 31, 2001, http://www.parl.gc.ca/committee/CommitteePublication.aspx?SourceId=55607.

44 Health Canada FAQ, see note 19.

45 Standing Committee report, see note 11 at 9.

46 This point is well put by Edgar Dahl (2004, 267) when he states: "In a society based on the idea of protecting each and everyone's right to life, liberty and property, the burden of proof always rests on the shoulders of the advocates of prohibition and legal coercion."

47 *R. v. Labaye,* 2005 SCC 80 at para. 35.

48 Ibid. at para. 14.

49 For example, in the event of a constitutional challenge (on the grounds that the legislation falls outside the legislative competence of the enacting body), the court's analysis will involve determining the "pith and substance" of impugned legislation. Part of the pith-and-substance analysis involves determining the "purpose and effects of the bill." Many statutes declare their own purpose, and the Assisted Human Reproduction Act is no exception (see s. 2(a-g) of the act). But, as Rand J. correctly notes in *Reference re Validity of s. 5(a) of Dairy Industry Act (Canada),* [1949] S.C.R. 1 at 48 *[Margarine Reference],* the declared purpose is not necessarily determinative. It is, rather, "a fact to be taken into account, the weight to be given to it depending on all the circumstances." If the true purpose of a statute is different from the declared purpose, the legislation is said to be "colourable," meaning that the enacting body has essentially dressed the wolf in sheep's clothing: it has presented a law that (possibly) exceeds its jurisdiction in such a way as to appear to fall within its legislative competence.

Courts can look at extrinsic aids such as parliamentary debates and committee reports in the process of determining the true purpose of a statute. As Ruth Sullivan (1997, 202) notes, extrinsic aids can be used as "indirect evidence of meaning or purpose."

50 *House of Commons Debates* (February 2, 1999) (Pauline Picard, MP), 17:35.

51 *House of Commons Debates* (February 27, 2003) at 4120 (James Lunney, MP).

References

BBC News. 2005. S Korea stem cell success "faked." BBC News, December 15, 2005. http://news.bbc.co.uk/2/hi/asia-pacific/4532128.stm.

Bernier, L., and D. Grégoire. 2001. Reproductive and therapeutic cloning, germline therapy, and purchase of gametes and embryos: Comments on Canadian legislation governing reproductive technologies. *Journal of Medical Ethics* 31 (6): 527-32.

Beyleveld, D., and R. Brownsword. 2001. *Human dignity in bioethics and biolaw.* Oxford: Oxford University Press.

Brownsword, R. 2003. Bioethics today, bioethics tomorrow: Stem cell research and the "Dignitarian Alliance." *Notre Dame Journal of Law, Ethics and Public Policy* 17 (1): 15-51.

–. 2004a. Regulating human genetics: New dilemmas for a new millennium. *Medical Law Review* 12 (1): 14-39.

–. 2004b. What the world needs now: Techno-regulation, human rights and human dignity. In *Human Rights,* ed. R. Brownsword, 203-34. Oxford: Hart.

Bubela, T., S. Debow, and T. Caulfield. 2005. Stem cells, politics and the progress paradigm. Paper presented at the annual general meeting of the Stem Cell Network, Calgary.

Campbell, A. 2002. A place for criminal law in the regulation of reproductive technologies. *Health Law Journal* 10: 77-101.

Caulfield, T. 2001. Clones, controversy and criminal law: A comment on the proposal for legislation governing assisted human reproduction. *Alberta Law Review* 39: 335-46.

–. 2003. Human cloning laws, human dignity and the poverty of the policy-making dialogue. *BMC Medical Ethics* 4: 3.

Caulfield, T., and R. Brownsword. 2006. Human dignity: A guide to policy making in the biotechnology era? *Nature Reviews Genetics* 7 (1): 72-76.

Caulfield, T., and T. Bubela. 2007. Why a criminal ban? Analyzing the arguments against somatic cell nuclear transfer in the Canadian parliamentary debate. *American Journal of Bioethics* 7 (2): 51-61.

Caulfield, T., and A. Chapman. 2005. Human dignity as a criterion for science policy. *PLoS Medicine* 2 (8): 101-3.

Caulfield, T., L. Knowles, and E.M. Meslin. 2004. Law and policy in the era of reproductive genetics. *Journal of Medical Ethics* 30 (4): 414-17.

Charo, R.A. 2004. Passing on the right: Conservative bioethics is closer than it appears. *Journal of Law, Medicine and Ethics* 32 (2): 307-14.

Childress, J. 2001. An ethical defence of federal funding for human embryonic stem cell research. *Yale Journal of Health Policy, Law and Ethics* 2 (1): 157-65.

Dahl, Edgar. 2004. The presumption in favour of liberty. *Reproductive BioMedicine* 8: 266-67.

Deckers, J. 2005. Why current UK legislation on embryo research is immoral: How the argument from lack of qualities and the argument from potentiality have been applied and why they should be rejected. *Bioethics* 19 (3): 251-71.

Devlin, Lord Patrick. 1965. *The enforcement of morals*. Oxford: Oxford University Press.

Devolder, K. 2005. Creating and sacrificing embryos for stem cells. *Journal of Medical Ethics* 31 (6): 366-70.

Dworkin, R. 2000. *Sovereign virtue: The theory and practice of equality.* Cambridge, MA: Harvard University Press.

Greenaway, Norma. 2002. Liberals reviving cloning, stem cell bill. *Edmonton Journal,* September 15.

Javitt, G. H., K. Suthers, and K. Hudson. 2005. *Cloning: A policy analysis*. Washington, DC: Genetics and Public Policy Center.

Macklin, R. 2003. Dignity is a useless concept. *BMJ* 327 (7429): 1419-20.

Magnus, D., and M.K. Cho. 2006. A commentary on oocyte donation for stem cell research in South Korea. *American Journal of Bioethics* 6 (1): W23-24.

Medical News Today. 2005. Royal Society comment on granting of therapeutic cloning license to Professor Ian Wilmut. http://www.medicalnewstoday.com/medicalnews.php?newsid=20004.

Minister of Health (David Dingwall). 1996. Message from the minister. http://www.hc-sc.gc.ca/dhp-mps/alt_formats/cmcd-dcmc/pdf/tech_reprod_e.pdf

Nelson, E. 2006. Comparative perspectives on the regulation of assisted reproductive technologies in the United Kingdom and Canada. *Alberta Law Review* 43 (4): 1023-48.

Nisbet, M. 2004. The polls—trends: Public opinion about stem cell research and human cloning. *Public Opinion Quarterly* 68 (1): 131-54.

Parry, S. 2003. The politics of cloning: Mapping the rhetorical convergence of embryos and stem cells in parliamentary debates. *New Genetics and Society* 22 (2): 145-68.

Pattinson, S. 2002. Reproductive cloning: Can cloning harm the clone? *Medical Law Review* 10 (3): 295-307.

Pattinson, S., and T. Caulfield. 2004. Variations and voids: The regulation of human cloning around the world. *BioMed Central Medical Ethics* 5: 9-16.

Shapiro, R. 2003. Legislative research bans on human cloning. *Cambridge Quarterly of Healthcare Ethics* 12 (4): 393-400.

Theise, N. 2003. Stem cell research: Elephants in the room. *Mayo Clinic Proceedings* 78: 1004-9.

WHO (World Health Organization). European Partnership on Patients' Rights and Citizens' Empowerment. 2003. Genetic databases: Assessing the benefits and the impact on human and patient rights. http://www.law.ed.ac.uk/ahrc/files/69_lauriewhoreport geneticdatabases03.pdf

Wilmut, I., A.E. Schnieke, J. McWhir, A.J. Kind, and K.H.S. Campbell. 1997. Viable offspring derived from fetal and adult mammalian cells. *Nature* 385: 810-13.

Wroe, D. Canberra. 2005. Embryo cloning gains backing. *The Age,* 20 December. http://www.theage.com.au/news/national/embryo-cloning-gains-backing/2005/12/19/1134840798458.html.

Young, A.H., and A. Wasunna. 1998. Wrestling with the limits of law: Regulating new reproductive technologies. *Health Law Journal* 6: 239-77.

Drugs–Up Close and Personal: Engaging Pharmacogenomics

10
Banking on Trust: Issues of Informed Consent in Pharmacogenetic Research

Rose Geransar

As the low-hanging fruit of the Human Genome Project are picked clean, researchers in industry and academia turn their attention to high-throughput strategies that can help us delve into the genome deeper than was possible before. The most modern genetic excursions include analysis of vast arrays of discrete variations imprinted in the human genome that tweak aspects of our experiences as modern beings. For a patient undergoing one or more pharmaceutical treatments, a slight break in rhythm of the biological "dance" caused by some genetic variations can make the difference between positive or poor treatment outcomes, and between life and death. These genetic imprints are called polymorphisms, and the "dance" that they influence is the process of drug metabolism.

Pharmacogenetics and pharmacogenomics are dedicated to studying the genetic basis of this dance in individuals and populations, respectively. Both areas share many of the features of other types of human research. They can best be compared to research on disease susceptibility, where the multifactorial nature of the research data compromises its clinical utility. What makes pharmacogenomic research different from other genetic research is that its primary focus is to develop treatments and help predict treatment outcomes, rather than to predict the risk of developing a particular disease; yet, this does not necessarily preclude the possibility of disease-related findings. This creates ambiguity as to how to assess the overall informational risks associated with such research results, which may potentially complicate the process of ethics review, disclosure of results, and informed consent.

This chapter focuses on how the new questions raised by pharmacogenomics challenge the current ethical frameworks for informed consent. These challenges are present in varying degrees depending on the research questions, contexts, and methodologies. They encompass research ranging from the elementary proof-of-concept stage to late-phase clinical trials, from targeted small-scale research to exploratory population-based studies. In Canada, the *Tri-Council Policy Statement* (TCPS) acts as the guidelines for ethical conduct in academic research; however, these guidelines only broadly address genetic research in general, failing to provide guidance on some of the issues unique to pharmacogenomic research (Canadian Institutes of Health Research et al. 1998). This leaves research ethics boards (REBs) a great deal of flexibility regarding issues such as disclosure and collaboration with other organizations or institutions. These gaps are also addressed in this chapter in the context of informed consent.

Background and Overview

There are two broad classes of consequential risks that can have implications for the informed consent process: physical risks and informational risks. Pharmacogenomic research poses little to no physical risks, since only a small volume of tissue or blood is required to extract DNA, and it can often be acquired through non-invasive and fairly convenient means, such as a buccal swab. The main class of risks often associated with genetic research is that of informational risks, which are the potential consequences of information flow to persons other than the authorized researchers (Anderson et al. 2002).

Since pharmacogenetic research directly concerns the genetic basis of drug response and not diseases, the Pharmacogenetics Working Group has proposed that the informational risks of such research are low relative to other areas of genetic research (Anderson et al. 2002). However, some pharmacogenomic gene markers convey secondary information about disease susceptibility, since genes related to drug metabolism may also be associated with disease risks (Lindpaintner 2002; Netzer and Biller-Andorno 2004; Knudson 2005). Disclosure of results to participants may induce psychological distress if, for example, a participant is found to be a non-responder and there is no other treatment available (Morely and Hall 2004). Because genetic data can also convey familial risks, disclosure to the patient's families may be problematic and potentially disruptive to family structures (Kissell 2005). Publication of information about certain ethnic or racial groups, communities, or clinical populations raises the issue of potential stigmatization of those groups, particularly if those groups are already struggling with societal stereotypes, as in the case of HIV patients (de Montgolfier et al. 2002). There is a two-way potential for discrimination if proper safeguards are not in place to prevent the flow of pharmacogenetic test results to insurers and employers; persons who are either susceptible to adverse drug reaction (the high-risk patients) or are non-responders to treatment (the difficult-to-treat patients) may find it harder to find affordable health insurance (Lipton 2003).

Biobanks for genetic research, including pharmacogenomic research, have recently become a growth industry. Since each gene often makes a relatively small contribution to a person's drug response, a large number of genetic and health profiles are needed to reveal correlations that will be relevant in predicting drug response (Kaiser 2002). Private banks may employ use of the samples in commercial research and development, or act as brokers of tissue to other institutions (Rothstein 2005; Wright-Clayton 2005). While some academic investigators collect deposits of tissues for future research purposes as part of their own REB-approved clinical research programs, others collaborate with public or private biobanks that may donate samples for use in research. Since private tissue banks are not required to adhere to the guidelines imposed by the TCPS, the role of REBs in ensuring informed consent for the use of privately supplied, retrospectively acquired tissue in academic research is unclear. (Burgess and colleagues [2003] describe an "implied regulatory bottom line" in Canada, characterized by the understanding that genetic information is shared, collected, or distributed only with informed consent and only as stipulated in the consent process.)

Informed Consent in Research

The notion of free and informed consent for research is highly valued in Canadian society within all levels of stakeholder groups, from publics to professionals (Einsiedel 2003; Pollara-Earnscliffe 2003). Yet, informed consent is a complicated concept to be denoted by such a simple phrase, and more so in pharmacogenomic research. What does it mean to be informed about the applications of pharmacogenomic research? Should it be possible to adopt a flexible, contextual interpretation of competence to expand the social inclusiveness of pharmacogenomic projects for research applications deemed to pose low risk by qualified REBs? And what if all potential applications cannot be defined at the outset of the project, as is often the case with pharmacogenomic research? Pharmacogenomic research applications challenge one-size-fits-all approaches to informed consent in genetic research, demanding a context-specific analysis of the issues. Thus, it remains a challenge as to how REBs can address the risks of pharmacogenomic research in a standardized way, given the variability in the scope, goals, and contexts of these studies.

The primary goal of free and informed consent is to engage the potential participants (or their guardians) in a meaningful exchange that enables them to voluntarily and knowingly decide whether or not to accept the potential risks and benefits that may be inherent to, or a potential outcome of, their (or their dependent's) contribution to any aspect of the research process (Canadian Institutes of Health Research et al. 1998). In short, the primary goal of informed consent is to act as a practical mechanism through which one shows respect for individual autonomy. The Pharmacogenetics Working Group has defined informed consent as the "document used to obtain consent as well as the process utilized to communicate the intended disclosures and to ensure accountabilities for the consequences of obtaining consent" (Anderson et al. 2002). The definition raises the question of what disclosures should be intended, by whom, and to whom. Various answers, discussed below, have been proposed in light of special considerations in pharmacogenomic research.

Scope of Consent

A challenge of informed consent in all genetic research is to strike a reasonable balance between the scope of consent and the clinical utility afforded by that consent. A key concern is that a broader scope of consent may imply less informed consent, more potential uses of the sample, and more informational risks. Table 10.1 provides a summary of consent models and their advantages and disadvantages. Pharmacogenomic research applications judged to be low risk may be better candidates for broader consent than other types of genetic research.

Traditional Informed Consent

According to the traditional model, free and informed consent is sought for a specific, well-defined research application. Recognizing that changes in the scope of the research can have ramifications on the potential benefits and risks of the research, TCPS requires re-consent for any such changes in research protocols involving identifiable tissue and genetic data (Canadian Institutes of Health Research et al. 1998, article 10.3a). This model of consent was originally developed in the context of traditional medical research to address the potential for harm to the participant as a result of the changes in the scope of the research question

This traditional model does not fit well with pharmacogenomic research, for several reasons. The only anticipated consequential risks are minimal informational risks that can usually be well managed with appropriate confidentiality measures. Changes in the questions, hypotheses, and scope of the research are the norm rather than the exception because of the rapid pace of pharmacogenomic scientific advances, and in some cases (such as in biobank research), the nature of the research is exploratory and makes it difficult to predict future uses. Re-consent can be cumbersome and problematic even when the institution doing the research has been involved in the collection process, let alone when the tissue has been donated by another institution. Restricting research to a defined set of gene markers severely limits the clinical utility of a given sample and may not make sense given that the scientific community may not yet be aware of all the genes involved in a drug response (Anderson et al. 2002; Buchanan et al. 2002; Caulfield, Upshur, and Daar 2003). That said, let us consider the opposite end of the consent spectrum.

Blanket Consent

A blanket or open-ended consent to any and all research solves the problems associated with the application of traditional consent frameworks in genetic research. Blanket consent is simple, easy to administer, and offers researchers maximum clinical utility. Furthermore, the approach can be consequentially unproblematic from an ethical perspective given (1) the appropriate oversight by a competent REB (Sheremeta 2003), (2) sufficient encryption of the samples such that individual donors may not be identified by the researchers (Knoppers 2004), and (3) handling of the research findings with due diligence (Canadian Institutes of Health Research et al. 1998).

However, blanket consent may introduce new problems. First, it is an all-or-none approach to consent that may not be appropriate since it offers participants no residual control over the broad range of potential applications for which the specimen may be used (Laurie 2003). Even if the sample were to be used exclusively for pharmacogenomic research, the scope of research within this area allows for applications with varying degrees of informational risk. As well, individuals' preferences may differ regarding donation of their sample to different types of research (Rothstein 2005). Second, although blanket consent may be voluntary consent, some authors insist that it cannot be informed consent since the nature of the research remains undefined (Hansson 2005). Gulcher and Stefansson (2000) argue that uninformed choice is consistent with autonomous choice: people should be free to take informational risks in the research context, as they do in the course of their clinical care. However, the primary goals and balance of interests are often different in clinical care than they are in research, rendering the comparison generally inappropriate.

Open-ended consent can be theoretically compatible with informed choice if a person has the opportunity and means to (1) be made aware of the research projects on an ongoing basis (perhaps through a continually updated website, or mail-in updates) (Mordini 2003), and (2) opt out of a particular research project, without withdrawing his or her sample from research altogether (Winickoff and Winickoff 2003). However, informed choice may be difficult to achieve in practice using blanket consent. Because of its one-size-fits-all approach to genetic research, TCPS guidelines would not support blanket consent for prospective collection of tissues for research, even for low-risk pharmacogenomic research applications, as the guidelines specifically require that the donors be informed about the potential uses of the tissue (Canadian Institutes of Health Research et al. 1998, article 10.2d). This leads to consideration of the next model of informed consent, which strives to reconcile the notion of informed choice with that of clinical utility.

Table 10.1

Some options for consent for the use of human blood/tissue in research

Option	Advantages	Disadvantages	Suggested applicability	Possible confidentiality features
Project-specific consent	• Consent required only once for initial collection and use of the sample for a specific use • Participants are presented with a well-defined protocol and informed of the duration of/specific use of their sample	• Restricts present use: research cannot shift away from the defined protocol • Secondary use requires re-consent	• Clinical trials, high-risk applications • Drug approval/regulatory applications • Smaller scale, specific research applications	• Identified, coded/double-coded samples
"Authorization" model/Dynamic consent	• Individuals can permit specific uses of their sample only, and specify the degree to which they wish to maintain authority over their sample/data	• May be difficult for researchers to anticipate all potential future uses • Logistically more complicated: may require re-contact	• Clinical trials, high- to moderate-risk applications • Biobanks	• Coded/double-coded samples
"Blanket" consent	• One-time consent • Researchers maintain flexibility to determine the scope of research	• Participants are not aware of the research applications that they are consenting to	• Population-based research • Low-risk research applications • Exploratory research	• Anonymous/Anonymized samples

Waiver of consent (no consent)	• Allows future use of tissue previously donated to research for a different purpose, for which re-consent may be difficult or impossible	• Is not suitable for prospective tissue collection	• Low-risk research applications	• Anonymous/ Anonymized samples (e.g., archival)
Community consent	• Opportunity for community consultation, consistent with respect for cultures and communities	• May conflict with individual consent	• Population-wide biobanks	• Part of a two-tiered consent model, may be used in conjunction with any of the aforementioned models

Sources: Sheremeta (2003); Burgess et al. (2003); Caulfield, Upshur, and Daar (2003); PWG (2001).

Authorization Model of Consent

The problems associated with the traditional model of informed consent and with blanket consent may be remedied in part by a modular approach to soliciting consent, which entails separate and independent consent to specific types of pharmacogenomic applications. This approach involves providing descriptions of specific areas of research in which the participant may choose to opt in or opt out. For example, a person may consent to storage of his or her sample for twenty-five years, but not to its use in creation of immortalized cell lines that can act as an inexhaustible source of DNA. Such an informed consent process need not present all potential uses of the sample: it may request permission for re-contact so that researchers could request consent for use of the sample in new and previously unforeseen areas of research. The model, supported by Caulfield, Upshur, and Daar (2003), respects the notion of informed autonomous choice, while leaving researchers with the flexibility to determine the particularities of future studies within the scope of the areas outlined in the consent form. Consistent with the idea of an authorization model is dynamic consent (Burgess et al. 2003). This concept shares many of the key features of the authorization model but may require more active and continuous involvement on the part of the participant.

Depending on its particularities, an authorization-like approach to informed consent may strike a good balance between scientific utility and the exercise of individual autonomy, and promote respect for differences in the personal and cultural values and beliefs of individuals (Caulfield, Upshur, and Daar 2002; Knoppers and Chadwick 2005). These qualities render the authorization approach to informed consent a good choice for many types of pharmacogenomic research. The key limitation of an authorization model, whether dynamic or not, is that it is inherently more complex in its logistics than either blanket or traditional consent.

Waiver of Consent

One of the most contentious issues in population biobanking has been the waiver of consent. Waiver of consent entails that persons are automatically enrolled in a research initiative, without having been informed about the project in question. In Canada, waiver of consent requires an REB to judge that consent is not necessary given that the research involves minimal risk, no therapeutic intervention, and could not be practically carried out without waiver (Canadian Institutes of Health Research et al. 1998, article 2.1).

In accordance to these stipulations, waiver of consent may be permitted for low-risk pharmacogenomic research using previously collected anonymized or anonymous samples (Canadian Institutes of Health Research et al. 1998, article 10.3b). Outside of this context, waiver of consent is incompatible with respect for persons and their right to be actively involved in making free and informed decisions. This conceptual feature renders waiver of consent unsuitable in the case of prospectively and retrospectively collected tissue in pharmacogenomic research (article 10.2).

Whose Consent?

Certain types of pharmacogenomic research can be associated with significant informational risks that may have ramifications for persons other than the individual participant. One of the unique features of genetic information is that it can be predictive of not only individual health-related susceptibilities but also familial susceptibilities (Mordini 2003), which can be a complicating factor when a guardian is required to act as a proxy for informed consent. Information on many families in a relatively closed community may also provide insight into future genetic trends in that community. These issues are compounded when the research participant is deemed both legally incompetent and a member of an identifiable population.

Public, Community, and Family Consent

Large-scale population studies can potentially be useful in identifying prevalent pharmacogenotypes, which can then be used to guide the investment of research funds toward the development of commercially viable drugs that will be of benefit to specific racial or clinical populations (Buchanan et al. 2002). A study by Chinese researchers, for example, shows that a genetic polymorphism in the mitochondrial aldehyde dehydrogenase-2 gene contributes to the lack of efficacy of glyceryl trinitrate, an angina and heart-failure drug in Chinese participants (Li et al. 2006). Biobanks may stratify individuals' profiles on the basis of shared genotypic characteristics and correlate these with shared phenotypic characteristics such as descent, geographic location, or ethnicity in order to discern these trends (Lunshof and de Wert 2004).

When susceptible subgroups of people constitute a minority of the population, they may be subject to collective informational risks, including the potential for prejudice arising from social stereotyping (Nuffield Council on Bioethics 2003; Gulcher and Stefansson 2000). The TCPS acknowledges that "misunderstanding or misuse of genetic testing has the potential to interfere with an individual's self identify . . . and to stigmatize the entire group to which that individual belongs" (Canadian Institutes of Health Research et al. 1998, section 8). For this reason, international bodies such as the Human Genome Organisation have recommended that informed consent be sought at the population or community level (HUGO 1996; UNESCO 2003). Community and family consent have since become a part of an emerging global trend in research ethics as a venue for achieving transparency and social accountability (UNESCO 2005; Knoppers and Chadwick 2005; Paul and Roses 2003).

In Canada, closed communities, such as certain Aboriginal communities, may be of particular scientific interest for pharmacogeneticists because of their relatively homogenous gene pools. In such cases, special considerations must be devoted to respecting and understanding cultural norms and worldviews of indigenous communities before proposing such research. A community's etiology of health and conceptualization of kinship, hierarchy, and inheritance may be very different from the scientific outlook of genetic researchers (Kissell 2005; Ermine, Sinclair, and Jeffery 2004). Consistent with this line of thinking, the TCPS gives special consideration to general research on Aboriginal communities and recommends a set of guidelines to ensure that the rights and interests of the community as a whole are respected (Canadian Institutes of Health Research et al. 1998, section 6). Among such considerations is the need for active involvement of indigenous authorities in the review process of research protocols (Ermine, Sinclair, and Jeffery 2004).

More broadly, the TCPS recommends that all the social structures affected by the genetic research be involved in the informed consent process, as far as possible, including the family and the community (Canadian Institutes of Health Research et al. 1998, article 8.1). However, even a two-tiered consent process does not address the question of whether and when a group or family's refusal to consent to research may trump an individual's right to participate. The only guidance the TCPS offers in this regard is that researchers must aim to enhance communication by offering a non-biased account of the risks and benefits, rather than supporting a particular view (Canadian Institutes of Health Research et al. 1998, article 8.1).

Research on Vulnerable Population

"Vulnerable persons" include persons whose diminished competence or decisional capacity make them particularly susceptible to harm (Canadian Institutes of Health Research et al. 1998). Vulnerable groups in the context of pharmacogenomic research include newborns, children, adolescents, and incompetent adults, and others that may be particularly susceptible to stigmatization as a result of the informational risks of genetic research. They may include cohorts of HIV patients (de Montgolfier et al. 2002), psychiatric patients (regardless of their legal competency) (Morely and Hall 2004), and even immigrants who are deeply embedded in their cultural heritage (Kissell 2005).

Coercion of participation is perhaps the most compelling issue in research involving vulnerable individuals. In the context of psychiatric PF research, Morely and Hall (2004) express concerns about the potential for coercion by family members who are in a position of decision-making authority over a vulnerable participant. The tri-council policy guidelines state that for a legally incompetent individual who "understands the nature and consequences of the research . . . the potential subject's dissent will preclude his or her participation" (Canadian Institutes of Health Research et al. 1998, article 2.7). Researchers must become sensitized so as to be able to discern a participant's reluctance to express dissent because of coercion induced by a proxy or other third party.

According to the TCPS, researchers must use legal competence as the one-size-fits-all measure to assess competence as it pertains to research participation (Canadian Institutes of Health Research et al. 1998, article 2.5b). This can exclude research participants whose decisional capacity may be attained gradually (e.g., an adolescent), lost gradually (e.g., an Alzheimer's patient), or present intermittently (e.g., a schizophrenic patient) (Knoppers et al. 2002). However, low-risk pharmacogenomic research applications may warrant consideration of a dynamic, contextual definition of competence that would require participants to be (1) *competent enough* to consent to the particular research study (i.e., not necessarily legally competent) and (2) competent *at the time* informed consent is sought. Using the latter criteria to assess competence, neither of the aforementioned participants would necessarily be precluded from providing informed consent to participate in pharmacogenomic research.

Consent to What?

What disclosures should be made in the informed consent process? Since the information provided as part of the informed consent process is likely to affect an individual's decision to participate, it is essential for the process to be as accessible, interactive, and transparent as possible. As with all studies, the informed consent process should explain both the general and specific purpose of the study, including its short- and long-term objectives.

Potential Risks and Benefits

The informed consent process should explain the nature of the potential benefits and risks. It should, for example, distinguish between individual, family, community, and societal benefits. Anticipated benefits should make reference to the current state of scientific knowledge and not be hyped or exaggerated (Anderson et al. 2002). Results from a questionnaire of a cohort of HIV-positive patients enrolled in a pharmacogenomic study indicated that the participants were under the false impression that the study would yield individual benefits in the form of a treatment tailored to their genetic makeup (de Montgolfier et al. 2002). Since false impressions can lead to future disappointment and even psychological harm, it is important to make the expectations clear at the beginning of the study.

Although the risks for pharmacogenomic research are generally considered to be minimal, the potential for information risks should be specified, particularly in research where the samples are not anonymized and are archived for future use.

Scientific versus Commercial Uses

The distinction between scientific use and commercial use of genetic information is directly related to the issue of ownership of that information. Although many of the specific uses of the sample may not be known or may be difficult to predict in pharmacogenomic research, investigators should indicate whether the development of proprietary applications or patents is a goal or possible product of the research, and if so, who will be the direct beneficiaries (Anderson et al. 2002). Disclosure of current and potential collaborating and funding partners must be included, which is important because some potential participants may not subscribe to the principles underlying commercial research and applications. This is a concern in pharmacogenetics, where a substantial portion of the funding is expected to come from industry, particularly in targeted intervention studies (Anderson et al. 2002). REBs should enforce the inclusion of this type of disclosure in informed consent documents, with an aim to help participants understand what commercial involvement could entail.

Confidentiality and Privacy Measures

The informed consent procedure should indicate the nature and volume of the sample to be taken from the participant; who will collect, handle, and process the specimen; and which, if any, of the procedures are a part of routine clinical care. Researchers should clearly indicate whether DNA will be collected from the sample, how long the sample will be stored, and whether the sample will be used to create immortalized cell lines (Canadian Institutes of Health Research et al. 1998, article 8.6). Most importantly, the informed consent process should clearly indicate what provisions will be made to safeguard the data.

The TCPS distinguishes between four types of tissue with respect to their identification status: (1) identifiable tissue, which can be linked to a particular individual; (2) traceable tissue, or codified tissue, whereby the code(s) link(s) to patient information in a database; (3) anonymous tissue, which is anonymous because of the "absence of tags or the passage of time"; or (4) anonymized tissue, which was originally identifiable but has been permanently stripped of its identifiers (Canadian Institutes of Health Research et al. 1998, section 10A). These categories are consistent with those presented by industry and may also be considered as referring to categories of genetic data (PWG 2001).

The four categories of genetic data each strike a different balance in terms of ethical trade-offs between confidentiality, clinical utility, and equity. At one extreme, identifiable data offer minimal confidentiality but maximum clinical utility, and opportunity for withdrawal from the study or return of the individual findings of study back to the participant (PWG 2001). Use of identifiable data is consistent with the view that genetic information is not fundamentally different from other health information and therefore requires no additional safeguards (Wertz 2003). Policy statements issued by many international bodies require the use of additional safeguards for genetic data such as sample coding (Canadian Institutes of Health Research et al. 1998; UNESCO 2003; HUGO 1996).

At the other extreme, anonymous or anonymized data offer maximum confidentiality but minimum clinical utility, and no opportunity for patients to benefit from the individual findings (if any) of the research or to withdraw from the study. Tissue that is not associated with a patient health profile has little utility in pharmacogenomic research because it cannot be used to study associations between genotypic and phenotypic (drug response) data (Caulfield, Upshur, and Daar 2003). However, anonymized or anonymous tissue could suffice in some large-scale exploratory research and may be regarded as a sufficient condition for the ethical application of open-ended consent in pharmacogenomic research consent (Knoppers 2004; Canadian Institutes of Health Research et al. 1998, article 10.3b).

Coding or double-coding of the data, if implemented properly, has the potential to maximize confidentiality and clinical utility while providing participants with the option of disclosure and withdrawal at least up to a certain predetermined time; such a model seems best suited for pharmacogenomic research. The TCPS guidelines make it incumbent on Canadian researchers to disclose the type of identifying information that will be attached to the tissue, potential risks to privacy, as well as safeguards to protect privacy. Samples donated by public or private institutions to academic research are (at the very least) codified, and the researcher does not have access to the code. Since such tissue is considered traceable, the TCPS requires that, where possible, researchers seek free and informed consent from donors or proxies for use of the tissue. However, this may be a problematic requirement because of unfavourable logistics and the undesirability of re-contacting participants after many years (Anderson et al. 2002; de Montgolfier et al. 2002). In the case of donated samples previously collected in a non-academic setting, such as a private tissue bank, the role of the REB remains unclear.

Primary and Secondary Uses

Where possible, any plans or conditions of distribution of genetic samples or data to other researchers for any purpose should be stipulated in the informed consent document (Lincinio and Wong 2002). Survey evidence implies that the Canadian public may be comfortable with secondary uses of their sample data after an initial consent process (Pollara-Earnscliffe 2003), and perhaps more so if the use is moderated by a competent REB. For example, de Montgolfier et al. (2002) recommend the use of a multidisciplinary, multi-stakeholder REB, with at least one knowledgeable patient group representative.

Withdrawal of Consent

The informed consent process should indicate whether and for how long it will be possible for the participant to withdraw his or her sample or data from the study. It is important to indicate whether researchers or sponsors of the study will be required to maintain records for verification in regulatory submissions. Consistent with the authorization model of consent, a distinction should be made between complete or partial withdrawal; for example, withdrawal from genetic research but not from the clinical trial of which it is a part. A patient must be free to choose not to participate in genetic research without jeopardizing his or her access to a required treatment (Anderson et al. 2002).

Disclosure

Researchers and REBs must learn to identify the kinds of circumstances in which a duty to inform may apply, and they must learn how to reconcile this duty, where necessary, with the duty to respect participants' preferences regarding whether and the degree to which they wish to be informed about their genetic findings (Lunshof and de Wert 2004; Anderson et al. 2002; de Montgolfier et al. 2002; Thomas 2001; Wertz 2003). Such judgment may be facilitated by an informed consent process that allows the participants to decide ahead of time whether and under what conditions they wish to be informed about their genetic results (UNESCO 2003, article 10). However, in pharmacogenomic research, a person's consent to the disclosure of genetic information does not make disclosure any less contentious. The kinds of test methods involved in genetic research are often not equivalent to those used in accredited clinical laboratories, and the full clinical implications of pharmacogenomic information on the individual, family, or community may not be known even at the end of a study (Anderson et al. 2002; Netzer and Biller-Andorno 2004). Furthermore, participants may not be able to foresee the kinds of information (and their implications) that may potentially arise as a result of their participation in a given project. In any case, the informed consent process should specify possible arrangements for disclosure, including how and by whom the information will be communicated.

When individual research outcomes may be clinically relevant to family members, the question arises as to whether researchers have a duty to inform family members. A complicated example is the apolipoprotein E gene marker, used to help predict a patient's response to treatment of Alzheimer's disease with cholinesterase inhibitors. The marker may also convey information about familial risk of the disease (Issa and Keyserlingk 2000), but professional organizations currently do not consider it reliable enough to be used in predictive testing. Disclosure of such results to surrogate decision makers may be necessary for researchers to obtain consent for a suitable course of treatment for the dependent participant.

The most compelling case for disclosure is when the results are based on a well-tested hypotheses, can be clinically validated, and can be used in the clinical management of disease for the participant or family members, and when investigators can make provisions for genetic counselling to help the participant put the information in context (de Montgolfier et al. 2002; Morely and Hall 2004). Conversely, disclosure of results is most controversial when the results of the study are preliminary and uncertain, the nature of the information is sensitive, the pharmacogenomic markers are known to convey a secondary risk of disease for which no effective treatment is available, genetic counselling is not available as part of the study, and family members are implicated (Rothstein 2005; Anderson et al. 2002).

The TCPS is consistent in international regulations in requiring that, *where appropriate,* researchers arrange for genetic counselling of participants (and their families) when they choose to offer the option of disclosure of genetic information (Canadian Institutes of Health Research et al. 1998, article 8.1; UNESCO 2003). However, ambiguities remain. The guidelines do not specify in which, if any, types of circumstances genetic counselling would be an imperative, and, therefore, this judgment is left to the discretion of researchers and REBs. Nor is there any consensus in the literature about whether individual results from pharmacogenomic studies should be released to participants, and, if so, in what manner (Morely and Hall 2004; Buchanan et al. 2002). These are among the many issues that remain open to debate.

Conclusion

Pharmacogenomic research, along with other emerging technological applications, is leading to the emergence of a more donor-centric approach to consent with a central notion of trust and responsibility (van Delden et al. 2004). Trade-offs among the scope of research and confidentiality, clinical utility, and respect for individual autonomy, and the dynamics among these trade-offs, demonstrate the interwoven nature of the ethical issues. In the wake of ethical complexities, trust has been and continues to be a major intermediary between REBs and researchers, and between the public and the academic enterprise. This trust is an important factor in public perception of science, research, academic institutions, and regulatory bodies (Einsiedel 2003; Hansson 2005; Kettis-Lindblad et al. 2005). Innovative and enhanced approaches to informed consent processes, such as active community involvement in the approval and design of the consent process, or use of authorization-type models implemented effectively over time, may support the development and maintenance of public trust in academic research (de Montgolfier et al. 2002).

A one-size-fits-all approach to informed consent is not likely to function better than a one-size-fits-all approach to drug therapy. Canada must work toward standardizing the capabilities of its REBs for evaluating the benefits and risks of pharmacogenomic research applications, for such assessments are the crucial premise of important subsequent judgments made regarding, for example, disclosure of results, involvement of children, and level of confidentiality. The development of guidelines to differentiate between high- and low-risk pharmacogenomic applications is an important step in avoiding the pitfalls of "genetic exceptionalism" and "genetic generalization" in pharmacogenomic research (Buchanan et al. 2002). The establishment of standards for disclosure, conditions for genetic counselling, and more fine-tuned evaluations of competence for participation in research are also relevant to the informed consent process. Until national and international guidelines are in place, the reality continues to be that donors are banking on trust: trust in the tissue banks, academic institutions, investigators, and the REBs that monitor their activities.

References

Anderson, D.C., B. Gornex-Mancilla, B.B. Spear, D.M. Barnes, K. Cheeseman, P.M. Shaw, and J. Friedman. 2002. Elements of informed consent for pharmacogenetic research: Perspectives of the Pharmacogenetics Working Group. *Pharmacogenomics Journal* 2 (5): 284-92.

Buchanan, A., A. Califano, J. Kahn, E. McPherson, J. Robertson, and B. Baruch. 2002. Pharmacogenetics: Ethical issues and policy options. *Kennedy Institute of Ethics Journal* 12 (1): 1-15.

Burgess, M.M., P. Lewis, P. Brornley, B. Kneen, and V. McCaffrey. 2003. Above and beyond: Industry innovation related to genetic privacy. In *Genomics, Health and Society: Emerging issues for public policy*, ed. B.M. Knoppers and C. Scriver, 157-97. Policy Research Initiative of the CIHR, CBAC, Health Canada, and Industry Canada.

Canadian Institutes of Health Research, Natural Sciences and Engineering Research Council of Canada, Social Sciences and Humanities Research Council of Canada. 1998 (with 2000, 2002, and 2005 amendments). *Tri-council policy statement: Ethical conduct for research involving humans.* http://pre.ethics.gc.ca/english/policystatement/policystatement.cfm.

Caulfield, T., R.E.G. Upshur, and A. Daar. 2003. DNA databanks and consent: A suggested policy option involving an authorization model. *BioMed Central Medical Ethics* 4 (January 3). http://www.biomedcentral.com/1472-6939/4/1.

de Montgolfier, S., G. Moutel, N. Duchange, I. Theodorou, C. Herve, C. Leport, and the APROCO Study Group. 2002. Ethical reflections on pharmacogenetics and DNA banking in a cohort of HIV-infected patients. *Pharmacogenetics* 12 (9): 667-75.

Einsiedel, E. 2003. *Whose genes, whose safe, how safe? Publics' and professionals' views of biobanks*. Ottawa: Canadian Biotechnology Advisory Committee.
Ermine, W., R. Sinclair, and B. Jeffery. 2004. *The ethics of research involving indigenous peoples: Report of the Indigenous People's Health Research Centre to the Interagency Advisory Panel on Research Ethics*. Regina: Indigenous People's Health Research Centre.
Gulcher, J.R., and K. Stefansson. 2000. The Icelandic Healthcare Database and informed consent. *New England Journal of Medicine* 342 (24): 1827-30.
Hansson, M.G. 2005. Building on relationships of trust in biobank research. *Journal of Medical Ethics* 31 (7): 415-18.
Hoeyer, K. 2003. "Science is really needed–that's all I know": Informed consent and the non-verbal practices of collecting blood for genetic research in northern Sweden. *New Genetics and Society* 22 (3): 229-44.
HUGO (Human Genome Organisation). 1996. *Statement on the principled conduct of genetic research*. HUGO Ethical, Legal, and Social Issue Committee Report to HUGO Council. http://www.eubios.info/HUGO.htm.
Issa, A.M., and E.W. Keyserlingk. 2000. Apolipoprotein E genotyping for pharmacogenetic purposes in Alzheimer's disease: Emerging ethical issues. *Canadian Journal of Psychiatry* 45 (10): 917-22.
Kaiser, J. 2002. Population databases boom, from Iceland to the United States. *Science* 298 (5596): 1158-61.
Kettis-Lindblad, A., L. Ring, E. Viberth, and M.G. Hansson. 2005. Genetic research and donation of tissue samples to biobanks: What do potential sample donors in the Swedish general public think? *European Journal of Public Health* 16 (4): 1-8.
Kissell, J.L. 2005. "Suspended animation," my mother's wife and cultural discernment: Considerations for genetic research among immigrants. *Theoretical Medicine and Bioethics* 26 (6): 515-28.
Knoppers, B.M. 2004. Biobanks: Simplifying consent. *Nature Reviews* 5 (7): 485.
Knoppers, B.M., D. Avard, G. Cardinal, and K.C. Glass. 2002. Children and incompetent adults in genetic research: Consent and safeguards. *Nature Reviews Genetics* 3 (3): 221-24.
Knoppers, B.M., and R. Chadwick. 2005. Human genetic research: Emerging trends in ethics. *Nature Reviews Genetics* 6 (1): 75-79.
Knudson, L. 2005. Global gene mining and the pharmaceutical industry. *Toxicology and Applied Pharmacology* 207 (2 Suppl.): S679-83.
Laurie, G. 2003. Privacy and property? Multi-level strategies for protecting personal interests in genetic material. In *Genomics, health and society: Emerging issues for public policy*, ed. B.M. Knoppers and C. Scriver, 83-98. Policy Research Initiative of the CIHR, CBAC, Health Canada, and Industry Canada.
Li, Y., D. Zhang, W. Jin, C. Shao, P. Yan, C. Xu, and H. Sheng. 2006. Mitochondrial aldehyde dehydrogenase-2 (ALDH2) Glu504Lys polymorphism contributes to the variation in efficacy of sublingual nitroglycerin. *Journal of Clinical Investigation* 116 (2): 506-11.
Lincinio, J., and M. Wong. 2002. Informed consent in pharmacogenomics. *Pharmacogenomics Journal* 2: 343-44.
Lindpaintner, K. 2002. The impact of pharmacogenetics and pharmacogenomics on drug discovery. *Nature Reviews Drug Discovery* 1: 463-69.
Lipton, P. 2003. Pharmacogenetics: The ethical issues. *Pharmacogenomics Journal* 3 (1): 14-16.
Lunshof, J., and G. de Wert. 2004. Pharmacogenomics, drug development, and ethics: Some points to consider. *Drug Development Research* 62 (2): 112-16.
Mordini, E. 2003. Ethical considerations on pharmacogenomics. *Pharmacological Research* 49 (4): 375-79.
Morely, K.I., and W.D. Hall. 2004. Using pharmacogenetics and pharmacogenomics in the treatment of psychiatric disorders: Some ethical and economic considerations. *Journal of Molecular Medicine* 82 (1): 21-30.
Netzer, C., and N. Biller-Andorno. 2004. Pharmacogenetic testing, informed consent and the problem of secondary information. *Bioethics* 18 (4): 344-60.

Nuffield Council on Bioethics. 2003. *Pharmacogenetics: Ethical issues.* http://www.nuffieldbioethics.org/go/ourwork/pharmacogenetics/introduction.

Paul, N.W., and A.D. Roses. 2003. Pharmacogenetics and pharmacogenomics: Recent developments, their clinical relevance and some ethical, social, and legal implications. *Journal of Molecular Medicine* 81 (3): 135-40.

Pollara-Earnscliffe. 2003. *Public opinion research into genetic privacy issues.* Prepared for the Biotechnology Assistant Deputy Minister Coordinating Committee, Government of Canada. http://biotech.gc.ca/epic/site/cbs-scb.nsf/vwapj/Wave_1_GPI_ExSum.pdf/$FILE/Wave_1_GPI_ExSum.pdf.

PWG (Pharmacogenetics Working Group). 2001. Terminology for sample collection in clinical genetic studies. *Pharmacogenetic Journal* 1: 101-3.

Rothstein, M.A. 2005. Expanding the ethical analysis of biobanks. *Journal of Law, Medicine, and Ethics* (Spring) 33 (1): 89-101.

Sheremeta, L. 2003. Population biobanking in Canada: Ethical, legal and social issues. Prepared for the Canadian Biotechnology Advisory Committee. http://cbac-cccb.ca/epic/site/cbac-cccb.nsf/vwapj/Research-2003_Sheremeta-Final_e.pdf/$FILE/Research-2003_Sheremeta-Final_e.pdf.

Thomas, S.M. 2001. Pharmacogenetics: The ethical context. *Pharmacogenomics Journal* 1 (4): 239-42.

UNESCO. 2003. *Draft international declaration on human genetic data.* http://unesdoc.unesco.org/images/0013/001312/131204e.pdf#page=27.

–. 2005. *Preliminary draft declaration on universal norms on bioethics.* http://portal.unesco.org/en/ev.php-URL_ID=17720&URL_DO=DO_TOPIC&URL_SECTION=201.html.

van Delden, J., I. Bolt, A. Kalis, J. Derijks, and H. Leufkens. 2004. Tailor-made pharmacotherapy: Future developments and ethical challenges in the field of pharmacogenomics. *Bioethics* 18 (4): 303-21.

Wertz, D.C. 2003. Ethical, social and legal issues in pharmacogenomics. *Pharmacogenomics Journal* 3 (4): 194-96.

Winickoff, D.E., and R.N. Winickoff. 2003. The charitable trust as a model for genomic biobanks. *New England Journal of Medicine* 349 (12): 1180-84.

Wright-Clayton, E. 2005. Informed consent and biobanks. *Journal of Law, Medicine, and Ethics* 33 (1): 15-21.

11
Pharmacogenomic Promises: Reflections on Semantics, Genohype, and Global Justice

Bryn Williams-Jones and Vural Ozdemir

Pharmacogenomics is an emerging medical specialty that investigates the role of genetic factors in drug response and adverse effects. It differs from its predecessor discipline pharmacogenetics by having a larger scope of inquiry: pharmacogenomics aims to characterize genetic differences among patients across the entire human genome. By contrast, pharmacogenetic studies typically involve investigations of single or a limited set of genes (Ozdemir and Lerer 2005). It is anticipated that a better understanding of genetic factors underlying individual differences in drug effects will help the customization of drug prescriptions.

The field of pharmacogenomics had its origins in the 1950s with the creation of human biochemical genetics and the discovery of single gene variations associated with enzyme deficiencies. These enzyme defects were responsible for certain unexpected adverse drug reactions, such as peripheral neuropathy in slow-acetylators of the anti-tuberculosis drug isoniazid. Twin studies in the 1970s subsequently confirmed the important role of genetics in drug disposition. These developments did not, however, permeate through mainstream medical research until the late 1990s, when DNA technologies spun off from the Human Genome Project became more widely available in clinical and research laboratories.

A key element of pharmacogenomic research is that it deals with questions of variability in drug effects. This approach runs counter to certain established and deterministic norms in medical practice and pharmaceutical industry drug development (Olivier et al. 2008). That is, drug researchers, policy makers, and industry representatives have traditionally approached drug efficacy and safety at a population level; rather little attention has been given to subpopulations or persons with a differential risk for treatment resistance or drug toxicity.

Much of the technical debate on pharmacogenomics and the promise of personalized medicine has centred on upstream applications that can facilitate research to discover new drug targets or the identification of patients with different molecular subtypes of disease. Although personalization of the choice of medicines or their doses is often highlighted in forward-looking statements on pharmacogenomics, little is mentioned about exactly how this may be achieved. Pharmacogenomics has also been advocated as a means of improving the efficiency (and thus reducing the cost) of drug discovery, clinical trials, and the drug approval process. But the very nature of the pharmacogenomics approach–its emphasis on individual variability in treatment outcomes–threatens the traditional business model associated with one-size-fits-all drug development and commercialization. To this end, reflections on the ethical and policy concerns associated with pharmacogenomic research have tended to focus on, for example, informed consent and banking of genetic material in clinical trials, privacy considerations, or potential for stigmatization (Tutton and Corrigan 2004; Weijer and Miller 2004).

As pharmacogenomic applications emerging from industry and academic laboratories start to enter the doctor's office (e.g., as genetic tests and individualized drug therapies), numerous other ethical concerns arise. In this chapter, we analyze three of these: (1) the semantic representations and significance of genetics language in public and policy discourse on drug efficacy and safety, (2) the implications of excessive genohype for the realistic application of pharmacogenomic technologies, and (3) the promise and challenges of applying pharmacogenomics for essential medicines used to treat diseases predominantly affecting people in developing countries.

Historical Context and Genealogy of Personalized Therapeutics

During the last century, and in particular the last thirty years, contemporary Western medicine has shifted its focus from the development of pharmaceutical drugs (whether prescription or over-the-counter) for rare medical conditions and infectious diseases to the development of treatments for common illnesses. Drugs are being used to alleviate both acute and chronic conditions and as prophylactic measures to reduce risk of long-term morbidity and mortality–in the West, drugs now constitute the primary treatment modality. Although drugs can be life-saving in some patients, there is growing concern over the marked uncertainty in drug toxicity or efficacy; these concerns relate both to new compounds in clinical trials and those already in clinical use.

A review of the published data on the efficacy of major drug classes being prescribed for common human diseases concluded that response rates vary substantially across various therapeutic areas. For example, while 80 percent of patients responded positively to pain medications such as COX-2 inhibitors, response rates dropped to 30 percent for treatments directed at Alzheimer's disease and 25 percent for cancer chemotherapies (Table 11.1). Although many drugs in the major therapeutic classes can be life-saving,

Table 11.1

Response rates of patients to major drug classes in selected therapeutic areas

Therapeutic area	Efficacy rate (%)
Alzheimer's	30
Analgesics (COX-2)	80
Asthma	60
Cardiac arrhythmias	60
Depression (SSRI)	62
Diabetes	57
HIV	47
Incontinence	40
Migraine (acute)	52
Migraine (prophylaxis)	50
Oncology	25
Osteoporosis	48
Rheumatoid arthritis	50
Schizophrenia	60

Source: Reproduced with permission from Spear, Heath-Chiozzi, and Huff (2001).

only about 50 percent of patients actually respond positively to their medications (Spear, Heath-Chiozzi, and Huff 2001). Of serious concern, then, are the remaining 50 percent of patients for whom their medication is either ineffective or even toxic.

A meta-analysis of prospective studies investigating drug safety in the United States indicated that 6.7 percent of hospitalized patients experience serious adverse drug reactions (ADRs), while 0.3 percent die from toxic drug effects. This translates into more than two million serious ADRs and an annual death rate of 106,000 patients. These estimates rank ADRs as the fourth leading cause of death in the United States (Lazarou, Pomeranz, and Corey 1998a). Subsequent extended analyses of thirty-two non-US studies from industrialized countries support the conclusion that fatal ADRs are a significant global public health concern (Lazarou, Pomeranz, and Corey 1998b). Interestingly, serious ADRs were observed during treatments with usual drug dosages, despite the exclusion of cases that were due to intentional or accidental overdose, human errors in drug administration, non-compliance, or drug abuse.

The patient and public health implications of non-response or toxic reaction to common medications, and the development of means to reduce these negative effects, have for decades been core research questions in the medical sciences (Reidenberg 2003). By better understanding the degree of and mechanisms governing the predictability of drug effects, biomedical scientists–and especially clinical pharmacologists–aim to rationalize the choice of drugs and customize dosages for individual patients and subpopulations; that is, the goal is personalized medicines (Sheiner 1997).

Some experts suggest that the origins of personalized medicines can be dated to about 510 BCE when Pythagoras, in Croton, in southern Italy, warned of "dangers of some, but not other, individuals who eat the fava bean" (discussed in Nebert 1999). The molecular basis of this historical observation was later found to be haemolytic anaemia attributable to glucose-6-phosphate dehydrogenase deficiency (Nebert 1999). Interest in the rational choice of therapies can be traced to the eighteenth century, when English naval surgeon James Lind demonstrated in 1747, in the first formal comparative trial of its kind, that scurvy could be cured by citrus juice but not by the other leading remedies of the day–cider, vinegar, sea water, or purgative mixtures (Sutton 2004).

Numerous recent policy initiatives have added support to the drive to develop personalized therapies (Kohn, Corrigan, and Donaldson 2000). Pharmaco-epidemiology studies have, however, highlighted important barriers to the rationalization of therapeutics, due in large part to the lack of reliable predictors for the marked inter-individual and population-to-population variability in drug treatment outcomes (Spear, Heath-Chiozzi, and Huff 2001).

One of the prominent research strategies for discovering and applying the requisite genetic predictors (i.e., biomarkers) needed to develop personalized medicines is pharmacogenomics. Employing a broad survey of the human genome made possible by recent advances in DNA sequencing and the identification of thousands of single nucleotide polymorphisms, pharmacogenomics studies the role of genetics on inter-individual and population-to-population variability in drug effects (Evans and McLeod 2003). As single nucleotide polymorphisms, genes or other biological markers are identified and associated with particular drug effects (therapeutic response, toxicity, or treatment resistance), it becomes possible to offer genetic testing to patients or population groups in order to individualize the selection of the type and/or dosage of a medication, thereby improving efficacy and reducing the risk of ADRs.

An important difference, then, between the present interest in pharmacogenomic-guided personalized medicine and previous attempts using more descriptive or demographic predictors (e.g., age, ethnicity, geographic origins) is that sufficient causal information now exists to allow for the development of DNA-based diagnostics and customized therapeutic interventions. This has led to the development of drugs such as Abacavir (for HIV/AIDS), Herceptin (for metastatic breast cancer) and Gleevec (for chronic myeloid leukemia). The hope is that such first generation pharmacogenomic applications will revolutionize medical practice by inaugurating the age of personalized medicines.

However, despite growing knowledge of the importance of genetics in drug effects, there is also an awareness that environmental and social factors play a significant role. Factors such as poor nutrition, low socio-economic status, lack of education, or inadequate health insurance contribute substantially to lack of efficacy or toxicity, a situation that applies to both mainstream and pharmacogenomic drugs (Corrigan 2002). Such a complex interaction of genetic, social, and environmental factors therefore requires more detailed and nuanced social and epidemiological research to better understand the etiology of drug effects. Consequently, although research into personalized medicine has become firmly placed on the social policy and science agendas and is a topic of public and media interest, this interest has been matched by increased scrutiny of the promises, timelines, and likely impact of pharmacogenomics on therapeutics and patient care (Corrigan 2005; Ozdemir and Lerer 2005).

Semantics Revisited: "Genetics" in "Biological Dogma" and "Risk Assessment"

The elucidation of the structure and functioning of DNA, and the growing understanding and application of genetic information in the biological sciences, has led to the deployment of powerful metaphorical language in scientific, public, and policy discussions. The human genome, for example, has variously been described as a "blueprint," an "instruction book," or, in the words of former US president Bill Clinton, "the language in which God created life." In drawing the map of the human genome, scientists are said to be building an "encyclopaedia," an orderly reference book that can be deciphered to predict health and well-being, individual behaviour, and personal destiny. Personalized medicines, gene therapies, genetically modified organisms, and genetic tests are "revolutionary" applications of knowledge from "the book of life," or instances of unacceptable hubris that "play God" and run the "risk" of unleashing "Frankenstein" creations (Hellsten 2005).

The use of metaphors and scientifically imprecise language is arguably an important means of translating, explaining, and simplifying complex ideas and is thus widespread in popular science and media discourses. But such language is not unproblematic (López 2004; Petersen, Anderson, and Allan 2005). Public representations of genomics research and applications often invoke simplistic, reductionist, or deterministic explanations for things that are significantly more complex. The mention of "genetics" or "genetically tailored drugs" creates expectations that highly effective and safe designer medicines will soon be available to the general public (Smart 2003). Similarly, when the diagnostics industry or academic groups seeking research funding or venture capital discuss the implications of their findings (e.g., in relation to variability in treatment outcomes), these are often attributed to "genetics" and thus tend to dismiss the role of non-genetic factors (Bowen, Battuello, and Raats 2005; Williams-Jones 2006).

Despite its frequent use, the term *genetics* is rarely contextualized in sufficient detail. Specifically, it is important to understand the different meanings of "genetics" in "biological dogma" and in "risk assessment" for medical outcomes such as drug response and disease susceptibility.

Biological dogma is a term that describes a fundamental process in cell biology: the unidirectional flow of biological information from the nucleotide sequences coded in individuals' genetic material (DNA), to messenger RNA (mRNA), through the process of gene transcription, and finally to proteins via translation of mRNA by cellular ribosomal machinery. Seen through the lens of biological dogma, then, DNA and genes are undoubtedly the indispensable currencies of all eukaryotic life forms (including humans) and, thus, a key focus for genomics research. Pharmacogenomics research, however, is about "risk assessment" and so poses a conceptually different question: Does human genetic *variation,* either in gene sequence or in the regulation of gene expression (transcription), explain person-to-person differences in drug efficacy or toxicity? Hence, while genes are an indispensable requirement in the context of the biological dogma, they are only one of many factors that contribute to the final composite risk for drug-related problems. The interchangeable use of the term *genetics* in either context–biological dogma or pharmacogenomics-based risk assessment–incorrectly implies a greater role for hereditary factors in pharmacogenomics.

Unfortunately, representations of "genetics" in the media or industry promotional press releases often erroneously refer to the biological dogma itself as the key focus of pharmacogenomics or human genetics research (Terwilliger and Weiss 2003). Rhetoric about the role of genes in sustaining "life" (i.e., more accurately, the biological dogma) is presented to support the importance of genetic factors for drug response. Such genetics talk is inherently misleading to non-specialist consumers of genomic technologies and may ultimately breach the public's trust in science and genetics research (Nelkin 2001). Thus, it is important to ensure that the public, as potential consumers of pharmacogenomic tests, and their physicians, as prescribers of personalized medicines, are able to adequately distinguish different meanings of "genetics" in risk assessment or discussions of the biological dogma.

Promissory Science and Genohype in the Evolution of Pharmacogenomics

The use of simplistic and reductionist language in public and policy discourse also links in important ways to concerns about the extent to which the promises of genomics and biotechnology will be actualized in clinical practice. Can pharmacogenomics deliver personalized medicines in a timely manner, or will it go the way of other genomics promises where "genohype" did not lead to "genoreality" (Ozdemir et al. 2005; Webster et al. 2004)?

The substantial historical–and continued–investment in genomics research by governments and corporations of developed and developing nations has been publicly justified on the grounds that this so-called new science would produce the necessary knowledge and derivative biotechnologies to solve many of the world's most challenging social, environmental, and economic problems. The Human Genome Project, for example, was billed as the means for identifying both rare and common disease susceptibility genes (e.g., for obesity, cancer, diabetes), information that could be quickly translated into cheap and accurate diagnostics to be followed soon after by gene therapies and personalized medicines. Genetics was (is) to be the future of medicine (Collins and McKusick 2001; United Kingdom, Department of Health 2003).

Yet, while the completion of the Human Genome Project may have allowed the "language" of DNA to become "readable," knowing the sequence of coding "letters" (base pairs) and deciphering some of the "words" (genes) has proven insufficient for a complete comprehension of the structure and function of the genome. The publication of the completed human genome map has thus been followed by other maps (e.g., the International HapMap Consortium)–genomics is only the beginning of what is likely to be a decades-long research endeavour to understand the complex interactions among genes, proteins, environment, socioeconomic, and cultural factors (Heymann et al. 2005; Terwilliger and Weiss 2003).

As numerous science, policy, and social science commentators have noted, the promissory fields of genomics and biotechnology have yet to deliver on most of the predicted "revolutionary" technologies (Ozdemir and Godard 2007). There have been important discoveries and developments, but these have tended to be incremental, following pre-existing and well-established lines of research (Nightingale and Martin 2004). For example, while a growing number of medical genetic tests have become available in the clinic and the marketplace, many of these tests were developed after decades of research involving detailed and extended family histories of hereditary disease. And for the most part, these tests are useful only for identifying or providing risk information about rare hereditary diseases; seldom are they sufficiently accurate for wider population screening (Baird 2000; British Medical Association 2005). Similarly, while many of the technical advances (e.g., DNA microarrays) can be represented as "engineering triumphs," they still require interpretation in the context of biology, clinical biomedicine, and commercial biotechnologies. The translation of genomics knowledge into scientifically and economically important biotechnologies has proven far more complicated than expected.

In the case of pharmacogenomics, proponents argue that the genetic customization of drugs will both radically improve medical outcomes by eliminating ADRs and develop new avenues for commercialization of pharmaceuticals and genetic technologies (Roses 2000). As with genetic testing, pharmacogenomics is being applied and is proving useful for some people (Dervieux, Meshkin, and Neri 2005). But despite high expectations, only a small number of pharmacogenomic products have actually entered clinical practice, and even these remain contentious (Woelderink et al. 2006).

For Herceptin, one of the first generation of personalized medicines, there has been some debate about whether or not this drug is in fact an instance of pharmacogenomics (Hedgecoe 2005; Lindpaintner et al. 2001). Herceptin is a monoclonal antibody directed at the human epidermal growth factor receptor 2 (HER2). It was approved by the US Food and Drug Administration in 1998 as an anti-cancer therapy for breast cancer patients who are HER2 positive. The Herceptin pharmacogenomic test involves the characterization of HER2 expression levels in tumour biopsy materials. A clearly discernible drug target, HER2 was a foreseeable candidate for diagnostic testing, the clinical adoption of which was facilitated by the relatively easy access to biopsy samples. In other medical specialties, such as psychiatry, obtaining target organ tissue samples for measurement of gene expression is not possible for obvious practical and ethical reasons. Instead, genetic tests using DNA from peripheral tissues (e.g., blood) remains the sole means of pharmacogenomic testing. Another notable aspect of the HER2 test is that it was not identified by a genome-wide search but rather by virtue of HER2 being a drug target whose expression levels may, understandably, influence treatment outcomes. In this light, HER2 is perhaps more appropriately framed as a product of classical pharmacogenetics rather than as an instance of a revolutionary pharmacogenomic technology.

Abacavir, a drug used to treat HIV-1, is associated with systemic hypersensitivity reactions in about 5 percent of patients. Genetic screening of these persons for the human major histocompatibility complex identified several susceptibility loci (Symonds et al. 2002). One particular allele, HLA-B*5701, was overrepresented among the Abacavir-hypersensitive patients in studies reported from Western Australia and North America and thus could be used as a pharmacogenomic test to identify at-risk persons who should not receive the drug (for an overview, see Quirk, McLeod, and Powderly 2004). Some would argue, however, that the Abacavir story is hyped and that the utility of pharmacogenomic testing is at best limited

(Lindpaintner 2002). The HLA-B*5701 allele does not account for all adverse reactions to Abacavir because people who do not carry this allele can also have a hypersensitivity reaction. Moreover, because exposure to Abacavir can lead to serious medical complications in hypersensitive patients, it is critical that the full complement of genetic and environmental risk factors be identified before pharmacogenomic testing can effectively guide prescription decisions.

Gleevec, a drug to treat persons with chronic myeloid leukemia who test positive for the Philadelphia chromosome (an abnormally short chromosome 22 identified in 1960), is likely a more clear-cut case of a functioning pharmacogenomic test and drug application. But this drug-test combination is based on a relatively simple cytogenetic test and would probably have been feasible without genomic technologies. An important point to note, then, is that the discovery of these drugs *did not* emerge from "genome-wide" searches per se, one of the long-advocated key deliverables of pharmacogenomics research. The attendant diagnostic tests were developed through more focused candidate gene studies (i.e., pharmacogenetics) or simply by using classical cytogenetic approaches. Further, the development of diagnostic tests for drug response appears to be an ad hoc effort rather than a prospective systematic approach in response to a serious adverse drug event or lack of efficacy.

An important goal of pharmacogenomics has been the identification of genes or biomarkers that are difficult to forecast from existing knowledge of a drug's chemistry or the disease pathophysiology. As with research into gene therapy, the enormous complexity of pharmacogenomics research and technology development means that much work remains before scientific discoveries will be readily translated into clinical applications (Freund and Wilfond 2002).

Investors, governments, and some in the pharmaceutical industry have also placed much hope on pharmacogenomics as the vehicle for re-energizing and reshaping the struggling pharmaceutical sector. Pharmaceutical companies have in the last ten years been finding it increasingly difficult to develop and market new blockbuster drugs–most new medications are "me too" drugs (Center for Drug Evaluation and Research 2005; Horrobin 2000). Further, a large number of blockbuster composition of matter patents are due to expire over the next several years, further threatening big pharma's economic model (Service 2004). In this context, upstream pharmacogenomic applications promise to help academic and industry researchers identify new disease-associated genes, some of which may serve as novel drug targets. Developments in pharmacogenomics may also enable companies to risk-proof their drug development pipelines by preventing catastrophic and extremely costly late-stage drug withdrawals (or class action lawsuits) because of unacceptable toxicity or lack of efficacy.

Pharmacogenomics may even revive, in specific subpopulations, patented and developed but uncommercialized drugs that failed during the clinical trial stage because of safety concerns or lack of efficacy in the broader patient population (Ozdemir and Lerer 2005).

The economic potential of pharmacogenomics, and the corresponding benefit for the pharmaceutical industry, may not however be so straightforward. The process of identifying genetic markers in individuals and communities creates pharmacological and disease subtypes that inevitably fragment both the disease and its treatment. That is, by personalizing medicines, pharmacogenomics undermines the traditional blockbuster model of drug development and marketing. No longer is it sufficient to develop a one-size-fits-all medication that will hopefully generate billions of dollars in revenues (i.e., be a blockbuster) (Danzon and Towse 2002); instead, a variety of personalized medications will be necessary for each molecular genetic subtype of disease or individual drug response. The upshot is a fragmented market and increased competition (or more complex collaborations) between drug manufacturers and biotechnology companies. Not surprisingly, then, individual pharmaceutical companies may have divergent views within their own constituents (e.g., among directors of marketing, genetics, or chemistry departments) about the appropriateness of pursing pharmacogenomics, at least in the short term (Williams-Jones and Corrigan 2003). It is very likely, however, that pharmacogenomics will make future predictions on drug efficacy and safety more reliable and reduce the risk for catastrophic and costly late-stage drug withdrawals from the market; but when and to whom these benefits will accrue is much harder to discern (Eisenberg 2002).

Such profound disconnects between the promises and realities of genomic knowledge and technologies have led many commentators to question the motivations of genomics advocates (whether they be academics, pharmaceutical or biotechnology industry representatives, or venture capital firms) and the veracity of their claims, and even to label these promises as "genohype" (Fleising 2001).

Hype may well be an important, even essential, part of the developmental phase of a new science or technology. Positive spin can facilitate collaboration and the uptake of new ideas; help in the acquisition of human, financial, and technical resources; and create a future-oriented consciousness about how the "new" will be an improvement over the "old" (Brown 2003; Hedgecoe and Martin 2003). However, while hype can be an effective means of achieving near-term objectives, it can also be counterproductive in the long run when not matched by tangible products.

Unfulfilled promises can severely undermine the credibility of stakeholders, be they scientists, technologists, companies, or governments. For example, when surrounded by academic, commercial, or other supporters of the promissory science, and in the face of hype and unrealistic public expectations, scientists genuinely committed to the attendant field of enquiry may become discouraged, feel pressured to pursue short-term goals centred on immediacy, or even decide to shift their research entirely to less hyped areas of scientific enquiry. By focusing on the short- to medium-term potentials of particular fields of enquiry or early-stage technologies, hype may also obscure the long-term potential of a new field of innovation (Caulfield 2000).

A more subtle and yet significant consequence of genohype is that it diverts attention from how the benefits of pharmacogenomics are to be distributed among the world's populations and the ways in which pharmacogenomic testing may shape healthcare delivery and treatment of diseases predominantly affecting developing countries.

Pharmacogenomics in the Developing World and the "90/10 Gap"

As discussed above, pharmacogenomics has traditionally been framed in the context of individualization of patented first-line drugs for illnesses such as type 2 diabetes or cancer, diseases that primarily affect the populations of developed and affluent countries. It is not surprising, then, that some would question the relevance of pharmacogenomics to the discovery and cost-effective delivery of drugs that will actually benefit patients in the developing world (Pang 2003).

Private and public investment in health research is estimated to be more than US$70 billion per year (Vijayanathan, Thomas, and Thomas 2002). However, less than 10 percent of global funding for research is directed toward diseases that affect more than 90 percent of the world's population– the so-called '90/10 gap' (Institute for OneWorld Health 2004). This situation is made worse by often inadequate research infrastructure and lack of trained personnel in developing countries who can conduct or advocate for research to tackle prevalent diseases such as malaria or tuberculosis. Because of wide disparities in wealth distribution and lack of a middle socio-economic class in resource-poor nations, efforts to develop local research capacity by training personnel abroad may also prove challenging. In particular, there may be uncertainty about who should have the opportunity to pursue academic training abroad (e.g., based on scientific and individual merit or socio-economic privilege?), not to mention difficulties in ensuring professional commitment and encouraging highly skilled scientists to return to their native countries (Hyder, Akhter, and Qayyum 2003).

In 1977, the World Health Organization (WHO) launched the Model List of Essential Medicines to identify and facilitate the provision of safe and effective treatments for communicable and chronic diseases affecting the vast majority of the world's population. This list is updated every two years by an expert committee to include drugs that will satisfy the priority healthcare needs of the population. The selection criterion takes into account public health relevance, evidence on efficacy and safety, and cost-effectiveness (Day et al. 2005). The concept of essential drugs is also a valuable measure to reduce irrational drug combinations or counterfeit products; the WHO donor programs use only the drugs listed in the model list.

Yet, of the 1,393 new drugs developed between 1975 and 1999, only 1.1 percent (16 drugs) were for the treatment or prevention of tropical and other diseases prevalent in developing countries (Trouiller et al. 2002). The commercial logic of contemporary drug development and marketing is such that developing countries are simply not a good market for first-line pharmaceuticals, nor are their diseases worth the research investment. Unlike drugs for chronic diseases that require multiple prescription refills (e.g., statins to lower cholesterol levels and reduce cardiovascular risk), and thus for which there is a substantial market in the West, antiviral drugs are not a profitable investment unless they also affect affluent countries (e.g., drug cocktails for "chronic" HIV infection or vaccines for avian influenza) because they are prescribed for a limited period.

A second critical element is ability to pay. Developing countries lack the large-scale public and private health-insurance programs common in North America and Europe that would cover the costs of prescription medications; nor do the peoples of developing countries have the personal wealth to afford costly drugs–one-third of the world's population lives on less than US$2 a day. Thus, while sub-Saharan Africa, for example, could provide a very large "volume" of clients for antiviral drugs or vaccines, they are too poor to pay even a reduced bulk price that might be offered by the big pharmas. Multinational regulation of trade and intellectual property rights, in particular the Agreement on Trade-Related Aspects of Intellectual Property Rights (TRIPS Agreement), may further hinder access to patented drugs in developing countries. Although the TRIPS Agreement, in the event of a public health emergency, allows for compulsory licensing by local governments, it is unclear whether this provides any help for countries with limited manufacturing infrastructure.

The high relative cost of pharmaceuticals in developing countries has a profound economic and social impact (Hale, Woo, and Lipton 2005). According to the WHO, "while spending on pharmaceuticals represents less than one-fifth of total public and private health spending in most developed countries, it represents 15 to 30 percent of health spending in transitional economies and 25 percent to 66 percent in developing countries. In most low-income countries pharmaceuticals are the largest public expenditure on health after personnel costs and the largest household health expenditure. And the expense of serious family illness, including drugs, is a major cause of household impoverishment" (WHO Medicines Policy and Standards 2007).

The result is global injustice and a widening gap in access to essential medicines between developed and resource-poor nations. Through no fault of their own–one's birthplace is obviously outside one's control–those people least able to afford basic (but costly) drugs live in poverty and in countries unable to provide affordable healthcare services. These people spend the largest percentage of their incomes acquiring essential medicines, whereas the majority of their wealthy neighbours in the North have access to a surplus of me-too drugs, a diversity of relatively affordable generic medicines, and a health insurance system to cover much of the costs. Given these profound social inequities, of what relevance is a developing but still not fully actualized and thus very expensive technology such as pharmacogenomics?

Given the upfront costs of genomics research and the development of pharmacogenomics technologies, in all likelihood drugs developed by pharmacogenomic guidance will be prohibitively costly for equitable access in developing countries. It is reasonable, then, to argue that other tested public health measures, such as the provision of clean water and equitable access to housing, would be a better use of the limited financial resources available for international development (Heymann et al. 2005). Thus, although genomics research or applied biotechnologies may benefit wealthier populations or individuals, it is not at all clear to many commentators how these technologies will address the serious issues of global public health and distributive justice (Dwyer 2005).

Despite these important critiques of pharmacogenomics (and other biotechnologies), there are two areas in which pharmacogenomics research might help with the equitable and timely provision of appropriate medicines in developing countries. The application of pharmacogenomics might (1) enable the provision of inexpensive generic drugs in subpopulations defined by pharmacogenomic tests wherein drugs display an optimal benefit-to-risk ratio, and (2) facilitate investment in the discovery of medicines for diseases such as malaria or parasitic infections.

It is conceivable, particularly given growing concerns about equitable access to safe and effective medications, that an internationally accepted and peer-reviewed essential biomarkers directory similar to the essential medicines library maintained by the WHO could be established. Such a biomarker directory would enable a broader utility for pharmacogenomic biomarkers by permitting the reintroduction of less costly second-line generic drugs with suboptimal safety as first-line treatments in subpopulations that demonstrate an improved benefit-risk ratio (Ozdemir et al. 2006). This possibility (i.e., a means of marketing patented but uncommercialized drugs) is also one of the hopes of pharmacogenomic advocates in the pharmaceutical industry, but for it to be applicable in the context of developing countries, it would have to be applied in concert with a different market model of drug provision. Specifically, drug developers would need to accept that the opportunity to market patented but shelved drugs brings with it a corollary obligation to market those drugs at a price affordable to developing countries. The payoff for drug companies would be profits from the sale (of relatively) large volumes of a medication that would otherwise be generating no revenues, as well as a means of enhancing their rather tarnished public images by seriously addressing the critical public health concerns of people in developing countries.

As part of their corporate social-responsibility programs, many big pharmas already donate drugs to developing countries; for example, Merck donates the drug ivermectin (Stromectol) for the treatment of river blindness in western and central Africa. However, these donations have had a limited effect on the provision of treatments for the broad range of diseases affecting developing countries. Another and more innovative approach is the development of public-private partnerships, wherein drug companies partner with not-for-profit foundations to engage in drug development for marginalized or orphan disease areas. A notable example is the Institute for OneWorld Health, a non-profit drug-development company that is partnering with big pharmas, small biotechs, and large public trusts such as the Bill and Melinda Gates Foundation to conduct clinical trials and develop a range of drugs for diseases affecting peoples in developing countries. For example, with a broad licence donation of a molecule owned by Celera Genomics, OneWorld Health is developing a treatment for Chagas' disease (Hale, Woo, and Lipton 2005).

Such a merging of innovative business strategies with developments in biotechnology and pharmacogenomics research may allow for more cost-effective identification of novel drug targets and pharmacological mechanisms for neglected diseases, and, it is hoped, lead to increased local access to affordable and effective drugs (Daar and Singer 2005). But although pharmacogenomic technologies may be valuable tools for improving global public health (in the long run), there are serious concerns over their representation as tools that will necessarily benefit science and the public interest in the near term. In comparison with other well-known social, cultural, or economic responses to the profound healthcare challenges facing developing nations or transitional economies, the potential and cost-effectiveness of pharmacogenomic technologies remains a contentious issue. Whether these technologies are applied as part of the standard big pharma business model or integrated in public-private partnerships around non-profit drug development, ultimately the question remains, who is going to fund the long-term research to actualize the potential of pharmacogenomics and translate knowledge into action (Pang, Gray, and Evans 2006)?

Conclusion: An Equitable Integration of Pharmacogenomics in Healthcare?

For pharmacogenomics "genohype" to become "genoreality," basic scientific and clinical discoveries associated with the genetics of drug effect and response must be converted into technologies that actually make a difference in the real world. Short of such translation of knowledge into product, the public as citizens and taxpayers may come to feel they have again been sold a bill of goods, so to speak. A related point is that pharmacogenomics, so far, has been largely an engineering triumph–the vast amount of information generated by genomics research is being transformed into pharmacogenomic tests that are helping researchers build more sophisticated understandings of individual and population variability in drug response. But translational biomarker research that transforms knowledge into functional pharmacogenomic tests and linked drugs will be crucial if this field is to also achieve a biological triumph. Only when the engineering triumph is matched with a biological triumph will the genohype become a genoreality. It is tempting to say, then, that part of the cause of genohype is the mismatch between these two very different types of triumph.

Developed and more affluent countries have a vested self-interest in advocating for global justice in access to essential medicines and provision of healthcare services in resource-poor countries. Recent world events such as the SARS outbreak in 2003, the AIDS pandemic, and the threat of avian influenza have shown the close relationships among national interests, global public health emergencies (regardless of geographical localization), and access to equitable healthcare services, including pharmacotherapy.

While reflecting on the promises of pharmacogenomics, it thus seems appropriate to pay attention to the ways in which new genomic technologies can benefit developing countries, for example, by the discovery of new drug targets for infectious diseases or treatment of resistant forms of tuberculosis. However, inadequate recognition of the strengths and limitations of pharmacogenomic technologies can widen the already existing gap in pharmaceutical research and access to equitable pharmacotherapies. In particular, the '90/10 gap' may become more pronounced unless the emerging pharmacogenomic technologies are appropriately evaluated regarding their implementation in populations with divergent socio-economic predicaments.

Finally, given the continued lack of access to essential medicines in developing countries, profound questions remain about how best to deploy limited financial resources, whether it be for the individualization of drug therapy or for other measures known to markedly improve health and prevent disease, such as access to clean water, education on sexually transmitted diseases (e.g., AIDS), adequate housing, and employment. Ultimately, however, it will still be important to evaluate both the social (environmental, economic, and cultural) and biological (genetic) factors, as well as their complex interactions, in order that biotechnologies and pharmacotherapies can collectively aid in the improvement of global public health (Heymann et al. 2005).

Acknowledgments
Williams-Jones is funded by grants from the Faculty of Medicine, University of Montreal, and the Canadian Institutes of Health Research (CIHR). Ozdemir is funded by an ethics operating seed research grant from the CIHR and a career scientist award in ethics, science, and society research from the Fonds de la recherche en santé du Québec (FRSQ).

References

Baird, P.A. 2000. Genes, health and science policy: Genetic technologies and achieving health for populations. *International Journal of Health Services* 30 (2): 407-24.

Bowen, D.J., K.M. Battuello, and M. Raats. 2005. Marketing genetic tests: Empowerment or snake oil? *Health Education and Behavior* 32 (5): 676-85.

British Medical Association. 2005. *Population screening and genetic testing: A briefing on current programmes and technologies*. London: British Medical Association.

Brown, N. 2003. Hope against hype: Accountability in biopasts, presents and futures. *Science Studies* 16 (2): 3-21.

Caulfield, T.A. 2000. Underwhelmed: Hyperbole, regulatory policy, and the genetic revolution. *McGill Law Journal* 45 (2): 437-60.

Center for Drug Evaluation and Research. 2005. NDAs approved in calendar years 1990-2004 by therapeutic potentials and chemical type. US Food and Drug Administration. http://www.fda.gov/cder/rdmt/pstable.htm.

Collins, F.S., and V.A. McKusick. 2001. Implications of the Human Genome Project for medical science. *Journal of the American Medical Association* 285 (5): 540-44.

Corrigan, O.P. 2002. A risky business: The detection of adverse drug reactions in clinical trials and post-marketing exercises. *Social Science and Medicine* 55 (3): 497-507.

–. 2005. Pharmacogenetics, ethical issues: Review of the Nuffield Council on Bioethics Report. *Journal of Medical Ethics* 31 (3): 144-48.

Daar, A.S., and P.A. Singer. 2005. Pharmacogenetics and geographical ancestry: Implications for drug development and global health. *Nature Reviews Genetics* 6 (3): 241-46.

Danzon, P., and A. Towse. 2002. The economics of gene therapy and of pharmacogenetics. *Value in Health* 5 (1): 5-13.

Day, R.O., D.J. Birkett, J. Miners, G.M. Shenfield, D.A. Henry, and J.P. Seale. 2005. Access to medicines and high-quality therapeutics: Global responsibilities for clinical pharmacology: A major theme of the 2004 World Congress of Clinical Pharmacology and Therapeutics was worldwide equity of access to medicines. *Medical Journal of Australia* 182 (7): 322-23.

Dervieux, T., B. Meshkin, and B. Neri. 2005. Pharmacogenetic testing: Proofs of principle and pharmaco-economic implications. *Mutation Research* 573 (1 and 2): 180-94.

Dwyer, J. 2005. Global health and justice. *Bioethics* 19 (5 and 6): 460-75.

Eisenberg, R.S. 2002. Will pharmacogenomics alter the role of patents in drug development? *Pharmacogenomics* 3 (5): 571-74.

Evans, W.E., and H.L. McLeod. 2003. Pharmacogenomics: Drug disposition, drug targets, and side effects. *New England Journal of Medicine* 348 (6): 538-49.

Fleising, U. 2001. In search of genohype: A content analysis of biotechnology company documents. *New Genetics and Society* 20 (3): 239-54.

Freund, C.L., and B.S. Wilfond. 2002. Emerging ethical issues in pharmacogenomics: From research to clinical practice. *American Journal of Pharmacogenomics* 2 (4): 273-81.

Hale, V.G., K. Woo, and H.L. Lipton. 2005. Oxymoron no more: The potential of non-profit drug companies to deliver on the promise of medicines for the developing world. *Health Affairs* 24 (4): 1057-63.

Hedgecoe, A. 2005. At the point at which you can do something about it, then it becomes more relevant: Informed consent in the pharmacogenetic clinic. *Social Science and Medicine* 61 (6): 1201-10.

Hedgecoe, A., and P. Martin. 2003. The drugs don't work: Expectations and the shaping of pharmacogenetics. *Social Studies of Science* 33 (3): 327-64.

Hellsten, I. 2005. From sequencing to annotating: Extending the metaphor of the book of life from genetics to genomics. *New Genetics and Society* 24 (3): 283-97.

Heymann, J., C. Hertzman, M.L. Barer, and R.G. Evans, eds. 2005. *Healthier societies: From analysis to action*. Oxford: Oxford University Press.

Horrobin, D.F. 2000. Innovation in the pharmaceutical industry. *Journal of the Royal Society of Medicine* 93 (7): 341-45.

Hyder, A.A., T. Akhter, and A. Qayyum. 2003. Capacity development for health research in Pakistan: The effects of doctoral training. *Health Policy and Planning* 18 (3): 338-43.

Institute for OneWorld Health. 2004. The global burden of infectious disease. http://www.oneworldhealth.org/global/global_burden.php.

Kohn, L., J. Corrigan, and M. Donaldson, eds. 2000. *To err is human: Building a safer health system*. Washington, DC: National Academies Press.

Lazarou, J., B.H. Pomeranz, and P.N. Corey. 1998a. Incidence of adverse drug reactions in hospitalized patients: A meta-analysis of prospective studies. *Journal of the American Medical Association* 279 (15): 1200-5.

–. 1998b. Reply: Adverse drug reactions in hospitalized patients. *Journal of the American Medical Association* 280 (20): 1743-44.

Lindpaintner, K. 2002. The importance of being modest: Reflections on the pharmacogenetics of Abacavir. *Pharmacogenomics* 3 (6): 835-38.

Lindpaintner, K., E. Foot, M. Caulfield, and I. Hall. 2001. Pharmacogenetics: Focus on pharmacodynamics. *International Journal of Pharmaceutical Medicine* 15 (2): 74-82.

López, J. 2004. Bridging the gaps: Science fiction in nanotechnology. *HYLE: International Journal for Philosophy of Chemistry* 10 (2): 129-52.

Nebert, D.W. 1999. Pharmacogenetics and pharmacogenomics: Why is this relevant to the clinical geneticist? *Clinical Genetics* 56 (4): 247-58.

Nelkin, D. 2001. Molecular metaphors: The gene in popular discourse. *Nature Reviews Genetics* 2 (7): 555-59.

Nightingale, P., and P. Martin. 2004. The myth of the biotech revolution. *Trends in Biotechnology* 22 (11): 564-69.

Olivier, C., B. Williams-Jones, B. Godard, B. Mikalsson, and V. Ozdemir. 2008. Personalized medicine, bioethics and social responsibilities: Re-thinking the pharmaceutical industry to remedy inequities in patient care and international health. *Current Pharmacogenomics and Personalized Medicine* 6 (2) (in press).

Ozdemir, V., E. Aklillu, S. Mee, L. Bertilsson, L.J. Albers, J.E. Graham, M. Caligiuri, J.B. Lohr, and C. Reist. 2006. Pharmacogenetics for off-patent antipsychotics: Reframing the risk for tardive dyskinesia and access to essential medicines. *Expert Opinion Pharmacotherapy* 7 (2): 119-33.

Ozdemir, V., and B. Godard. 2007. Evidence based management of nutrigenomics expectations and ELSIs. *Pharmacogenomics* 8 (8): 1051-62.

Ozdemir, V., W. Kalow, L. Tothfalusi, L. Bertilsson, L. Endrenyi, and J.E. Graham. 2005. Multigenic control of drug response and regulatory decision-making in pharmacogenomics: The need for an upper-bound estimate of genetic contributions. *Current Pharmacogenomics* 3 (1): 53-71.

Ozdemir, V., and B. Lerer. 2005. Pharmacogenomics and the promise of personalized medicine. In *Pharmacogenomics*, 2nd ed., ed. W. Kalow, U.A. Meyer, and R.F. Tyndale, 13-50. New York: Taylor and Francis.

Pang, T. 2003. Impact of pharmacogenomics on neglected diseases of the developing world. *American Journal of Pharmacogenomics* 3 (6): 393-98.

Pang, T., M. Gray, and T. Evans. 2006. A 15th grand challenge for global public health. *The Lancet* 367 (9507): 284-86.

Petersen, A., A. Anderson, and S. Allan. 2005. Science fiction/science fact: Medical genetics in news stories. *New Genetics and Society* 24 (3): 337-53.

Quirk, E., H. McLeod, and W. Powderly. 2004. The pharmacogenetics of antiretroviral therapy: A review of studies to date. *Clinical Infectious Diseases* 39 (1): 98-106.

Reidenberg, M.M. 2003. Evolving ways that drug therapy is individualized. *Clinical Pharmacology and Therapeutics* 74 (3): 197-202.

Roses, A.D. 2000. Pharmacogenetics and future drug development and delivery. *The Lancet* 355 (9212): 1358-61.

Service, R.F. 2004. Surviving the blockbuster syndrome. *Science* 303 (5665): 1796-99.

Sheiner, L.B. 1997. Learning versus confirming in clinical drug development. *Clinical Pharmacology and Therapeutics* 61 (3): 275-91.

Smart, A. 2003. Reporting the dawn of the post-genomic era: Who wants to live forever? *Sociology of Health and Illness* 25 (1): 24-49.

Spear, B.B., M. Heath-Chiozzi, and J. Huff. 2001. Clinical application of pharmacogenetics. *Trends in Molecular Medicine* 7 (5): 201-4.

Sutton, G. 2004. James Lind aboard *Salisbury*. The Library and Information Services Department of the the Royal College of Physicians of Edinburgh. http://www.jameslindlibrary.org.

Symonds, W., A. Cutrell, M. Edwards, H. Steel, B. Spreen, G. Powell, S. McGuirk, and S. Hetherington. 2002. Risk factor analysis of hypersensitivity reactions to Abacavir. *Clinical Therapeutics* 24 (4): 565-73.

Terwilliger, J.D., and K.M. Weiss. 2003. Confounding, ascertainment bias, and the blind quest for a genetic "fountain of youth." *Annals of Medicine* 35 (7): 532-44.

Trouiller, P., P. Olliaro, E. Torreele, J. Orbinski, R. Laing, and N. Ford. 2002. Drug development for neglected diseases: A deficient market and a public-health policy failure. *The Lancet* 359 (9324): 2188-94.

Tutton, R., and O.P. Corrigan, eds. 2004. *Genetic databases: Socio-ethical issues in the collection and use of DNA*. London: Routledge.

United Kingdom, Department of Health. 2003. *Our inheritance, our future: Realising the potential of genetics in the NHS*. London: Department of Health.

Vijayanathan, V., T. Thomas, and T.J. Thomas. 2002. DNA nanoparticles and development of DNA delivery vehicles for gene therapy. *Biochemistry* 41 (48): 14085-94.

Webster, A., P. Martin, G. Lewis, and A. Smart. 2004. Integrating pharmacogenetics into society: In search of a model. *Nature Reviews Genetics* 5 (9): 663-69.

Weijer, C., and P.B. Miller. 2004. Protecting communities in pharmacogenetic and pharmacogenomic research. *Pharmacogenomics Journal* 4 (1): 9-16.

WHO Medicines Policy and Standards. 2007. Technical cooperation for essential drugs and traditional medicine. http://www.who.int/medicines/services/essmedicines_def/en/.

Williams-Jones, B. 2006. "Be ready against cancer, now": Direct-to-consumer advertising for genetic testing. *New Genetics and Society* 25 (1): 89-107.

Williams-Jones, B., and O.P. Corrigan. 2003. Rhetoric and hype: Where's the "ethics" in pharmacogenomics? *American Journal of Pharmacogenomics* 3 (6): 375-83.

Woelderink, A., D. Ibarreta, M.M. Hopkins, and E. Rodriguez-Cerezo. 2006. The current clinical practice of pharmacogenetic testing in Europe: TPMT and HER2 as case studies. *Pharmacogenomics Journal* 6 (1): 3-7.

12
Envisioning Race and Medicine: BiDil and the Insufficient Match between Social Groups and Genotypes

Benjamin R. Bates

On June 16, 2005, the US Food and Drug Administration's (FDA) Cardiovascular and Renal Drugs Advisory Committee held hearings on a new drug. This drug was a combination of two previously approved ingredients, isosorbide dinitrate and hydralazine hydrochloride. The company submitting the application, NitroMed, was seeking to have a fixed combination of these ingredients approved to treat heart failure. The submission of previously approved drugs as a combination drug is not unusual and may, according to section 300.50 of the 2005 federal regulation *Fixed-combination prescription drugs for humans*, be approved if "the combination is safe and effective for a significant patient population requiring such concurrent therapy as defined in the labelling for the drug."

What made this application different was that NitroMed defined its significant patient population as self-identified black patients. If the committee recommended approval of this indication, it would be the first time that a drug was specifically indicated for a particular race to the exclusion of others. Following the hearings, BiDil was approved and was indicated for treating heart failure in self-identified black patients. At the hearing, Shomarka Keita stated that he was concerned that this indication would label the drug as a black drug, a labelling he thought "will invariably happen and has already happened" (*Hearings of the Cardiovascular and Renal Drugs Advisory Committee* 2005, 221; references to the hearings hereafter cited as *Hearings* 2005). Indeed, when news of BiDil's approval became imminent, it was labelled the "black pill" (Usborne 2004) and "the country's first race-based medicine" (Howatt 2005, 14A).

As news about BiDil spread, the drug was framed as the first step to an era of pharmacogenomics medicine, a method of prescription that assigns drugs based on a patient's genetic makeup. Drawing on the FDA summaries of the hearings, several newspapers reported that BiDil was "a step toward the promise of personalized medicine" (e.g., Stein 2005b, A15; Sternberg 2005, 6A; Zone 2005, A4). Although some writers saw BiDil as a "rudimentary first step" toward pharmacogenomic medication (Rowland 2005, E1), others argued that BiDil was the first pharmacogenomic medication (*Omaha World-Herald* 2005; Stein 2005a). One writer went so far as to say that "if there ever was a positive way to practice racial profiling, this is it," because BiDil treated people based on genetic difference (*Minneapolis Star-Tribune* 2005, 18A). Only "politically correct dogma" prevented people from recognizing genetic differences between black and white Americans, and failure to approve BiDil for self-identified blacks was named "invidious racial discrimination" (Bailey 2005, B2).

Other reports were less optimistic. Because the BiDil trials employed "crude simplifying labels" rather than genetic markers, calling BiDil a pharmacogenomic medication could detract from future medical research (Highfield 2004, 13). NitroMed had offered a double shortcut to considering pharmacogenomically significant differences. First, the trials employed a "crude shortcut for genetic typing" by employing a phenotype instead of looking to actual genetic difference (Saul 2005c, C2). Second, the committee's decision to consider this application as a reflection of sound science would "shortcut truly genetics-based medical research," as this decision made the conflation of socially defined race, phenotype, and genotype appear to be scientifically acceptable (Nesmith 2005, 1C). Because socially defined race is "poorly connected to the group's underlying genetics," this shortcut to pharmacogenomic claims is questionable (Saul 2005a, 1A).

The FDA's approval of BiDil as a (quasi)pharmacogenomic medication offers an opportunity to examine questions related to the scientific, social, and policy implications of race-based pharmacogenomic technologies. To explore these questions, I use the approval of BiDil as a case study in the negotiation of pharmacogenomic science, social understandings of racial difference, and the implications of approving race-based medicines. I begin by outlining the current debate over the sufficiency of race as a genomically significant category. The chapter then provides a close reading of the BiDil approval hearings. Hearings can be an invaluable body of discourse to explore how public questions are negotiated; they allow multiple voices to interact and inform each other, provide a record of the give and take of public argument, and outline how decisions are reached when multiple perspectives interact (Bates 2003; Filler 2001). In this reading, three themes emerge. First, there is a debate over whether BiDil is a pharmacogenomic medication, a debate in which it appears that BiDil is found to rely on "clear enough" genomic difference. Second, there is a discussion of the sufficiency of race as a proxy for genomic differences, a debate that similarly finds a "good enough" correlation. Third, the symbolic implications of BiDil become a tie-breaker for approving BiDil as a drug for blacks only. I conclude the chapter by outlining some ramifications that approving BiDil as a pharmacogenomic medication has for future drug development and public health needs.

Race as a (Non)biological Category

Many scientists and medical professionals believe that the completion of the Human Genome Project will encourage a shift in drug development. Rather than seeking drugs that work equally well for as many people as possible, pharmacogenomic research may encourage drug developers to find drugs that work based on a person's genome (Bamshad et al. 2004; Burchard et al. 2003; Cooper, Kaufman, and Ward 2003; Phimister 2003). Because pharmacogenomic research seeks to assign the best drug to a person, rather than trying to find generic drugs for all people, pharmacogenomic medications may become more clinically useful than general drugs (Goldstein, Tate, and Sisodiya 2003; Schmith et al. 2003). Genetic testing for these differences, however, is not commonly employed. Tests are rarely available in the clinician's office and are often too expensive for the individual patient (Bates et al. 2004). Although access to these technologies may expand as costs decrease and frontline providers become more familiar with genetics, stop-gap solutions based on clinically assessable phenotypes have been proposed.

"Race" has been used as one of these stop-gap solutions. Perhaps the most significant challenge to using race as a clinically significant biological category is that there is no common definition of race (Collins 2004; Foster and Sharp 2002; Keita et al. 2004; Long and Kittles 2003; Sankar and Cho 2002; Tishkoff and Kidd 2004). Race is, in part, defined by skin colour, which is, in part, dependent on a person's genetics (Sturm, Teasdale, and Box 2001). Skin colour may not be an adequate marker of race: when people are grouped by perceived race, they often end up in categories different from when they are grouped by shared genetic makeup (Braun 2002; Lee, Mountain, and Koenig 2001). Therefore, when a clinician attempts to use a patient's perceived race to choose a drug, he or she may assign a drug that is not best for that person's genotype. In addition to skin colour and assumed ancestry, race is defined by a person's culture, history, socio-economic status, political status, and other factors (Dubriwny, Bates, and Bevan 2004). Thus, when patients are asked to identify their race, they may answer based on a social definition, not a biological one. Because of the difficulty of defining race, there is no consensus on the pharmacological importance of race to drug response (Burchard et al. 2003; Schwartz 2001).

Race as a biological category is also challenged by the lack of discrete markers between races. There are some patterns of genetic variation that correspond to continental origin (Bamshad et al. 2003; Rosenberg et al. 2002; Rotimi 2004). Despite these correspondences, only a small amount of genetic difference can be assigned to continental origin (Jorde and Wooding 2004). Rather than discrete differences between continental groupings, genetic variation is both continuous across groupings and heterogeneous within groupings, complicating attempts to find racial markers (Cooper, Kaufman, and Ward 2003; Schwartz 2001; Tate and Goldstein 2004; Tishkoff and Kidd 2004). In addition, the search for genetically distinct races is challenged by the overwhelming genetic similarities among people. Humans differ, on average, at only one of every five hundred to thousand nucleotides between chromosomes (Chakravarti 2001; Sachidanandam et al. 2001; Schneider et al. 2003). The shared sequence between individuals and between continental groupings may overwhelm the few differences (which are, in themselves, shared) among groupings (Bamshad et al. 2003; Rosenberg et al. 2002). A patient's particular genome may not correspond to that predominate in people with a similar continental origin or in the social category with which he or she is identified. The widely shared genetic patterns among all humans may undermine attempts to sort them into meaningful groups.

However problematic the category of race may be, drug trials generally sort people into racial groupings for analysis. Not only is this grouping required by the FDA (2005), but several researchers have called explicitly for race-based research because there may be underlying, yet to be discovered, genetic differences (Risch et al. 2002; Wood 2001). Based on trial results and clinical experience, there are already some uses of race in diagnosing and treating conditions (Exner et al. 2001; McLeod 2001). Race appears to matter in some cases because, on a group level, some "races" seem to respond better to some drugs than do others. BiDil is one of these drugs.

BiDil as a (Non)Pharmacogenomic Medication in the Hearings

Exactly how BiDil functions is unclear. Nevertheless, it does appear that, in a self-identified black population, BiDil, in conjunction with standard therapies, works better than a placebo to improve patient survival, to delay hospitalization for heart failure, and to improve quality of life. It has been posited that BiDil functions to increase available nitric oxide in the patient's blood. Because nitric oxide is, on average, less available in black patients than it is, on average, in white patients, BiDil may be more likely to improve heart function in black patients than in white patients (*Hearings* 2005, 30).

Although BiDil may be more efficacious for black patients than for white patients, BiDil does not work based on the colour of a patient's skin. A better way to assess whether BiDil is appropriate for a particular patient may be to determine whether the patient has a nitric deficiency rather than asking what his or her "race" is. As Steven Nissen, the chair of the committee, put it, "Obviously, if you could actually have a direct marker of who would benefit maybe there are some Caucasian or white Americans that would have this trait and would fall into the same group" (*Hearings* 2005, 98). There are non-black patients who have nitric oxide deficiencies and are candidates for the drug. As yet, BiDil has not been evaluated directly for efficacy in patients with nitric oxide deficiencies as compared with those without a deficiency. Although there are data from the studies cited by the applicants indicating that persons who use alcohol to excess also have, on average, lower levels of available nitric oxide, and, on average, benefit from BiDil, the applicants asked that BiDil be indicated based on membership in a racialized group, not on nitric oxide deficiency.

Because BiDil was not proposed as a universal medication, but one based on individual patient differences, BiDil was seen by some advocates as a pharmacogenomic medication. This view was expressed most clearly by Donna Christensen, who said that it was "critical that we continue the kind of research that was inherent in the promise of the decoding of the human genome whereby we move closer and closer to identifying targeted treatments and more precise measures than race for determining the effectiveness of a treatment" (*Hearings* 2005, 209). As genetic science improves and genetic testing becomes more common, people can be tested for the version of the genes they have and medicine can become more precise. In the interim, however, cruder categorizations, such as those offered by skin colour, may allow the best medical treatment.

Christensen drew on the mystique of the Human Genome Project and treated BiDil as a product of this research. Other speakers at the hearing were concerned that individualized medicine would be undermined if race-based medicine was considered a kind of pharmacogenomics. Basil Halliday testified that, although Christensen's argument was "the way it is playing out in the newspapers, that BiDil some day is going to lead us towards individualized medicine," he disagreed (*Hearings* 2005, 242). Because the BiDil studies evaluated differences among people at the group level and not the individual level, how, Halliday asked, can we say that BiDil is evaluating individual genetic differences "when, indeed, we are doing the very opposite? We are using group as a definition for the people that BiDil would be effective for" (*Hearings* 2005, 242). Gwynn Kendrick agreed, holding that NitroMed's "designation of race/ethnicity categories as sociocultural rather than anthropologic, while politically correct, weakens the utility of genetically influenced differences between populations" (*Hearings* 2005, 254). Had there been a categorization of people based on their genome, rather than a socio-cultural categorization, the BiDil study may have been a step toward pharmacogenomics. However, because no genetic tests were performed in the study, calling the BiDil studies a step toward individualized patient treatment may be a stretch. Yet, because of growing evidence indicating that there are differences in drug response between different groups, and evidence that there are genetic differences between these groups, race may serve, in some cases, as a legitimate proxy for genetics.

This (partial) overlap between the social definition of race and biological differences among groups foregrounds several concerns about the relationship between race and pharmacogenomics. Although the mandate of the FDA is to approve drugs based on their safety, efficacy, and security, the FDA also evaluates the study's design, the statistical procedures used in analysis, and the public health impacts of the drug alongside the reported results of the clinical trial. The committee considered the definition of race employed in the BiDil study to evaluate the study design. Moreover, because of the inconsistency of the statistical support, the committee chose to add points to BiDil's score because of the differential impacts that heart disease has on blacks in the United States and because of the active inclusion of blacks in all levels of the study design. In other words, BiDil was not approved because it is the best drug for blacks pharmacogenomically, but because pharmacogenomics justifications allowed the designation of BiDil as a drug for blacks.

BiDil's Use of Race as a Proxy

One crucial topic in the BiDil discussion is the definition of race. As explained by one of the applicants, Anne Taylor, the first level of inclusion/exclusion criteria employed in the study was race, followed by medical criteria. She noted that "patients were eligible for randomization if they self-identified as African American; had symptomatically stable New York Heart Association Class II-IV heart failure; and were on standard background heart failure medications" (*Hearings* 2005, 60-61). Although it may seem self-evident whether a person is black or not, the discussion of this criterion reveals that NitroMed's definition was a poor one. Charmaine Royal asked the committee, "Who is African American? How are we going to identify African Americans? How is the decision going to be made about who is black? . . . The question about identity is one that is critical here" (*Hearings* 2005, 250). As Royal explained, the lack of a precise definition made it unclear who was in the study population and who would benefit. Without national standardized criteria for identifying blacks, some physicians might use their perceptions of a person's race, others might allow the patient to self-identify, and still others might not use the drug at all. Because race is a socio-cultural label rather than a biological one, Halliday agreed that the definition of race might lack clinical utility. He posited that, without an agreed-on definition, "we are going to exacerbate that whole social phenomenon, that group identity is confused with ancestry and that African Americans have multiple ancestry, and we must consider that when we are talking about biology" (*Hearings* 2005, 242-43).

To avoid this issue of defining people for the study, Taylor and other NitroMed representatives emphasized that the inclusion criterion was *self-identified* race. The flaw in this definition was revealed by an interaction between Taylor and Vivian Ota Wang, a member of the approval committee. Ota Wang, a person whom most would classify as being of Asian descent, pushed Taylor to recognize the slipperiness of self-identification as a standard.

> *Ota Wang:* So, if I presented myself and said I am black, would you allow me to participate in the study?
> *Taylor:* Yes. Self-identification was the criterion.
> *Ota Wang:* So, there is a possibility that there are people who look like me or other people around the table who were included in the study based on their assumption of who they are and not what you are presuming they should be.
> *Taylor:* Yes. (*Hearings* 2005, 96)

Although most members of the study population are likely to be people whom most would classify as black, this brief dialogue does show the insufficiency of the boundaries of self-identification. The scenario described by Ota Wang does not rely on a person's choice to "be" black for a particular study. Indeed, studies have shown that self-identification is variable, with up to one-third of the population changing their self-identification within a two-year period (Cho and Sankar 2004). Moreover, the growing admixture in the population is such that many people may have multiple ancestry of which they are not aware, meaning that a person's self-identification may correspond poorly to their actual heritage (Tishkoff and Kidd 2004; Cooper, Kaufman, and Ward 2003; Schwartz 2001).

The insufficiency of self-identification as an inclusion criterion calls BiDil's indication for black patients only into doubt. Ota Wang argues that, in defining race, "people use skin tone and the racial categories are a proxy for skin tone, and I don't think skin tone is necessarily a great proxy for a biological sort of trait" (*Hearings* 2005, 97). Indeed, as presented by some speakers to the committee, racial classification was distinctly unscientific. Gary Puckrein noted that "we all recognize that the race and ethnic categories that we are currently using are not anthropological, meaning they are not scientifically based. Those categories described the sociocultural construct of our society" (*Hearings* 2005, 213). Although data from clinical trials can be sorted into nominal racial categories, these distinctions may not have great scientific power. When assessing the efficacy of a drug, medications do not depend on a person's self-identification alone but also on biological factors, environmental conditions, and behaviours. Keitá testified that "medications work at the levels of pathophysiology, clinical phenotypes and individuals and not on sociodemographic categories, groups or mystical identities" (*Hearings* 2005, 222). Moreover, self-identification does not serve as a clear proxy for biological differences. As Keita continued to say, "The African American group does not consist of uniform individuals or biologically the same, due to genealogical uniformity, or even environment insult. The race concept does not apply to modern humans" (*Hearings* 2005, 222).

Along with the several challenges to the definition of race offered to the committee, and broad agreement that self-identification was suspect, the argument that, absent widespread genetic testing and additional association studies, race was the best proxy for a person's genome in general was also presented. Moreover, if there were benefits to BiDil that tracked along racial lines, then race was an acceptable proxy for this particular study.

Consistent with NitroMed's definition, the majority of the committee agreed that race was an acceptable proxy. Indeed, the hearings concluded that, because there was no scientific alternative, social definitions had to be held sufficient. Puckrein argued that "some geneticists and social scientists denounce the combination as unscientific, but they cannot offer an immediate alternative to identify this population" (*Hearings* 2005, 203-4). Although critics of socially defined race may be scientifically correct, the lack of an immediate alternative meant that waiting for better definitions would ignore the substantiated clinical utility of BiDil. As Puckrein put it, "Our position is that we cannot allow people not to have their medications. It is important that this new medication be made available to all" (*Hearings* 2005, 203-4). The "all" in Puckrein's statement, however, was equivalent to "all persons who self-identify as black." Because the clinical utility of BiDil had not been demonstrated for the whole population, it could not be approved for all users but only for those who had been members of the study population. This indication, however, might bar prescribing BiDil to non-blacks who had genotypes contributing to nitric oxide deficiencies. Nissen recognized that this gap existed, and, although he "wish[ed] we could do it on a genetic basis . . . in the absence of that, we have some information that suggests that African Americans–we know that African Americans, self-identified, get a pretty robust response to the drug" (*Hearings* 2005, 355-56). Nissen's statement recognizes that direct genetic evidence would be preferable, yet he changes "some information" into a definitive statement of "we know." The committee's findings were recognized by committee members to be a blurring of social definitions with scientific definition, a hardening of the boundaries between racialized groups, and a transformation of a proxy into a direct indicator. Although the scientization of race is questionable, and most committee members evinced discomfort with the conflation of social race and biological difference, the committee nonetheless approved the drug.

Social Differences as Warrants for Approval of BiDil

Although BiDil was ultimately approved as a drug indicated for black patients only, the committee recognized that self-identified race was not the same as genetic categorization. This apparent inconsistency of a scientific determination that relies on unscientific categories appears, at first, to be surprising. The inconsistency is, however, partially resolved by the FDA's secondary mandate to promote public health. Two non-genetic differences between socially designated races may have encouraged the committee to recommend BiDil's approval. The committee found the existence of epidemiological differences in heart disease between blacks and the majority population and differences in participation in medical research between blacks and the majority population. If BiDil were approved based on race, the approval would send two messages to the black community. The first is that the federal government was seeking to build a more responsive healthcare system for blacks. The second was that the inclusion of blacks in clinical trials is desirable. Although neither of these messages is predicated on the safety or efficacy of BiDil in a strict, logical sense, BiDil became a symbol of responsiveness and inclusion and, thus, was given additional reasons for approval.

BiDil as a Symbol of Responsiveness

The medical system in the United States has not responded as well to the needs of blacks as it has to the needs of the majority population (Smedley, Stith, and Nelson 2003). Blacks consistently perceive that they receive less care and lower quality of care than the non-black population, and, in some cases, these perceptions are accurate. Moreover, many common preventable diseases occur at higher rates and occur earlier in blacks than in the majority population. Within this environment, there is a need to develop innovations that take blacks' medical concerns seriously and to promote these innovations. BiDil was framed as a symbol of this kind of responsiveness. As Christensen put it, the committee was given "an unprecedented opportunity to significantly reduce one of the major health disparities in the African American community and, in doing so, to begin a process that will bring some degree of equity and justice to the American healthcare system" (*Hearings* 2005, 203-4). Although BiDil was not a panacea for heart disease, the possibility that BiDil was a response, however flawed, to the severity of heart disease's impact on blacks became one significant reason to approve the drug.

The US federal government has known for a decade that heart disease occurs earlier in blacks, that blacks experience a greater incidence of the risk factors for heart disease, and that blacks have benefited less, on average, than the majority population from innovations in prevention and treatment (Lenfant et al. 1994). Clyde Yancy, for instance, argued that "heart failure is a pressing cardiovascular illness and BiDil represents a new treatment for heart failure as it affects African Americans. African Americans experience heart failure at a greater frequency and have an unusual natural history" (*Hearings* 2005, 110). Specifically, Yancy noted, heart disease occurs earlier in blacks, results in more advanced left ventricular dysfunction in blacks, and has a different etiology in blacks. Other advocates emphasized the quantitative burden heart disease imposed on blacks. Waine Kong noted that "heart failure affects approximately five million Americans and more than 750,000 of them are African American. Between the ages of 45 and 64, African Americans suffer from heart failure 2.5 times more than whites" (*Hearings* 2005, 216). Kong also claimed that black patients were younger and died sooner than white patients.

Given these differences, a responsive medical system had to find additional strategies for treating heart disease in a black population. Lucille Perez argued, "Given that you are convinced that . . . cardiovascular disease [has a disproportionate impact] on African Americans, anything short of approval of BiDil for use in this population cannot be justified" (*Hearings* 2005, 258). In Perez's view, BiDil's absolute efficacy is not the issue; the severity of heart disease for blacks is. Likewise, Puckrein concluded that he did not support BiDil because it was efficacious for blacks on a pharmacogenomic level. Indeed, he posited that there is "no absolute or implied correlation between social, race, genetic type and the efficacy of BiDil"; rather, he supported BiDil "because it will extend the life of many Americans with heart failure" (*Hearings* 2005, 211).

The separation of BiDil from pharmacogenomics reasoning also allowed the committee to separate the causes of heart disease in the black population from its treatment. That is, the fact that BiDil was a response became one ground for approving the drug. This separation worried some advocates. For instance, Keita warned that ignoring the presumed genetic warrant for BiDil's efficacy would lead to "equating a social designation with a particular medicine as if ontologically connected" and imply that genetic difference was the reason blacks suffered a disproportionate impact from heart disease (*Hearings* 2005, 223). By neglecting to differentiate between genetic and social causes, Keita worried that the committee would endorse the view that "the developmental and later environmental causes of disease–namely social inequality, the biology and the poverty–will persist" (*Hearings* 2005, 223). Nevertheless, because BiDil was a response, its perceived responsiveness to a need felt by blacks became the overriding concern. Ronald Portman, for example, stated, "I don't know whether those differences are genetic or whether they are social or whether they are economic or whether health delivery-related and I don't think that particular issue is germane here" (*Hearings* 2005, 353). In other words, although approving BiDil would not address the causes of heart disease, the reasons underlying heart disease in the black population were excluded from consideration. The germane issue, in Portman's view, was whether there was evidence that BiDil was a potential solution to these impacts on the black community.

This separation of the causes of heart disease from the responsiveness of the drug was generalized to the whole of the US healthcare system. Robert Samuels posited that "we may have to adjust slightly to accommodate the need of this very hard-pressed segment of our population that suffers disproportionately from cancer, from heart disease, from diabetes. I mean, there are just so many health issues that affect this community" (*Hearings* 2005, 310). This slight adjustment meant that the central concern of the committee was not the effect of drugs but the effect of diseases on the black population. And, if the disease was severe enough, the existence of a drug for the affected population could serve as an indicator that the committee took the health issues of the black population seriously. The logic of approving a drug based on the symbolism of BiDil was articulated most clearly by Halliday when he argued that the United States had neglected minority needs. Halliday reasoned: "This overlook then forces a healthcare system that is unresponsive to the needs of the people that it is supposed to be serving. This perceived lack of responsiveness is then perceived as a lack of caring which then affects trust. You mix all this together and what you end up with is the stuff of health disparities. I submit to you that with approval of BiDil we can at least begin to break this cycle" (*Hearings* 2005, 237).

BiDil would be perceived as a response to the needs of blacks. This perceived response would then be perceived as an act of caring about the black community's health. This perceived caring would promote trust in the healthcare system. This trust would then help resolve health disparities. Halliday's mixture of perceptions of responsiveness with the materiality of health disparities may muddle the distinction between BiDil's pharmacological effectiveness in the patient body and its social effectiveness in promoting trust in the healthcare system. As a warrant for approving the drug, however, Halliday's argument was persuasive. As Nissen put it, "When you get information that is potentially very valuable and informative about a group that can be very difficult to treat, you have to give a sponsor some points for going after that" (*Hearings* 2005, 307). Although these "points" were not premised on science, they did lend support to BiDil's approval. Because BiDil responded to blacks' social needs that manifested themselves on a biological level, the equation of social self-identification to biological differences became more acceptable to much of the committee.

BiDil as a Symbol of Inclusion

The committee also noted that the BiDil study would promote a politics of inclusion that would encourage black participation in medical research. Many blacks are reluctant to participate in medical research because of a history of exploitative research in the black population. These concerns, however, can be allayed if blacks play a significant role in the design of medical research and if the research is touted as addressing the needs of the black community when participants are recruited to the study (Bates and Harris 2004). If BiDil was model research, endorsing BiDil could encourage other researchers to also include blacks in their studies. As such, the inclusiveness of NitroMed's research design became a second significant reason to approve the drug.

Blacks are significantly underrepresented in medical trials. The percentage of blacks enrolled in clinical trials declined by half to only 6 percent between 1995 and 1999 (Evelyn et al. 2001) and this percentage has continued to decline (Murthy, Krumholz, and Gross 2004). The reasons behind this disparity are manifold. One significant contributor, however, may be that drug manufacturers and allied researchers do not make significant efforts to recruit black populations to their studies. Halliday told the committee that he had asked drug companies why they do not include more blacks. He said drug company executives told him, "'We get our drugs approved anyway, so why bother?' The fact of the matter is when you put all this together we have that minority participation in clinical trials averages less than five percent in trials supporting drug safety and efficacy, and it is time that we changed that" (*Hearings* 2005, 235-36). Indeed, most medications are indicated without regard for socially designated race, even though very few persons of colour are enrolled in clinical trials. Even as FDA advisory committees regularly note the low participation by persons of colour in a drug's trial, these committees also regularly approve drugs with predominantly white populations.

The members of the committee recognized that NitroMed's trial was a significant deviation from the usual practice of testing in an (effectively) white-only population. As the patient advocate on the committee, Susanna Cunningham, put it, "Usually I am here asking for representation of ethnic groups and gender, and I think in this case we do have data and that is unusual and refreshing in some ways" (*Hearings* 2005, 399). NitroMed's decision to conduct a trial in an exclusively black population provided data specifically to represent a minority group, and the high percentage (approximately 40 percent) of black women enrolled in the study indicated that NitroMed wanted to include, and was able to include, blacks in the study population. This atypical recruitment decision meant that the population was different from that usually considered in FDA committees. Although concerns were raised that the recruitment of blacks meant only that the indication for the drug would be race-based, NitroMed's strategy helped answer social concerns about including persons of colour in medical trials. Nissen noted that the committee has "sat around this table and complained many times about getting trials where we don't have enough women and we don't have enough minorities to come to some conclusions" (*Hearings* 2005, 189). Moreover, NitroMed's recruitment strategy could be seen as a response to the FDA's desire for more inclusive medical trials. Although NitroMed recruited only blacks, Nissen claimed that "one of the prices that you have to pay for getting that information is to accept that this kind of exploration is desirable" (*Hearings* 2005, 189). That is, because of FDA calls for more persons of colour to participate in trials, trials that enrolled only persons of colour had to be accepted as proper protocol.

The data for the BiDil application were assigned support because so many blacks had been involved as participants. Royal argued that this study design made blacks feel more involved in the research process. Upon hearing the results, he claimed, "Many people in the African American community were thinking finally we have our drug; something for us. Somebody is paying attention to us, the whole issue of inclusion being part of the process" (*Hearings* 2005, 247). The design of a drug to address black concerns through extensive testing with blacks may make black patients more likely to use the drug if it is prescribed to them. In addition to involving blacks as research participants, blacks assumed other roles in NitroMed's study, further reinforcing BiDil as a symbol of inclusion in the research process. Kong, for instance, emphasized the multiple roles blacks had played in the study. According to him, members of the Association of Black Cardiologists "felt that by direct trial participation, including principal investigator and subject recruitment, we would be able to confirm the data's validity for the medical community, particularly those caring for African American heart failure patients" (*Hearings* 2005, 216). In addition to ensuring that blacks would be looking out for black interests in medical research on blacks, Kong explained that this trial participation on multiple levels might also make black patients more likely to trust the results of the research. Similarly, Halliday stated that, as a person of colour, she would "ask that we become informed and consent to participate in the process not only as patients but also as investigators and also as advisory panel members. We need to become our own experts" (*Hearings* 2005, 239). The opportunity to become one's own expert was offered by NitroMed.

Because blacks were involved both as participants and as investigators, the NitroMed study departed from standard medical research conducted (mainly) by white researchers on (mainly) white participants. This difference in design allowed for a more inclusive study, even if that study did not depend on biological differences. Although this element of the study design was unusual and foreclosed any comparisons between a black population and some other reference population, the committee chair, Nissen, claimed, "We have to overcompensate in order to make people comfortable in minority groups with participating in clinical trials . . . These folks were able to pull it off and I am going to give them some points for that" (*Hearings* 2005, 309). As with the arguments that BiDil was a symbol of responsiveness, the points given to BiDil for serving as a symbol of inclusion were not premised on science but nevertheless offered further support to BiDil's approval. Because BiDil included blacks as a significant population, social self-identification within a racial group became acceptable to the committee as a legitimate part of the study design.

Implications of BiDil's Approval for Pharmacogenomics

BiDil was approved by the committee, though exactly what was approved was unclear. The lead advocate for NitroMed, Manual Worcel, proposed that BiDil be "indicated for the treatment of heart failure as an adjunct to standard therapy in black patients to improve survival, prolong time to hospitalization for heart failure, and improve quality of life" (*Hearings* 2005, 14). Following the hearing, however, Nissen stated that the committee had been "asked to opine on whether V-HeFT I, V-HeFT II and A-HeFT adequately support a claim that BiDil, hydralazine plus isosorbide dinitrate, improves outcome in patients with heart failure" (*Hearings* 2005, 273). Significantly, the modifier "black" was omitted from the issue put to a committee vote. As the committee members explained their votes, only two members explicitly stated that "black" should be included in the approved indication, and only two members explicitly stated that the drug should be approved for all patients (*Hearings* 2005). The desired label reading for the other six participants on the committee was not provided. This confusion over exactly what had been decided was replicated in the media coverage of this decision, with the *New York Times* emphasizing that the drug had been race-labelled despite two members' concerns about including the term *black* (Saul 2005b) and the *Atlanta Journal-Constitution* reporting that the drug had been approved by all members, with just two recommending that the term *black* be included (Nesmith 2005).

Ultimately, the package insert for BiDil repeated NitroMed's desired language, and BiDil was fixed as a drug for blacks only (NitroMed 2005). The naming of BiDil as a drug for blacks was not fixed by the committee. Rather, the silence of those six members of the committee about the race label helped the committee avoid the propriety of racial categorization in medical research. This silence, and NitroMed's ability to use this silence as a warrant for labelling BiDil a black drug, indicates that the social definition of race became a "good enough" proxy for pharmacologically significant genetic differences. The committee assumed that, because there was a statistically significant improvement in cardiac function when taking

BiDil instead of a placebo in a study conducted in a black-only population, others with a similar genotype as represented by skin colour and named by self-identification would similarly benefit. Rather than conflating social race and genetic difference, the committee compressed an imprecise proxy into a direct, associative marker. The debate over whether race labels are sufficient is made irrelevant, and the reader of the BiDil indication–be they a patient or a physician–is encouraged to read "race" as a shorthand for genetic difference. If one asks about the reasoning for BiDil's "blacks only" indication, the compression of social categories into biological difference can be explained, but the shorthand, by its very nature, discourages the unpacking of this debate. Moreover, the approved race-based label encourages the reader to assume that there is a uniform category of "black" in which all blacks have the same genotype and to assume that non-blacks do not share this genotype. The approval of a race-based label through a (primarily) scientific committee, then, allowed a socially defined category to become a pharmacogenomic and biological category.

In this compression of the social and the biological, the committee's statements and the testimony given to the committee emphasized pharmacological solutions to broad challenges in the public health arena. By foregrounding a causality premised on genetics and proposing BiDil as a solution to the significant impact that heart disease has on the American black population, the multifactorial challenges to public health could be set aside in favour of a biology-only model. The concerns articulated to the committee about inclusive medicine as a way to promote trust and participation in biomedical research were generated by a particular sociohistorical experience. The approval of a new drug does not, in itself, counter that historical experience. Moreover, to address suspicion of the medical system as a biologically-based concern may not only ignore the social forces that contribute to this suspicion but also reinforce it. If blacks are concerned that the medical system treats them differently, and medicines are approved on the fundament of biological difference, they may actually become more suspicious of a system that is attempting to enact inclusiveness. Likewise, the concerns about responsiveness of the medical system to black needs were generated by a confluence of environmental, behavioural, and biological factors that cause higher rates and earlier incidence of cardiovascular disease for black Americans. Although BiDil may be an appropriate allopathic response to the development of heart disease, the foregrounding of a biological treatment to a multifactorial etiology may ignore preventive measures that could lower the incidence of heart disease and sidestep social justice concerns, such as economics and healthcare access, which could provide a more comprehensive solution. The emphasis on biological differences grounded in assumed genetic differences may be part of a larger pattern of pharmacogenomic solutions to common medical ailments. If differences between socially defined populations in terms of medical outcomes or use of the medical system are ascribed to membership in a racialized group, social contributors to these differences that correspond better to socially ascribed groupings may be de-emphasized in favour of a pharmacological solution that is less efficacious for these outcome and usage differences.

Finally, using race as a stop-gap form of pharmacogenomic research may undermine actual pharmacogenomic research in the future. In approving BiDil, racial classification served as a double proxy for pharmacogenomic research; self-identification into a racial category (a social definition) was a proxy for skin colour (a phenotypic distinction) and, in turn, skin colour was a proxy for genotypic differences leading to nitric oxide deficiency. There was no demand that NitroMed extend its search into identifying actual genetic contributions to nitric oxide deficiency or to identify non-black persons with nitric oxide deficiencies that could also benefit from BiDil. Instead, race was "good enough" for indicating BiDil for blacks only. Recent evidence (Lapu-Bula et al. 2005) suggests that, in humans, variations in the NOS3 gene cluster may contribute to lower levels of available nitric oxide in the blood and contribute to left ventricular dysfunction. Other evidence suggests that variations in NOS2 in rat and AGTR2 in mouse have a similar impact. A better understanding of variants in these, and potentially other, gene clusters would provide a more direct indicator of who is likely to suffer from heart disease. Nitric oxide deficiency was the factor NitroMed posited as the reason BiDil worked, on average, better in black patients than a placebo. These genetic variants, however, are not exclusive to the black population. Assigning BiDil to patients with these variants may improve patient survival, delay hospitalization for heart failure, and improve their quality of life. Unless associations between these gene variants and BiDil are tested, the actual pharmacogenomic of BiDil will go unfulfilled. If race-based medicine is used as a proxy for pharmacogenomics, the promise of patient treatment will go unmet and the potential of the Human Genome Project to improve patient and public health will be unsatisfied.

References

Bailey, R. 2005. When medicine that discriminates is good. *Chicago Sun Times*, December 18.

Bamshad, M., S. Wooding, B.A. Salisbury, and J.C. Stephens. 2004. Deconstructing the relationship between genetics and race. *Nature Reviews Genetics* 5 (8): 598-609.

Bamshad, M.J., S. Wooding, W.S. Watkins, C.T. Ostler, M.A. Batzer, and L.B. Jorde. 2003. Human population genetic structure and inference of group membership. *American Journal of Human Genetics* 72 (3): 578-89.

Bates, B.R. 2003. Ashcroft among the senators: Justification, strategy and tactics in the 2001 attorney general hearings. *Argumentation and Advocacy* 39 (4): 254-73.

Bates, B.R., and T.M. Harris. 2004. The Tuskegee study of untreated syphilis and public perceptions of biomedical research: A focus group study. *Journal of the National Medical Association* 96 (8): 1051-64.

Bates, B.R., K. Poirot, T.M. Harris, C.M. Condit, and P.J. Achter. 2004. Evaluating direct-to-consumer marketing of race-based pharmacogenomics: A focus group study of public understandings of applied genomic medication. *Journal of Health Communication* 9 (6): 541-59.

Braun, L. 2002. Race, ethnicity, and health: Can genetics explain disparities? *Perspectives in Biology and Medicine* 45 (2): 159-74.

Burchard, E.G., E. Ziv, N. Coyle, S.L. Gomez, H. Tang, A.J. Karter, J.L. Mountain, E.J. Perez-Stable, D. Sheppard, and N. Risch. 2003. The importance of race and ethnic background in biomedical research and clinical practice. *New England Journal of Medicine* 348 (5): 1170-75.

Chakravarti, A. 2001. Single nucleotide polymorphisms: To a future of genetic medicine. *Nature* 409 (6822): 822-23.

Cho, M.K., and P. Sankar. 2004. Forensic genetics and ethnical, legal and social implications beyond the clinic. *Nature Reviews Genetics* 36 (11): S8-12.

Collins, F.S. 2004. What we do and don't know about "race," "ethnicity," genetics and health at the dawn of the genome era. *Nature Reviews Genetics* 36 (11): S13-15.

Cooper, R.S., J.S. Kaufman, and R. Ward. 2003. Race and genomics. *New England Journal of Medicine* 348 (25): 1166-70.

Dubriwny, T.N., B.R. Bates, and J.L. Bevan. 2004. Lay understandings of race: Cultural and genetic definitions. *Community Genetics* 7 (4): 185-95.

Evelyn, B., T. Toigo, D. Banks, D. Pohl, K. Gray, B. Robins, and J. Ernat. 2001. Participation of racial/ethnic groups in clinical trials and race-related labelling: A review of new molecular entities approved, 1995-1999. *Journal of the National Medical Association* 93 (12): S18-24.

Exner, D.V., D.L. Dried, M.J. Domanski, and J.N. Cohn. 2001. Lesser response to angiotensin-converting-enyzme inhibitor therapy in black as compared with white patients with left ventricular dysfunction. *New England Journal of Medicine* 344 (18): 1351-57.

FDA (US Food and Drug Administration). 2005. *Guidance for industry: Collection of race and ethnicity data in clinical trials*. Rockville, MD: Food and Drug Administration, US Department of Health and Human Services.

Filler, D.M. 2001. Making the case for Megan's Law: A study in legislative rhetoric. *Indiana Law Journal* 76 (2): 315-65.

Foster, M.W., and R.R. Sharp. 2002. Social classifications as proxies of biological heterogeneity. *Genome Research* 12 (6): 844-50.

Goldstein, D.B., S.K. Tate, and S.M. Sisodiya. 2003. Pharmacogenetics goes genomic. *Nature Reviews Genetics* 4 (12): 937-47.

Hearings of the Cardiovascular and Renal Drugs Advisory Committee, Volume II. 2005. (NDA) 20-727. Gaithersburg, MD: Center for Drug Evaluation and Research, US Food and Drug Administration.

Highfield, R. 2004. "Ethnic drug" raises fears over race and genetics. *London Daily Telegraph,* November 1.

Howatt, G. 2005. FDA is close to approving medicine to treat heart failure in black patients. *Minneapolis Star-Tribune,* June 16.

Jorde, L.B., and S.P. Wooding. 2004. Genetic variation, classification and "race." *Nature Reviews Genetics* 36 (11): S28-33.

Keita, S.O.Y., R.A. Kittles, C.D.M. Royal, G.E. Bonney, P. Furbert-Harris, G.M. Dunston, and C.N. Rotimi. 2004. Conceptualizing human variation. *Nature Reviews Genetics* 36 (11): S17-20.

Lapu-Bula, R., A. Quarshie, D. Lyn, A. Oduwole, C. Pack, J. Morgan, S. Nkemdiche. 2005. The 894T allele of endothelial nitric oxide synthase gene is related to left ventricular mass in African Americans with high-normal blood pressure. *Journal of the National Medical Association* 97 (2): 197-205.

Lee, S., J. Mountain, and B. Koenig. 2001. The meanings of "race" in the new genomics. *Yale Journal of Health Policy, Law and Ethics* 1 (1): 33-75.

Lenfant, C., C.K. Francis, A.O. Grant, D.M. Becker, R.O. Cannon III, E.S. Cooper, V. Fuster. 1994. *Report of the Working Group on Research in Coronary Heart Disease in Blacks.*

Bethesda, MD: National Heart, Lung and Blood Institute, National Institutes of Health, Public Health Service, Department of Health and Human Services.

Long, J.C., and R.A. Kittles. 2003. Human genetic diversity and the non-existence of biological races. *Human Biology* 75 (4): 449-71.

McLeod, H.L. 2001. Pharmacogenomics: More than skin deep. *Nature Genetics* 29 (3): 247-48.

Minneapolis Star-Tribune. 2004. "Black pill" a good case of racial profiling. November 12.

Murthy, V.H., H.M. Krumholz, and C.P. Gross. 2004. Participation in cancer clinical trials: Race-, sex-, and age-based disparities. *Journal of the American Medical Association* 291 (22): 2720-26.

Nesmith, J. 2005. FDA advisors endorse drug for blacks only. *Atlanta Journal-Constitution*, June 17.

NitroMed. 2005. BiDil, official package insert, rev. August 2005. http://www.bidil.com/pdf/pi_hcp.pdf.

Omaha World-Herald. 2005. Personalized treatment. July 11.

Phimister, E.G. 2003. Medicine and the racial divide. *New England Journal of Medicine* 248 (12): 1081-82.

Risch, N., E. Burchard, E. Ziv, and H. Tang. 2002. Categorization of humans in biomedical research: Genes, race and disease. *Genome Biology* 3 (7): 2007.1-2007.12.

Rosenberg, N.A., J.K. Pritchard, J.L. Weber, H.M. Cann, K.K. Kidd, L.A. Zhivotovsky, and M.W. Feldman. 2002. Genetic structure of human populations. *Science* 298 (5602): 2381-85.

Rotimi, C.N. 2004. Are medical and nonmedical uses of large-scale genomic markers conflating genetics and "race"? *Nature Reviews Genetics* 36: S43-47.

Rowland, C. 2005. Panel backs drug for blacks. *Boston Globe*, June 17.

Sachidanandam, R., D. Weissman, S.C. Schmidt, J.M. Kakol, L.D. Stein, G. Marth, S. Sherry. 2001. A map of human genome sequence variation containing 1.42 million single nucleotide polymorphisms. *Nature* 409 (6822): 928-33.

Sankar, P., and M.K. Cho. 2002. Genetics: Toward a new vocabulary of human genetic variation. *Science* 298 (5597): 1337-38.

Saul, S. 2005a. US to review drug intended for one race. *New York Times*, June 13.

–. 2005b. FDA panel approves heart remedy for blacks. *New York Times*, June 17.

–. 2005c. FDA approves a heart drug for African Americans. *New York Times*, June 24.

Schmith, V.D., D.A. Campbell, S. Sehgal, W.H. Anderson, D.K. Burns, L.T. Middleton, and A.D. Roses. 2003. Pharmacogenetics and disease genetics of complex diseases. *Cellular and Molecular Life Sciences* 60 (8): 1636-46.

Schneider, J.A., M.S. Pungliya, J.Y. Choi, R. Jiang, X.J. Sun, B.A. Salisbury, and J.C. Stephens. 2003. DNA variability of human genes. *Mechanisms of Ageing and Development* 124 (1): 17-25.

Schwartz, R.S. 2001. Racial profiling in medical research. *New England Journal of Medicine* 344 (18): 1392-93.

Smedley, B.D., A.Y. Stith, and A.R. Nelson. 2003. *Unequal treatment: Confronting racial and ethnic disparities in health care*. Washington, DC: National Academies Press.

Stein, R. 2005a. Heart drug for blacks endorsed. *Washington Post*, June 17.

–. 2005b. FDA approves controversial heart medication for blacks. *Washington Post,* June 24.

Sternberg, S. 2005. Heart drug for blacks gets OK. *USA Today*, June 24.

Sturm, R.A., R.D. Teasdale, and N.F. Box. 2001. Human pigmentation genes: Identification, structure and consequences of polymorphic variation. *Gene* 277: 49-62.

Tate, S., and D. Goldstein. 2004. Will tomorrow's medicines work for everyone? *Nature Genetics* 36 (Suppl.): S34-S42.

Tishkoff, S.A., and K.K. Kidd. 2004. Implications of biogeography of human populations for "race" and medicine. *Nature Reviews Genetics* 36 (11): S21-S27.

Usborne, D. 2004. Controversy over heart drug for blacks only. *London Independent*, November 10.

Wood, A.J.J. 2001. Racial differences in the response to drugs: Pointers to genetic differences. *New England Journal of Medicine* 344 (18): 1394-96.

Zone, R. 2005. Heart drug OK'd for use by blacks. *Seattle Times*, June 24.

Is Small Really Beautiful? Does Size Matter? Nanotechnologies

13
Nanotechnology and Human Imagination

Susanna Hornig Priest

The Project on Emerging Nanotechnologies at the United States' Woodrow Wilson International Center for Scholars so far has identified over six hundred consumer products using–or claiming to use–nanotechnology.[1] These range from the mundane and now-familiar stain-resistant pants to more exotic items, including a recently announced spray-on antibacterial condom for use by women developed in China. One thing is clear: browsing this inventory of advertising hype should convince anyone that from an industry perspective the "nano" designation seems to be more of a selling point than a consumer turn-off. While nanotechnology is widely (or is that wildly?) heralded as the next new breakthrough technology and regularly compared, often in the next breath, to biotechnology, the climate of public opinion for nano is still evolving.

A Google search requiring the words *promise* and *nanotechnology* yielded 868,000 results for me when I was in the process of developing this chapter. Not a very scientific approach, to be sure, but a rather dramatic demonstration of the association of one concept with the other. While many other relevant but somewhat more generic combinations yield even more hits ("public understanding of science" yielded 115 million, for example, and "science literacy" 35.7 million), the idea that the nascent technology exploiting the nanoscale should hold so much promise for the future in resolving problems in health, energy, environment, even material abundance, is apparently well rooted. And I choose the verb *exploiting* deliberately. When Richard Feynman made his famous "plenty of room at the bottom" statement, laying out the vision of nanotech's future at the California Institute of Technology in 1959, the cultural resonance with the attraction of unexplored and unexploited frontiers was unmistakable, and it remains so today. Although the West has been won (or, depending on how you look at it, already lost), the nanoscale is yet to be conquered.

Yet "nanotechnology defined" yielded even more hits, 938,000 to be specific, suggesting that public confusion continues over just what this term encompasses. According to the US National Nanotechnology Initiative website at nano.gov, nanotechnology is "the understanding and control of matter at dimensions of roughly 1 to 100 nanometres, *where unique phenomena enable novel applications*" (emphasis added). (A nanometre is one billionth of a meter.) It is not just about being tiny, in other words, but about applications that exploit the atomic properties of tininess and novel instrumentation that permits nanoscale observation by human beings. In part because of the substantial government and industry funding available for work in nanotechnology and nanoscience in anticipation of its revolutionary impacts, more and more work is described in these terms. As the defining characteristic is one of scale rather than of the field of inquiry or the scientific theory involved, this expansion undoubtedly encompasses applications that, were it not for the availability of funding, would have been described in other terms. It almost seems that the nanohype in the academic world rivals that in the world of consumer advertising.

Debate also continues between a vision of nanotechnology known as molecular manufacturing (7.73 million hits), which some seem to imagine might enable the on-demand creation of almost any materially or biologically desirable structure and which others see as threatening to unleash hoards of uncontrolled nanobots on the world, versus more conservative views. It is the molecular manufacturing scenario that once raised fears of the world reducing itself to a pile of "grey goo" as rampant nanobots destroy all other material in their effort to reproduce only themselves. While most of the scientific community dismiss these unsettling projections as strictly science-fiction scenarios designed to instil fear in the hearts of the scientifically illiterate masses, they do seem to fear a popular backlash occurring that would parallel the public reaction to genetically modified foods. (Perhaps with some justification, they also fear involvement of some of the same activist interest groups.) At least in the United States, one result of this concern has been a substantial public investment in studying societal implications, improving related public education and outreach, and sponsoring public engagement in discussions of nanotechnology's future.

Whether nano will ultimately yield a Star Trek-inspired brand of molecular magic in which free filet mignon dinners and clean human arteries alike are manufactured on demand is an issue well beyond the ability of my own imagination to resolve, and I will certainly not attempt to go beyond this limitation to try to predict what seems a very uncertain future. In the more mundane world in which we actually live, however, nanotechnology is more commonly seen as promising a variety of more down-to-earth advances, as varied as radically reduced-scale electronics, stronger materials, new medical treatments for disease, and a range of advances in disease prevention, more efficient energy production, and sensors capable of detecting individual atoms, not to mention wrinkle-free pants, improved automobile tires, and nano-based sunscreens and cosmetics that more easily penetrate the human body (with health effects that are not yet completely certain). What unites such diverse developments is, again, simply the scale at which the engineering research takes place. Regulators are scurrying to develop schemes to monitor the environmental and health effects of a wave of technologies connected more by their foundational size than necessarily by their actual properties or the processes used to develop them. A great deal of contemporary concern surrounds the potential of nanoparticles to penetrate and disrupt tissues (such as lung tissue) that did not evolve to resist this artificially fine-grained form of matter. It now seems likely that workers in nanotechnology industries may be those exposed to the greatest health risks.

In this world of confused and competing visions, both positive and negative, combining fact, fantasy, and science fiction, and all difficult to grasp because of their intimate connection to matters too small to be seen or (generally) even "accurately" imagined,[2] the visions imposed on nanotechnology by diverse publics are naturally very dependent on pre-existing attitudes toward science and technology. It is through analogy with earlier technologies, their regulation, their positive and negative effects on everyday life, and their more problematic dimensions that ordinary people find the means to grapple with the newest ones. Levels of trust in government, trust in industry, and trust in science help determine whether people who are not themselves scientific or engineering experts will have trust in current emerging technologies; these levels of trust are determined in large part by previous experience. Levels of trust in the critics of these emerging technologies also matter.

Public responses in pluralistic, multicultural societies vary widely, and these variations produce dynamic climates of public opinion that can be difficult to conceptualize or predict, especially for newly emerging technologies that rest on complex science. In such cases, opinion is often seen as volatile, easily swayed by partial evidence or exposure to only one side of a complicated issue. However, I argue here that it is easier and more accurate to conceptualize public opinion in this area as being formed on rather different foundations by different segments of the so-called general public. Seen from this perspective, ordinary people appear less "volatile" or "irrational" as they do responsive to different sets of influences, depending on their own values, interests, and beliefs. Several distinct attitudinal groups of interest can be identified in this context, based on available North American survey data that provide windows (however limited) into popular thinking and its many variations.[3]

And when we consider the visions of nanotechnology such groups might create, it is important to remember that, on the one hand, nano is in many respects a completely blank slate, with expressed hopes for it therefore seeming to reflect fundamental human values more than the promises articulated for this particular set of technologies, even as broadly defined as these have been. When we are presented with a set of technologies that seem to promise the potential to provide us with almost everything, responses about what is most sought from them are a window on what we most desire. On the other hand, ordinary people understand new technologies primarily in terms of their experiences with older ones, experiences that provide most of the basis of public trust or distrust in both technology's managers and their critics.

Who Are the Publics for Nano?

Using opinion poll data, it is possible to discern the parameters that may separate different publics for technology, including nanotechnology. By a "public" I mean a group characterized by a particular constellation of interests and perspectives vis-à-vis this (or any other) new constellation of technologies. To be a member of a public should imply having an interest and some kind of reasoned perspective on an issue or set of issues, but the foundations on which interests and reasoning are based varies from group to group. This concept is in some ways very similar to that proposed in the literature on "interpretive communities," reflecting the recognition that a variety of cultural factors influence the understandings of new information and that members of an interpretive community share common elements that may significantly influence that understanding.

Public opinion researchers (and, on occasion, informed public relations specialists as well) remind us that thinking in terms of a single generic public can be misleading and suggest this language of "publics" instead. In slightly different contexts, it might be appropriate to refer to different stakeholders (those with particular stakes or investments in the outcome of an issue or issue set) or audiences (those with particular information and media consumption patterns of a sort that might pique the interest of both media professionals and marketers) to identify other, related divisions within the so-called general public. Those trying to sell products to consumers may make parallel distinctions but are more likely to refer to them as market segments. Clearly, the concept of "consumer"–while it has elements in common with the concept of "public," implying "citizen"–conjures up an image of someone we intend to manipulate, as opposed to someone we might consult about a social decision.

In fact, each of these different terms has its own set of connotations, but each is based on a similar recognition that it is useful to think beyond a single mass public, or of public opinion as a single entity. The bottom line is that it is more useful, and more accurate, to think in terms of a range of publics (or communities, or sets of interests) than a more faceless and uniform general public, particularly when thinking about reception for information about specialized areas such as science and technology, religion and ethics, or public or international affairs. People not only have different levels and kinds of knowledge, but they listen to different types of leaders and believe that scientific controversies are best settled by different kinds of evidence and on different grounds. While this recognition certainly challenges the assumption that public dialogue and public engagement efforts will readily produce consensus, it should help those trying to introduce new technologies into contemporary societies to anticipate the range of reactions that will occur and to understand why. The concept of publics is not reducible to demographic distributions; however, the demographic characteristics of particular publics do vary in what are presumably nonrandom ways.

Previous analysis (Priest 2005a, 2005b) has established that about five publics can be usefully identified for emerging technologies. While these groupings are the result of inductive analysis of available survey data and therefore have some statistical foundation, it is important not to reify these particular divisions. Nevertheless, these categories not only emerged on analyzing available survey data (always an impoverished reflection of actual opinion) but also have some resonance to observers of public opinion formation. They are as follows: *true believers* who assume science and technology are generally benevolent; *utilitarians* who want experts to examine risks as well as benefits in weighing new options–by some measures the largest group in both Canada and the United States; *moral authoritarians* who seek leadership and guidance on the ethical issues and possibly also other socio-political factors they see associated with developments in science and technology; *democratic pragmatists* who want people to be empowered to make up their own minds about risks and benefits; and *ethical populists* who want people to be empowered to make up their own minds but primarily on moral grounds.

This analysis elaborates substantially on Miller's well-established suggestion (1986) that there are attentive and interested publics for science who keep up with the news in this area. Although this observation by Miller has been very useful to many studying public understanding of science, the real picture is more complex. Persons who spend statistically disproportionate amounts of time consuming information on scientific topics–often true believers, in terms of the analysis above, or the "attentive public" in Miller's terms–should not be entirely discounted as potential audiences for information about these topics. Other publics are relevant as well. People who want to promote science need to reach these alternative audiences, not just the "attentive" audience or the true believers. In particular, the scientific community wanting to anticipate and manage public criticism should pay special attention to a variety of publics, including those who are critical as well as supportive in their responses to science and technology. The era in which science policy was made behind closed doors is behind us.

Again, the specific list of publics resulting from this broader exploratory analysis should not itself be reified. A different set of survey questions could yield a slightly different set of publics. This is a realm of no particular right answers, but the existence of a range of perspectives–potentially irreconcilable perspectives–certainly seems to be supported by the available evidence. While true believers accord high credibility to scientific voices but tend to discount environmentalist ones, democratic pragmatists find environmentalist voices most credible; moral authoritarians and ethical populists both listen to religious and political voices, not as much to science; utilitarians weigh both scientific and environmentalist arguments (Priest 2005a). No wonder the debates about biotechnology and nanotechnology (different in some ways, alike in others) seem to go on forever without resolution.

The crucial point about the introduction of nanotechnology (itself a concept with many variations) is that although this is in some ways a sort of blank-slate technology that promises–it seems–almost everything, it is a slate drawn upon by more than one type of visionary artist. Like a Rorschach drawing, the vagueness of nanotechnology's definition and the as-of-today incompleteness of public understanding–we might say its emergent character–invite the application of rich imagination. Competing visions of technology's general promise, and competing ideas about how policy decisions are best approached, characterize modern pluralistic societies, and nanotechnology–in which so much has been invested–becomes the most recent exemplar. The challenges of governance in such circumstances are substantial. Nanotechnology would not occupy centre stage in contemporary debates about the deployment of technology if there were not already a substantial financial commitment from developed economies around the world in this area. While crying out for public engagement to establish its legitimacy, the nanotechnology "enterprise" in fact presents itself as a *fait accompli* to most observers. Under these circumstances, public engagement is difficult to distinguish from market research.

Upstream Engagement in Defining the "World of Small"

In the recent past, Western societies operating within strongly democratic traditions did not always accommodate emerging technologies easily. Genetically modified foods struck people as unnatural; stem cell technology struck people as immoral; cloning struck people as simply unimaginable. Engaging the public earlier in the process–"upstream" from status quo decision making–has become the panacea solution to the "problem" of public resistance. Ironically, this is a solution that seems to have been proposed, in the first instance, primarily by somewhat anti-establishment intellectuals, yet it is one that in many ways undeniably serves establishment interests. The whole point of so-called upstream engagement is to get ordinary people to have a say in technologies' development and deployment before it is too late for them to influence outcomes–thus converting them to supporters of these self-same technologies in which it is assumed they will see their own values mirrored. So the end result is broadly conceived to be a public less resistant to the technologies in question. For establishment science, the purpose may be less to improve democracy as to quiet criticism. Politics does indeed make strange bedfellows.

Nanotechnology opinion data suggest that this broad range of defined publics of science and technology is less likely to believe nanotechnology is being produced by those who keep their (that is, ordinary peoples') interests, values, and beliefs in mind than it is to believe that biotechnology is being produced by such people (Priest 2006). Nanotechnology's promoters should pay attention to the implications of this kind of evidence. In other words, despite the public controversy over stem cell research and human cloning, nanotechnology is seen as existing in a social realm yet more distant from people's interests, values, and beliefs than is biotechnology. On whose behalf are all these millions and billions of dollars being invested, then? And for what? Perhaps this perception is only because nano is new and not yet especially familiar.

Upstream public engagement in a context in which nanotechnology is only vaguely understood but is broadly imagined, and promoted, as providing whatever one wants it to provide (material goods, economic prosperity, improved health and longevity) will largely reflect human social values, rather than choices among those elements that nanotech is actually capable of providing. Research on what people want from nanotechnology is interesting, then, in part because it provides a window on what people imagine technology might provide; that is, it provides a window on what they want from technology (and science) in general. But it is hardly a reality-grounded foundation for deciding in what directions to take nanotechnology, in the end. Nanotechnology (or any technology) is likely to be able to provide some things, but not other things–most certainly not all things. Ordinary people have experienced this before–every technological fix has its finite limitations.

Were I to be invited to some sort of upstream engagement exercise, no doubt incognito, I think I would ask whether nanotechnology could reduce global warming (in my own lifetime), halt the spread of HIV/AIDS infection, poverty, and malnutrition in the developing world (in my own lifetime), or improve my own First World health and longevity (very clearly, in my own lifetime). Can nanotechnology do any of these things?

Will nanotech's vague promises about new forms of energy production and new cures for cancer actually respond to these hopes? No, or perhaps, not necessarily. Then (I would have to ask) why are we pursuing it at all? Because frankly it seems we will all be dead or poisoned by the time we reap any of its more nuanced benefits. Yet, serious scientists do hope that nanotechnology can make a contribution to most, if not all, of these problems. If so, what investment is too great to make? The uncertainty of the promise is difficult to grasp, a poor foundation on which to act–or not act. What if we reject nano, when it might possibly be capable of all these things? What opportunity costs are attached to investment in one form of science and technology and not another?

If upstream engagement does not entertain discussion on whether the social distribution of benefits from these technologies in which we are investing so many resources will actually improve the current maldistribution of wealth in the world–particularly, in my own myopic view, in the United States, as well as between the United States and the rest of the world–then what is the point? If nano does not solve our dependence on types of energy sources that exacerbate global climate change (and some in the field doubt that it can do so anytime soon), then where will it be able to take us? Empowering ordinary people to inscribe on nanotechnology's future their particular parochial visions of what they would like–better health, more jobs, improved material conditions–simply does not go far enough. I have complete and total sympathy for these aspirations and share them myself. But they will be impossible to achieve unless we can address the bigger issues such as poverty, epidemic disease, and global warming that threaten our very existence. In this context, upstream engagement seems like child's play–and perhaps it is. "Public understanding of science" and "technological literacy" also need to mean teaching people to connect the dots between the development and exploitation of particular technologies and the future (or extinction) of human civilization.

Cultural Mythology and the Pursuit of Values

Earlier in this chapter I raised the issue of the cultural myth, completely familiar to all North American media audiences and visibly resonant in the "room at the bottom" rhetoric, of exploiting the frontier. The Western genre television and movies I grew up with were all about this theme, as was *Star Trek* and its various spin-offs, including *Star Trek: The Next Generation,* which took the old mythology and wrote it larger, scribbled freehand across the entire universe: Thou shalt exploit the frontier albeit on behalf of democracy and freedom! Let me not be misunderstood: I am unambiguously in favour of democracy and freedom, and I am also generally in favour of science and technology–although I occasionally have my doubts in particular instances. I do worry about the deployment of our frontier mythology to justify the narrow pursuit of nanotechnology, or any other technology afforded such power to both revitalize our economy and restore our health, in the face of so many immediate threats to our very existence brought on in many cases by that same technology. And if it is only for better sunscreen and more wrinkle-resistant pants that we are investing so many of our precious spare dollars for research and development, we are falling short of the mark. We have other things to worry about.

But let me regain my composure and shelve my personal concerns about priorities, just for the moment. The pursuit of nanotechnology and its promises certainly resonates with US cultural mythology about conquering the frontier, but the data suggest that North American publics–whether inclined to embrace science or to question it–nevertheless remain alienated from the realm of all things nano. Across all the publics–from true believers to moral populists–nanotechnology's developers are not seen as representing people's own interest, values, and beliefs so much as is biotechnology, the object of such considerable controversy but still also the carrier of so much optimism when it comes to health, food, heredity, and human promise (Priest 2006). North American culture is largely built on this kind of technological optimism. Nanotech needs to demonstrate its ability and willingness to live up to this promise, or at least to contribute to our imagining of a promising future expanded–not circumscribed–by what nano actually proposes to do for us. Both the utilitarian public that seeks to balance perceived risk and perceived benefit and the moral authoritarians who seek ethical foundations for public policy decisions might actually agree on this point.

Preliminary focus group results from six discussion groups conducted in three US cities during the summer of 2005 and three in Canada suggest that the most common public reaction to nanotechnology is that it is helpful, making life easier and providing additional convenience.[4] People are not especially afraid. But the second most common type of comment is about the idea that nanotechnology is harmful or has risks; in addition, participants in these groups were aware that too much is unknown to fully evaluate these risks. The implication is that ordinary people would like to weigh risks against benefits and engage in this actively when invited to consider new technological possibilities. This is consistent with research on the publics for technology in North America, which suggests that utilitarian perspectives asking about the balance of risks and benefits predominate. Participants were uncertain what kinds of risks, and benefits, nanotech might actually provide, but concerns over the distribution of benefits (differential access to health care, who gains jobs and who loses them) are much more commonly expressed than concerns over nanobots destroying the world.

If public engagement is a serious undertaking, people's real concerns need to be taken seriously. A significant risk for science, however, is overselling technology as a solution to everything. Nanotechnology is a good case in point in this regard.

Conclusion

Public perceptions of nanotechnology can reasonably be proposed as emblematic of all North American responses to technology. We are often asked, after the fact, to embrace technologies that have been created in response to perceived scientific and market opportunities and that, we are told, will also be certain to help us to achieve our vision of a good life. It has undoubtedly often been the case that these technologies *have* improved our lives by lowering the prevalence of many diseases and vastly increasing our material wealth.

In addition, our receptivity to such arguments is rooted in several centuries of cultural commitment to the enabling and liberating potential of a variety of technologies; to the Jeffersonian vision of science and technology as progress; to the Deweyan conception of the rational human being making reasonable, informed, and intelligent choices; to the agrarian roots of our contemporary vision of the good life. Generally positive public expectations for nanotechnology are no doubt rooted in this longstanding cultural proclivity, despite the public controversies associated in our recent history with various controversial technologies, ranging from biotechnology to nuclear technology.

But our technologies have also caused a host of new and often unanticipated environmental and health problems, and early indications are that nanotechnology will be no exception. When asked to think about new technologies and their impact, ordinary people remember their own past experiences with technological risk and ask themselves questions about whether they can trust the regulators. Nuclear and chemical technologies have raised new issues of the social distribution of risk. Biotechnology has also challenged cultural notions of individuality and mortality in ways that nanotechnology does not but may soon.

The current upstream public engagement movement–if it is more than a marketing ploy to enlist the public as supporters by capitalizing on and reinforcing our historical good-faith receptivity to technology–must find some political purpose beyond simply allowing people to express their hopes and fears. Such expressions may have inherent value but will not of themselves generate consensus or provide a path forward *despite* this lack, given the existence of a variety of publics for new science and technology. Those who make public policy in this arena have a complicated challenge.

Because nanotechnology promises everything, it seems to serve as a *tabla rasa* that invites speculation about what "everything" might actually consist of. The result could well be an unusual opportunity to speculate on the relationship between visions for technology and visions for what broader American values might ask of it. This speculation is a valuable opportunity to clarify our values but should not blind us to the fact that technological fixes may not always be available to save us from our contemporary shortcomings. Significant humanitarian and environmental challenges lie ahead, and the technologies we should embrace are those that address them.

Engaging the public–whether upstream or downstream or midstream–does not let us off the hook on this point. Nor will it necessarily unite those who seek ethical technologies and those who seek technologies that provide material benefits. Perhaps it will ultimately draw our attention to the necessity of finding and developing those technologies that contribute to a good life that is both economically abundant and socially just.

Notes

1 The database can be browsed or searched at http://www.nanotechproject.org/inventories/consumer/.

2 Debate also rages over the "proper" visual representation of these things that are too tiny to be actually seen by human—or any other biological—eyes.

3 That is, based on 2004 and 2005 surveys in Canada and the United States funded and implemented by the Canadian Biotechnology Secretariat, whose support of the present analysis is gratefully acknowledged.

4 Results based on groups in Columbia, South Carolina; Chicago, Illinois; and Phoenix, Arizona; research funded by the National Science Foundation, Science and Society Program, under NSF Grant Number 0523433 to the University of South Carolina. The Canadian groups were funded by the Canadian Biotechnology Secretariat through a grant to the University of Calgary.

References

Miller, Jon D. 1986. Reaching the attentive and interested publics for science. In *Scientists and journalists: Reporting science as news,* ed. Sharon Friedman, Sharon Dunwoody, and Carol Rogers, 55-69. New York: Free Press.

Priest, Susanna H. 2005a. International audiences for news of emerging technologies: Canadian and US responses to bio- and nanotechnologies. In *First impressions: Understanding public views on emerging technologies,* ed. E. Einsiedel, 20-42. Report prepared by the University of Calgary Genome Prairie GE^{3}LS team for the Canadian Biotechnology Secretariat. http://www.biostrategy.gc.ca/CMFiles/CBS_Report_FINAL_ENGLISH249SFD-9222005-5696.pdf.

–. 2005b. The public opinion climate for gene technologies in Canada and the United States: Competing voices, contrasting frames. *Public Understanding of Science* 15 (1): 55-71.

–. 2006. The North American opinion climate for nanotechnology and its products: Opportunities and challenges. *Journal of Nanoparticle Research* 8 (5): 563-68.

14 Nanotechnology: The Policy Challenges

Lorraine Sheremeta

Nanoscience is the study of phenomena and the manipulation of materials at the atomic, molecular, and macromolecular scales. At this scale (approximately 1-100 nm), certain properties of matter–including those once thought immutable (including melting point, boiling point, and electrical conductivity)–can differ significantly from those observed at the macro-scale. Nanotechnology is the application of nanoscience in the design, characterization, production, and application of materials, structures, devices, and systems at the nanoscale. Nanotechnology is inherently interdisciplinary in that it has come to exist through evolutionary developments in physics, chemistry, biology, engineering, and medicine as researchers probe progressively smaller size scales. It is evident that advances in nanoscience and nanotechnology are bringing about revolutionary changes in our understanding of, and ability to manipulate, the physical world. The societal implications of these new technologies promise to be profound.

In 1999, Nobel laureate Richard Smalley predicted that

> the impact of nanotechnology on health, wealth, and the standard of living for people will be at least the equivalent of the combined influences of microelectronics, medical imaging, computer-aided engineering, and man-made polymers in this century. (Smalley 1999, 2)

The purpose of this chapter is to identify the main policy challenges that nanotechnology and, in particular, engineered nanomaterials will raise. These challenges include the following: (1) definitional and measurement challenges, (2) risk to human health and the environment, (3) the complexity of the regulatory regimes that are relevant to the consideration of nanotechnology, and (4) the need to maintain public trust through responsible stewardship. It will also demonstrate that funding is urgently required to enable human health and environmental effects research as

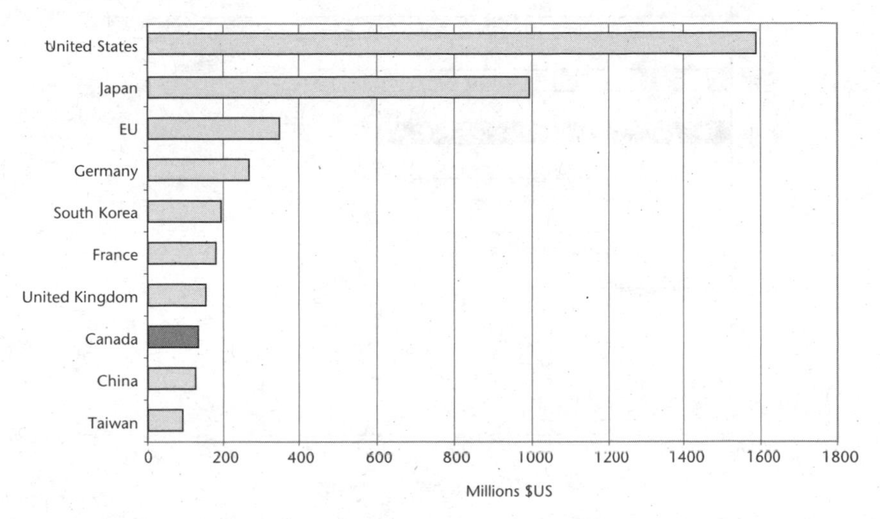

Figure 14.1 Worldwide government nanotechnology funding (2004)
Source: Office of the National Science Advisor (2005), data used with permission

well as large-scale interdisciplinary research, including research into the ethical, environmental, economic, legal, and social issues.

The Promise and Perils of Nanotechnology

The development of nanotechnology will have implications for every economic sector and "holds great promise for revolutionizing how we manufacture products, communicate with each other, and treat disease" (Goldman and Coussens 2005, xi). Indeed, *Science* magazine named the construction of functional nanoscale circuits able to perform logical operations the breakthrough of the year for 2001 (Kennedy 2001). The Woodrow Wilson International Center for Scholars hosts an online nanotechnology consumer products inventory that currently lists nearly four hundred marketed products, including, among other things, paints, sunscreens, antibacterial wound dressings, and natural health products.[1]

Whether exaggerated or not, the potential benefits of nanotechnology are likely to be significant and are inspiring governments around the world, including the Canadian government, to invest heavily in nanoscience and nanotechnology research and development (Roco 2003; Royal Society and the Royal Academy of Engineering 2004; Office of the National Science Advisor 2007). The 2008 budget request for the US National Nanotechnology Initiative (NNI) was in excess of US$1.44 billion (NSET 2007).

In terms of per capita spending on nanotechnology, Canada slipped

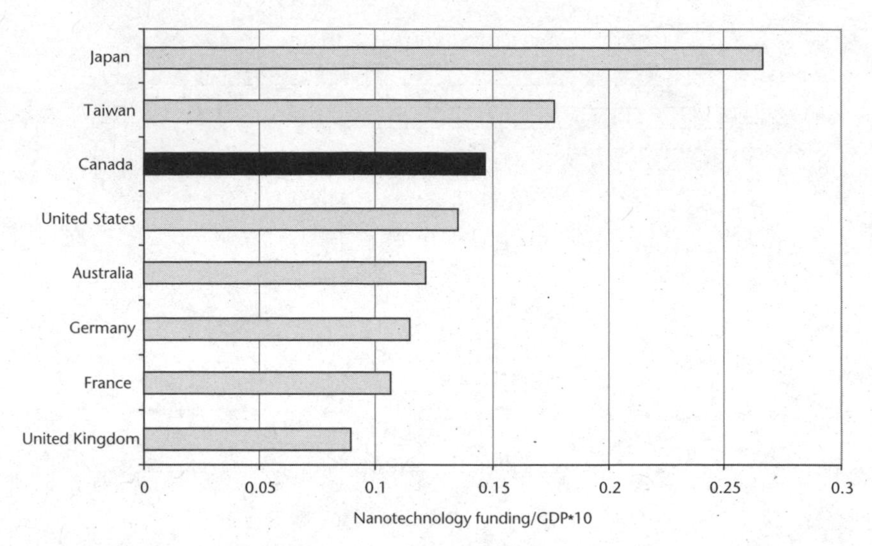

Figure 14.2 Nanotechnology funding relative to GDP (2004)
Source: Office of the National Science Advisor (2005), data used with permission

from third in 2004 to eighth place in 2005, behind Japan, Germany, the United States, Taiwan, the United Kingdom, France, and Australia (in decreasing order of spending (see Office of the National Science Advisor 2007, Figure 2). Analysis by the Office of the National Science Advisor and Statistics Canada estimates that total investments in nanotechnology research reached Cdn$246 million in 2005, a 33 percent increase from 2004. About a fifth of this was from private sources, with the federal government as the major source of funding through a wide array of programs, including research funding to universities through granting councils and direct funding for business through economic development research support and assistance.

A commercial research firm reported funding for nanotechnology worldwide in 2006 as totalling US$11.8 billion, an increase of 13 percent from 2005 (Lux Research 2006). In the area of government spending, according to this report, the European Union outspent the United States on nanotech research and development in 2006.

Given the overt and aggressive national and international initiatives to develop nanotechnology products, it is instructive to reflect on the historical parallels of the efforts to develop biotechnology products and the plethora of problems that have subsequently been encountered in that sphere. It is trite, but unfortunately necessary, to say that nanotechnology policy should not be developed on the basis of wildly inflated potential benefits nor paralyzed on the basis of irrational fear of unknown risks.

Sound policy decisions about nanotechnology must be based on the best available scientific information and in combination with a concerted effort to predict and proactively address the ethical, environmental, economic, legal, and social issues that will inevitably arise.

Policy Challenges

Defining and Measuring Nanotechnology

Nanoscience and nanotechnology are characterized by the unique elements of size and scale, as well as of complexity. These elements lead to profound definitional, metrological, and regulatory challenges. It follows that, from a policy perspective, that which is inherently hard to characterize and measure will be difficult to regulate. Definitional challenges have already led to significant difficulties in the identification of parameters to understand research and development expenditures. Measurement challenges are a particular problem for international coordination and for the development of appropriate occupational health and safety norms.

The most frequently cited definition of nanotechnology is that of the US National Nanotechnology Initiative, which defines nanotechnology as the understanding and control of matter at dimensions of roughly 1 to 100 nanometres, where unique phenomena enable novel applications. Encompassing nanoscale science, engineering, and technology, nanotechnology involves imaging, measuring, modelling, and manipulating matter at this length scale. At the nanoscale, the physical, chemical, and biological properties of materials differ in fundamental and valuable ways from the properties of individual atoms and molecules or bulk matter. Nanotechnology research and development is directed toward understanding and creating improved materials, devices, and systems that exploit these new properties.[2]

The US National Nanotechnology Initiative distinguishes nanotechnology research and development from other types of scientific research that simply operate at the nanometre scale but do not exploit novel properties. The distinction between what is, and what is not, considered nanotechnology is particularly challenging at the intersection of nanotechnology and biotechnology. In light of this, the US National Institutes of Health has clarified the situation as follows:

> While much of biology is grounded in nanoscale phenomena, NIH has not re-classified most of its basic research portfolio as nanotechnology. Only those studies that use nanotechnology tools and concepts to study biology; that propose to engineer biological molecules towards functions very different from those they have in nature; or that manipulate biological systems by methods more precise than can be done by using

> molecular biological, synthetic chemical, or biochemical approaches that have been used for years in the biology research community are classified as nanotechnology projects.[3]

In a similar vein, Canada's National Research Council defines nanotechnology as

> the application of science and engineering at the atomic scale. It facilitates the construction of new materials and devices by manipulating individual atoms and molecules, the building blocks of nature. Nanotechnology enables the atom-by-atom design and fabrication of tiny structures that are very small, typically 1-100 nanometres, and which have new properties and powerful application in medicine and biotechnology, in energy and the environment, and in computing and telecommunications.[4]

In its first request for nanomedicine funding proposals, the Canadian Institutes of Health Research (CIHR) defined nanotechnology as "the design, synthesis or application of materials, devices or technologies in the nanometre scale for the basic understanding, diagnosis, and/or treatment of disease" (CIHR 2005). However, "many current research initiatives in the development of novel techniques and methodologies relevant to biomedical research and clinical practice [did] not necessarily fit within this strict definition" (CIHR 2005). For purposes of the second announcement, the CIHR recognized that microscale technologies, though not meeting the strict size definition, are relevant for nanomedicine and regenerative medicine (i.e., cellular imaging, biophotonics, drug delivery and targeting, and molecular characterization of cellular processes) and were potentially fundable. The expanded definition also includes the application *to health* of existing technologies and methodologies not traditionally associated with the life sciences. Relevant disciplines could include, for example, mathematics, computational sciences, chemistry, physics, and engineering and applied sciences.

In addition to creating administrative challenges for research funding agencies, the definitional issues described above create challenges for governments, market analysts, and others who seek to quantify how much nanotechnology research and development is being done in a given region and what is the likely commercial potential of the developments associated with nanoscience research. Given the acutely competitive nature of nanotechnology research and development, there will inevitably be inconsistencies (intentional and unintentional) in the definition and/or the application of the definition of nanotechnology when attempting to measure current investment and to attract future investment. A common definition applied in a standardized fashion by all funding agencies and investors would facilitate accurate comparisons of relative research and development expenditures both within and between nations. The issue has arisen in Canada, for example, where two provinces have sought to quantify their nanotechnology research and development capacity using markedly different definitions.

Related to the definitional challenge is the need to develop common terminology and formalisms that will appropriately describe nanotechnology to enable scientists from diverse disciplines to communicate and work together. Nomenclature conventions are necessary to eliminate ambiguity when communicating differences between nanomaterials and bulk materials and in reporting for regulatory purposes. To that end, Canada is leading the International Organization for Standardization (ISO) working group on terminology and nomenclature (ISO/TC299). Two additional working groups have also been created under TC299: metrology and characterization (Japan); and health, safety, and the environment (United States). ISO/TC229 will produce standards for classification, terminology and nomenclature, basic metrology, calibration and certification, and environmental issues related to nanotechnology. It will also develop standardized test methods that will focus on physical, chemical, structural, and biological properties of materials or devices whose performance is critically dependent on one or more dimension of less than 100 nanometres (ISO 2005; Hatto 2006). Other organizations are also working toward developing a common nomenclature for nanotechnology. These include, among others, the American National Standards Institute, ASTM International (formerly the American Society for Testing and Materials), and the Chemical Abstracts Services.

In a recent report published by the Project on Emerging Technologies at the Woodrow Wilson International Center for Scholars, J. Clarence Davies (2006) questions whether it makes sense to regulate nanotechnology on the basis of size and whether a definition can be found that will facilitate certainty between manufacturers and regulators. These questions remain to be answered and highlight a critical issue for policy makers in Canada and elsewhere, given that the ability to regulate this area will depend on the definition of nanotechnology, the standards that are developed, and the ability of regulators to apply them (Davies 2006).

Nanotechnology Risks: Human Health and the Environment

The Institute of Medicine and others have noted that nanotechnology holds much promise through new applications directed toward human health and the environment. A concurrent development of the knowledge base about the risks associated with nanomaterials and products that contain them is also needed. By knowing about the risks and ways to address them, environmental health scientists will be better able to serve the public (Goldman and Coussens 2005, xii).

Nanotechnology spans a vast area of scientific endeavour that is rapidly developing and evolving. Not surprisingly, new risks will accompany these technological advances. Appropriate safeguards to protect human health and the environment will need to be developed as the technology matures and enters widespread use. The emerging body of evidence about the adverse effects of nanomaterials is nascent, but it is sufficient to raise genuine concern about potential harm to human health and the environment (Colvin 2003; Maynard and Kuempel 2005; Oberdörster 2004; Hoet, Nemmar, and Nemery 2004; Shvedova et al. 2003; IRSST 2006).

Engineered nanomaterials–"materials designed and produced to have structural features with at least one dimension of 100 nanometres or less" (Oberdörster et al. 2005, 2)–can possess size-related properties that make them ideal for use in a variety of applications.[5] It is these very properties that may lead to "nanostructure-dependent biological activity that differs from and is not predicted by the bulk properties of the constituent chemicals and compounds" (Oberdörster et al. 2005, 2). The US Environmental Protection Agency proposes that nanomaterials be organized into four basic types (EPA 2005) (see Table 14.1).

The state of knowledge about the effects of nanomaterials on human health and the environment is incomplete (President's Council of Advisors on Science and Technology 2005). To date, relatively few studies that evaluate the effects of nanomaterials on human health or the environment have been published (Oberdörster et al. 2005; Maynard and Kuempel 2005). Concerns have been raised, however, by numerous groups that nanoparticles, because of their size and surface characteristics, can behave differently *in vivo* than larger particles of the same material (Warheit et al. 2004; Maynard et al 2006). Preliminary studies in fish suggest that C60, a carbon molecule in the conformation of a truncated icosahedron consisting of precisely sixty carbon atoms (Kroto et al. 1985; Smalley 1996) may be able to translocate to the brain and cross the blood brain barrier (Oberdörster 2004). In response to these growing concerns, the new field of nanotoxicology has emerged to study the effects of nanomaterials in biologic systems (Oberdörster, Oberdörster, and Oberdörster 2005; Oberdörster et al. 2005).

Engineered nanoparticles, including metal oxides and carbon nanotubes, are currently used in a variety of applications, including sunscreens, cosmetics, sports equipment, resins, paints, and other surface coatings.[6] Products employing nanomaterials have been launched on the market following a relatively short research and development phase, and this may be cause for concern. It is of particular concern to insurers and re-insurers

Table 14.1

Four categories of nanomaterials

Category	Description
Carbon-based materials	Nanoparticles composed entirely of carbon and taking the form of a sphere (buckyball, fullerene, nano-onion), ellipsoid, or nanotube (single or multi-walled). These particles have many potential applications, including improved films and coatings and stronger, lighter materials.
Metal-based materials	These materials include metal oxides such as titanium dioxide, quantum dots, nano-gold, and nano-silver.
Dendrimers	Nano-sized polymers built up from branched units called monomers. Technically, a dendrimer is a branched polymer. The surface of a dendrimer has numerous chain ends that can be tailored to perform specific chemical functions. Dendrimers are also characterized by cavities that may be useful for drug delivery or other delivery applications.
Nanocomposites	Materials that combine nanoparticles with other nanoparticles and/or other bulk materials to yield new materials with enhanced properties (i.e., enhanced mechanical, thermal, barrier, or flame-retardant properties).

who, because of a lack of exposure and toxicity data, are unable to assess the associated risks to human health and the environment. Increasingly, this lack of data is creating uncertainty among industry experts (Swiss Re 2004) and public action groups (ETC Group 2003; Arnall 2003).

Nanomaterials pose various challenges to those seeking to clarify their toxicological profiles. For example, there is a hugely diverse complement of materials that are commercially available at present or that are under development. Frequently, a given nanomaterial can be produced by several different processes that lead to different versions of the same type of material. The EPA notes that single-walled carbon nanotubes, for example, can be mass-produced by employing four different processes, each of which produces carbon nanotubes with a specific size, shape, and composition. These differences may lead to different toxicological profiles for each species of nanotube produced (EPA 2005). It cannot be predicted, at least until more data are amassed, whether toxicity data pertaining to one type of source material will be relevant to similar source materials produced by a different method. What is clear is that test methods to determine the toxicity of engineered nanomaterials in a timely and efficient manner are needed so that accurate risk assessment information can be developed for nanomaterials that are currently in use–especially those that have already entered trade and commerce.

Given the stage of technological development, it appears that in the short term the greatest potential for human exposure to engineered nanomaterials will be in the workplace–both in public and private laboratories (Royal Society and Royal Academy of Engineering 2004). However, sunscreens and cosmetics also pose a potential source of exposure to the general public. Standards for measuring and monitoring workplace exposures need to be examined and developed on an urgent basis. In the meantime, it is not clear what protective measures are appropriate for those working with nanoparticles (Maynard and Kuempel 2005; Maynard et al. 2006). Current data suggest that for insoluble particulates, toxicity increases as particle size decreases (and relative surface area increases). These findings challenge the standard mass-based risk assessment approach and suggest that standard protective measures commonly used in laboratories may require modification. The Royal Society and the Royal Academy of Engineering (2004, 43) assert that "given previous experience with asbestos, we believe that nanotubes deserve special toxicological attention" and have gone so far as to assert that until proven otherwise, nanoparticles should be presumed hazardous. In the United States, the Department of Health and Human Services, the Centers for Disease Control and Prevention, and the National Institute for Occupational Safety and Health have implemented an information exchange that aims to ensure the advancement and sharing of knowledge that will lead to the safe handling of nanoparticles in the workplace (DHHS, CDCP, and NIOSH 2006).

Nanomaterials and the Environment

Nanotechnology raises significant issues for environmental protection. The US Environmental Protection Agency (2005) questions how we can simultaneously "allow full realization of the societal benefits of nanotechnology, while identifying and minimizing any adverse impacts to humans or ecosystems from exposure to nanomaterials" (14). This same question is now being asked by regulatory agencies around the world.

It is anticipated that nanotechnology will provide tools that will enhance environmental quality and sustainability through pollution prevention, treatment, and remediation (Zhang 2003; Masciangioli and Zhang 2003; Glazier et al. 2003). Research is ongoing to determine potential new applications for nanotechnology in the environmental arena. Nanotechnology applications also could be used, for example, to create benign substances that will replace toxic substances that are currently in use. Non-toxic, energy-efficient computer monitors are replacing older models that employ cathode ray tubes. New LCD monitors are smaller, do not contain lead, and consume significantly less power than previous generation monitors (Masciangioli and Zhang 2003).

In situ field research in contaminated sites has shown promising results using nanoscale bimetallic particles for rapidly remediating various common environmental contaminants, such as chlorinated organic solvents and hexavalent chromium in groundwater (Glazier et al. 2003). Field studies undertaken in New Jersey and North Carolina have shown that nanoscale iron particles provide great flexibility in the treatment of environmental contamination. They can be injected into the ground subsurface or deployed in off-site slurry reactors for treating contaminated soil, sediment, and solid wastes. Alternatively, they can be mounted on solid matrices such as activated carbon for treatment of water, waste water, or gaseous process streams (Glazier et al. 2003).

Recently, the use of nanoscale iron particles for Ocean Iron Fertilization (OIF) to reduce atmospheric $C0_2$ has created much controversy in the international arena (Buesseler et al. 2008). The process, involving the stimulation of phytoplankton growth by releasing iron into the ocean, has been tested in small-scale experiments since the early 1990s. Private firms are now driving toward large-scale use of the technology (Climos Inc. 2008). Greenpeace published a scientific critique of OIF in 2007 and noted that the concerns are not new but have recently become widely debated. In 2007 the Scientific Groups to the London Convention and London Protocol (Convention on the Prevention of Marine Pollution by Dumping of Wastes and Other Matter) issued a statement of concern, which set out their view that large-scale ocean OIF operations are not justified by current knowledge on the effectiveness in sequestering carbon and the potential negative impacts on environment and human health (Allsopp et al. 2007). The Ottawa-based ETC Group is concerned that large-scale geo-engineering "schemes" to fix climate change, including seeding the upper atmosphere with nanoparticles and OIF, without understanding the potential effects is irresponsible (ETC Group 2007). At the end of May 2008, at the ninth meeting of the UN Convention on Biological Diversity, a general agreement was reached on the need for a "de-facto moratorium" on ocean fertilization activities (ETC Group 2007; Climos Inc. 2008).

Although nanotechnology is generally expected to offer substantial benefits in the environmental arena, very little is known about the potential impact of nanomaterials on the environment. The increasing manufacture of nanoparticles for use in industrial and medical applications will lead to an increased potential that nanoparticles will be released, as pollution, into the air, water, soil, and groundwater as a manufacturing by-product or as products are disposed of at the end of their useful life (Dreher 2004). More knowledge about the potential environmental impact of nanomaterials must be sought (Maynard et al. 2006; ICF International 2006; Kreyling, Semmler-Behnke, and Möller 2006).

The UK Royal Society and the Royal Academy of Engineering (2004, 32) have recommended that life cycle assessments be undertaken to demonstrate a net savings in consumption for nanotechnology applications and products "to ensure that savings in resource consumption during the use of the product are not offset by increased consumption during manufacture and disposal."[7]

Regulatory Challenges

Both health and the environment are diffuse subject matters that cut across many areas of constitutional responsibility, some federal, some provincial.[8] Both areas may be dealt with by valid federal or provincial legislation, depending on the circumstances and nature and scope of the problem in question. Accordingly, there are many laws and regulations, some overlapping, promulgated by both the federal and provincial governments that are currently relevant to the regulation of nanotechnology (see Table 14.2).

To date, no comprehensive analysis has been performed in Canada to evaluate the overall legislative/regulatory framework for nanotechnology. It is, therefore, unclear whether current Canadian laws and regulations concerning human health and the environment can adequately protect the public from potential risks that nanomaterials and products containing nanomaterials may pose. A preliminary review of the main federal departments, statutes, and regulations applicable to the governance of nanotechnology in Canada, as outlined in Table 14.2, has been commenced (Sheremeta 2006).

National and international governance systems, including Canada's, reflect the scientific understanding of bulk and macro-scale materials. Such systems may, therefore, be inappropriate for the regulation of nanoscale materials. We can expect that new procedures and assessment tools will need to be developed to appropriately assess the products of nanoscience and nanotechnology. Indeed, this is precisely what commentators in the United States, for example, have suggested (American Bar Association 2006a, 2006b, 2006c, 2006d, 2006e, 2006f, 2006g; Davies 2006; Bergeson and Auerbach 2004; Wardak 2003). In 2004, the UK Royal Society and the Royal Academy of Engineering (2004, 77) recommended that "all relevant regulatory bodies [in that country] consider whether existing regulations are appropriate to protect humans and the environment from the hazards outlined in this report and publish their review and details of how they will address any regulatory gaps." A follow-up report commissioned by

Table 14.2

Main Canadian statutes and corresponding regulations relevant to nanotechnology (federal department)

Statute	Regulation
Environment Canada	
Canadian Environmental Protection Act, R.S.C. 1999, c. 33	• New Substances Notification Regulations (Chemicals and Polymers), S.O.R./2005-247 • Persistence and Bioaccumulation Regulations, S.O.R./2000-107
Fisheries Act, R.S.C. 1985, c. F-14	
Agricultural Products Act, R.S.C. 1985, c. 20	
Feeds Act, R.S. 1985, c. F-9	
Fertilizers Act, R.S. 1985, c. F-10	
Pest Control Products Act, R.S.C. 1985, c. P-9	
Oceans Act, S.C. 1996, c. 31	
Arctic Waters Pollution Prevention Act, R.S.C. 1985, c. A-12	
Health Canada	
Food and Drugs Act, R.S., c. F-27	• Food and Drugs Regulations, C.R.C., c. 870 • Medical Devices Regulations, S.O.R./98-282 • Cosmetics Regulations, C.R.C., c. 869 • Natural Health Products Regulations, S.O.R./2003-196
Hazardous Products Act, R.S.C. 1985, c. H-3	• Controlled Products Regulations, S.O.R./88-66 • Ingredient Disclosure List, S.O.R./88-64
Workplace and Public Safety Programme	• Work Hazardous Materials Information System (WHMIS)
Employment and Immigration	
Canada Labour Code, R.S.C. 1985, c. L-2	• Canada Occupational Health and Safety Regulations, S.O.R./86-304
Provincial Labour Codes and Occupational Health and Safety Codes	• Work Hazardous Materials Information System (WHMIS)

the Department for Environment, Food and Rural Affairs (DEFRA) and published online concludes that

> it is likely that the lack of understanding about the potential impacts of NMs [nanomaterials] on human and environmental health, coupled with limited validated methods for monitoring their levels in the environment, will significantly affect the ability of operators and regulators to act in this area. A substantial body of work will be required to reduce the uncertainties already clearly expressed in the Royal Society/Royal Academy of Engineering report published in 2004. There is a need for setting clear definitions for nanotechnologies and NMs, and categorising different types of NMs into *new* (or different form) or *existing* substances, as this will have a major bearing on the appropriateness and applicability of a number of current and future legislation. (Chaudhry et al. 2006, 94)

A recent report by J. Clarence Davies (2006) of the Woodrow Wilson International Center for Scholars in the United States suggests that a new nanotechnology law is needed in that country. The US Environmental Protection Agency argues that, in the United States, federal authority to regulate nanomaterials pursuant to the Toxic Substances Control Act and other relevant statutes is sufficient to regulate nanotechnology. The American Bar Association's Section of Environment, Energy, and Resources recently published a comprehensive series of papers which, when taken together, comprise a regulatory gap analysis of core US federal environmental statutes vis-à-vis nanotechnology (American Bar Association 2006a, 2006b, 2006c, 2006d, 2006e, 2006f). In general, the papers concur with the Environmental Protection Agency's position and conclude that the agency possesses sufficient legal authority under the statutes to address the challenges it is likely to encounter as it assesses the risks and benefits associated with nanotechnology. A similar analysis of Canadian laws and regulations is required.

In the Canadian context, government departments need to foster an innovative interdepartmental/interdisciplinary approach to deal with the challenges nanotechnology poses. Although a new law may not be needed, regulatory clarification will almost certainly be required. In the development of new regulations or guidelines, critical expertise from various parties must be sought and utilized (Bergeson and Auerbach 2004). Necessary expertise may reside in the minds of government scientists or policy makers, or experts from industry, academia, or the publics at large. Collaboration between all constituents is necessary if sound policy responses are the desired outcome. So far, Environment Canada alone has been exploring its regulatory options in the context of new chemical substances (Environment Canada and Health Canada 2007).

In the meantime, regulatory uncertainty will become increasingly costly to industry and may be fatal to some private firms. We have moved into a world which is, as David Rejeski states, "dominated by rapid improvements in products, processes and organizations, all moving at rates that exceed the ability of our traditional governing institutions to adapt or shape outcomes." He warns, "If you think that any existing regulatory framework can keep pace with this rate of change, think again (David Rejeski, quoted in Davies 2006).

The UK government has recently implemented a "Voluntary Reporting Scheme" for engineered nanoscale materials. The scheme, scheduled to run for at least two years, from September 2006 to September 2008, is aimed at gathering evidence that will help it move toward the development of appropriate regulatory controls. It will run parallel to a government research program and will lead to "a better understanding of the properties and characteristics of different engineered nanoscale materials, so enabling potential hazard, exposure, and risk to be considered" (United Kingdom, Department for Environment, Food and Rural Affairs 2006). The deliberate development of an evidence base will allow for a more informed and reasoned debate about possible regulatory solutions. At present, no similar voluntary reporting scheme exists in the United States (Service 2005).

Maintaining Public Trust

It is well recognized by those in industry, academia, and government (and increasingly by the broader publics) that, given the potential impact of nanotechnology, it is important to proactively consider the ethical, environmental, economic, legal, and social issues in conjunction with the science and the development of emerging technologies (Royal Society and Royal Academy of Engineering 2004; Roco 2003; Arnall 2003; ETC Group 2003; Sheremeta and Daar 2003). The urgent need to engage in this dialogue is heightened as potential risks to human health and the environment from engineered nanomaterials have been identified (Maynard et al. 2006). This highlights a clear need for close collaboration between scientists, clinicians, engineers, and social scientists to address the concerns and to forge a scientifically, ethically, and legally sound approach to nanotechnology in Canada and throughout the world. The question remains as to how this dialogue can best incorporate and respond to rational public concern and limit the attention paid to unfounded (and often extreme) views.

The relevance of public attitudes in the realization of the potential of technological advances was crystallized in the debate over genetically modified crops and food in the United Kingdom, and abroad. Whether we are ready or not, nanotechnology is poised to emerge as yet another public issue. The state of public opinion on nanotechnology is, however, muddied. Most people know very little about nanotechnology (Cobb and

Macoubrie 2005; Scheufele and Lewenstein 2005; Priest 2006; Waldron, Spencer, and Batt 2006), the actual risks associated with nanotechnology are unknown or disputed and are difficult to assess in an objective manner (IRGC 2006), and portrayals of nanotechnology in the popular press (e.g., Michael Crichton's bestselling novel *Prey*) are effective at blurring the boundary between fact and fiction. The challenges are high.

Recent results of quantitative and qualitative public opinion research in the United States, United Kingdom, and Canada have shown that public knowledge about nanotechnology is low (Canadian Biotechnology Secretariat 2005; Priest 2006). Few people can offer a definition of nanotechnology. Having said that, most people think nanotechnology will improve their lives; only a small percentage thinks it will make their lives worse (Canadian Biotechnology Secretariat 2005). In all jurisdictions, intelligent concerns are raised about the potential long-term side effects of medical applications; for example, whether nanomaterials are biodegradable and whether nanotechnology might get into the wrong hands and be used in weaponry by rogue states and terrorist groups. The public's view of nanotechnology, however, will depend on what it is used for. For now, people remain tentative. The public views nanotechnology, correctly, as being untried but expects to see the potential benefits and drawbacks to become clear over time (Canadian Biotechnology Secretariat 2005).

Furthermore, Canadian public opinion data clearly suggest that people want stricter regulatory controls to safeguard human health and the environment from potential risks associated with new technologies, including nanotechnology. Focus group discussions held in Canada consistently turn to concerns over the drug approval process (Canadian Biotechnology Secretariat 2005). The Vioxx fiasco is frequently cited as one example of how current regulatory systems are flawed. If the government cannot get the drug approval process right, how can we expect that it will regulate nanotechnology appropriately?

The UK Royal Society and the Royal Academy of Engineering (2004) views public open-mindedness as an opportunity to begin a proactive dialogue (using various methods, led by numerous parties, and adequately funded) about the future of nanotechnology. They recommend that the research councils in that nation fund an extensive and sustained research program focusing on public attitudes to nanotechnologies. Importantly, "whether the public accepts the new technology and sees in it advantages for itself–or rejects it–will largely depend on how well informed it is and to what degree it is able to make objective judgments" (44). Gaskell et al. (2005, 1909) note that

> the public expect and want science and technology to solve problems but they also want a say in deciding which problems are worth solving. This is

> not a matter of attracting public support for an agenda already established by science and scientists, but rather of seeing the public as participants in science policy with whom a shared vision of socially viable science and technological vision can be achieved.

Careful thought must be given to the development of a bundle of strategies that will yield the best possible understanding of public and stakeholder sentiment *for the Canadian context.* Appropriate and sustained levels of funding must be secured for these purposes (Sheremeta and Daar 2003).

The Role of Public Action Groups

The ETC Group's recommendation that a moratorium be imposed on the development and release of new nanomaterials was a strategically effective way to set the agenda for ongoing discussion about the regulation of nanotechnology (ETC Group 2003). To the horror of industry players, the ETC Group forced the consideration of a moratorium as a potential policy option on governments around the world. Although the idea of a moratorium has been refuted as a realistic option, what is clear is that industry, academia, and governments (federal, provincial, and local) must be proactive in the consideration of ethical, environmental, economic, legal, and social issues. As Greenpeace (Arnall 2003) has aptly noted, failure to do so could result in a *self-imposed* moratorium. This possibility is disconcerting in light of the potential benefits that nanoscience and nanotechnology are expected to yield for both the developed and the developing world (Mnyusiwalla, Daar, Singer 2003).

Extreme views are an unavoidable part of the nanotechnology debate. Such views play an important role to the extent that they challenge the status quo and force a more thorough consideration of available alternatives. Insofar as they hijack the debate and limit the consideration of reasoned alternatives, they are problematic. Not all public action groups and nongovernmental organizations are created equal. They have vastly different objectives and modes of operation. If we are not cautious, society runs a risk of inappropriately equating the loudest voice with a correct opinion. The role of public action groups in the abortion debate and the debate over global warming provide a contextual backdrop from which we stand to learn.

Conclusions and Future Priorities

The speed at which nanotechnology research is proceeding means that the reality of the daily use of nanotechnology-enabled products and the potential societal risks (and benefits) of these applications are near. It is expected that nanotechnology will enable the development of a broad array of innovations that will affect all economic sectors. Although we expect nanotech innovations to lead to tangible improvements in the quality of human life, risks are inevitable. Accordingly, the main objective of Canadian policy makers must be the legitimate maintenance of public trust through responsible stewardship of nanotechnology. This will necessarily involve maximizing the potential benefits of nanotechnology while managing the risks of harm to human health, to the environment, and to society. Strategically planned multidisciplinary and interdisciplinary efforts are needed to support this important policy objective. For example:

- Scientific efforts must be directed toward defining and measuring nanomaterials and to understanding human health and environmental effects of nanomaterials.
- Policy makers and legal experts need to consider the applicability and appropriateness of existing statutes and regulations to nanotechnology.
- Social scientists must gauge and consider opinion (expert and lay) about various aspects of nanotechnology and develop tools to effectively communicate information about risks and benefits to various interested publics.
- Educators must consider the development of strategies that will facilitate the expansion of awareness about nanoscience and nanotechnology (general public, K-12, and university level).

Although widely heralded as an essential part of a successful nanotechnology strategy, public engagement is fraught with complexity. Past experiences may not be appropriate or applicable in the context of nanotechnology; there exists no formulaic approach to engagement that can be readily adopted. As noted, few Canadians know anything about nanotechnology. Relatively few scientists claim expertise in nanoscience. Few policy makers and lawyers understand the legal regime in which nanomaterials and products incorporating them will fall and the nuanced issues that nanotechnology will raise. Collaborative strategies and careful planning must be directed toward the development of a bundle of strategies that will yield the best possible understanding of public and stakeholder sentiment over time in the Canadian context.

Sound policy decisions about nanotechnology must be based on the best available scientific information and in combination with a concerted effort to predict and proactively address the ethical, environmental, economic, legal, and social issues that will inevitably arise. In spite of the increasing expenditures on nanotechnology research and development, it is now a matter of societal urgency that insufficient funds are being dedicated to examining the risks and benefits of nanotechnology to the environment, human health, and society writ large. Failure to act quickly to fill this gap will certainly undermine public trust in the scientific enterprise and may well spell disaster for the broad development and introduction of nanotechnology products in the long term.

Notes

I would like to thank the National Institute for Nanotechnology, Genome Canada, and Health Canada for their research funding support, and Randy Reichardt for his research assistance.

1 Woodrow Wilson International Center for Scholars, Project on Emerging Nanotechnologies, http://www.nanotechproject.org/index.php?id=44.
2 National Nanotechnology Initiative, http://www.nano.gov/html/facts/whatIsNano.html.
3 US National Institutes of Health Bioengineering Consortium, http://www.becon.nih.gov/nano.htm.
4 National Institute for Nanotechnology. National Research Council of Canada, http://nint-innt.nrc-cnrc.gc.ca/nano/index_e.html.
5 Chemical properties that are important in the characterization of chemical substances include, but are not limited to, molecular weight, melting point, boiling point, vapour pressure, octanol-water partition coefficient, and water solubility, reactivity, and stability. Chemical properties such as those listed above may be important for nanomaterials, but other properties such as particle size and distribution, surface/volume ratio, magnetic properties, coatings, and conductivity are expected to be more important for the majority of nanoparticles (EPA 2007, 34).
6 Woodrow Wilson International Center for Scholars, Project on Emerging Nanotechnologies, http://www.nanotechproject.org/consumerproducts.
7 Life cycle analysis (LCA) contemplates the entire life cycle of a product, process, or activity. The objective of LCA is the evaluation of environmental impacts through raw material acquisition, manufacture, use, and disposal. Use of the LCA is gradually being viewed by environmental managers and decision makers as an important element in achieving environmental sustainability, and it can help industry and government avoid the unintended substitution of one environmental problem for another.
8 *R. v. Hydro Québec*, [1997] 3 S.C.R. 213.

References

Allsopp, Michelle, David Santillo, and Paul Johnston. 2007. A scientific critique of ocean iron fertilization as a climate change mitigation strategy. Greenpeace Research Laboratories Technical Note 07/2007. http://www.greenpeace.de/fileadmin/gpd/user_upload/themen/klima/CBD/iron_fertilisation_critique.pdf.

American Bar Association. 2006a. ABA SEER [Section of Environment, Energy, and Resources] CAA [Clean Air Act] nanotechnology briefing paper. Chicago: American Bar Association. http://www.abanet.org/environ/nanotech/pdf/CAA.pdf.

–. 2006b. CERCLA [Comprehensive Environmental Response, Compensation and Liability Act] nanotechnology issues. Chicago: American Bar Association. http://www.abanet.org/environ/nanotech/pdf/CERCLA.pdf.

–. 2006c. Nanotechnology briefing paper Clean Water Act. Chicago: American Bar Association. http://www.abanet.org/environ/nanotech/pdf/CWA.pdf.

–. 2006d. EMS/innovative regulatory approaches. Chicago: American Bar Association. http://www.abanet.org/environ/nanotech/pdf/EMS.pdf.

–. 2006e. The adequacy of FIFRA [Federal Insecticide, Fungicide, and Rodenticide Act] to regulate nanotechnology-based pesticides. Chicago: American Bar Association. http://www.abanet.org/environ/nanotech/pdf/FIFRA.pdf.

–. 2006f. RCRA [Resource Conservation and Recovery Act] regulation of wastes from the production, use, and disposal of nanomaterials. Chicago: American Bar Association. http://www.abanet.org/environ/nanotech/pdf/RCRA.pdf.

–. 2006g. Regulation of nanomaterials under the Toxic Substances Control Act. Chicago: American Bar Association. http://www.abanet.org/environ/nanotech/pdf/TSCA.pdf.

Arnall, A.H. 2003. *Future technologies, today's choices*. London: Greenpeace Environmental Trust.

Bergeson L.L., and B. Auerbach. 2004. The environmental regulatory implications of nanotechnology. *Daily Environment Report* 71 (BNA 4-14-04): B1-B7. http://www.lawbc.com/other_pdfs/The%20Environmental%20Regulatory%20Implications%20of%20Nanotechnology.pdf.

Buesseler, Ken O., Scott C. Doney, David M. Karl, Philip W. Boyd, Ken Caldeira, Fei Chai, Kenneth H. Coale, Hein J. W. de Baar, Paul G. Falkowski, Kenneth S. Johnson, Richard S. Lampitt, Anthony F. Michaels, S.W.A. Naqvi, Victor Smetacek, Shigenobu Takeda, and Andrew J. Watson. 2008. Ocean iron fertilization–Moving forward in a sea of uncertainty. *Science* 319 (January 11): 162.

Canadian Biotechnology Secretariat. 2005. First impressions: Understanding public views on emerging technologies. Government of Canada BioPortal. http://www.biostrategy.gc.ca/CMFiles/CBS_Report_FINAL_ENGLISH249SFD-9222005-5696.pdf.

Chaudhry, Q., J. Blackburn, P. Floyd, C. George, T. Nwaogu, A. Boxall, and R. Aitken. 2006. *A regulatory gaps study for the products and applications of nanotechnology: Final report*. Department for Environment Food and Rural Affairs, UK Government. http://www2.defra.gov.uk/research/project_data/More.asp?I=CB01075&M=KWS&V=Nanotech&SUBMIT1=Search&SCOPE=0.

CIHR (Canadian Institutes of Health Research). 2005. *Regenerative medicine and nanomedicine: Innovative approaches in health research*. http://www.cihr-irsc.gc.ca/e/22842.html#4.

Climos Inc. 2008. Statement on recent activity regarding ocean iron fertilization. May 30. http://www.climos.com/.

Cobb, M.D., and J. Macoubrie. 2005. Public perceptions about nanotechnology: Risks, benefits, and trust. *Journal of Nanoparticle Research* 6 (4): 395-405.

Colvin, V. 2003. The potential environmental impact of engineered nanomaterials. *Nature Biotechnology* 21: 1166.

Davies, J.C. 2006. Managing the effects of nanotechnology, project on emerging technologies. Woodrow Wilson International Center for Scholars. http://www.nanotechproject.org/index.php?id=39.

DHHS, CDCP, and NIOSH (Department of Health and Human Services, the Centers for Disease Control and Prevention, and the National Institute for Occupational Safety and Health). 2006. *Approaches to safe nanotechnology: An information exchange with NIOSH*. http://www.cdc.gov/niosh/topics/nanotech/safenano/pdfs/approaches_to_safe_nanotechnology_28november2006_updated.pdf.

Dreher, K.L. 2004. Health and environmental impact of nanotechnology: Toxicological assessment of manufactured nanoparticles. *Toxicological Sciences* 77 (1): 3-5.

Environment Canada and Health Canada. 2007. Proposed regulatory framework for nanomaterials under the Canadian Environmental Protection Act, 1999. September 10. http://www.ec.gc.ca/substances/nsb/eng/nanoproposition_e.shtml.

EPA (Environmental Protection Agency). 2007. Nanotechnology Workgroup. Nanotechnology white paper, February 15. Environmental Protection Agency. http://www.epa.gov/osa/nanotech.htm.

ETC Group. 2003. *From genomes to atoms: The big down; Atomtech; technologies converging on the nano-scale*. Winnipeg: ETC Group.

–. 2007. Geoengineers to foul Galapagos Seas—Defying climate panel warning. Press Release, May 3. ETC Group. http://www.etcgroup.org/en/materials/publications.html?pub_id=617.

Gaskell G., E. Einsiedel, W.Hallman, S. Hornig Priest, J. Jackson, and J.Olsthoorn. 2005. Social values and the governance of science. *Science* 310 (5756): 1908-9.

Glazier, R., R. Venkatakrishnan, F. Gheorghiu, L. Walata, R. Nash, and W. Zhang. 2003. Nanotechnology takes root. *Environmental Science and Technology* 73 (5): 64-69. http://oaspub.epa.gov/eims/eimsapi.dispdetail?deid=71568.

Goldman, Lynn, and Christine Coussens, eds. 2005. *Implications of nanotechnology for environmental health research*. Washington, DC: National Academies Press.

Hatto, Peter. 2006. The role of standardization in risk governance of nanotechnologies. Paper presented at IRGC Workshop on Risk Governance for Nanotechnology, July 6-7, Ruschikon, Switzerland. http://www.irgc.org/IMG/pdf/Peter_Hatto_The_Role_of_Standardization_in_Risk_Governance_of_Nanotechnologies_-2.pdf.

Hoet P.H., A. Nemmar, and B. Nemery. 2004. Health impact of nanomaterials? *Nature Biotechnology* 22: 19.

ICF International. 2006. *Characterizing the environmental, health, and safety implications of nanotechnology: Where should the federal government go from here?* Fairfax, VA: ICF International. http://www.icfi.com/markets/environment/doc_files/nanotechnology.pdf.

IRGC (International Risk Governance Council). 2006. *White paper on nanotechnology risk governance.* Zurich: International Risk Governance Council.

IRSST (Institut de recherché Robert-Sauvé en santé et en securité du travail). 2006. Health effects of nanoparticles. Report R-469. http://www.irsst.qc.ca/en/_publication-irsst_100209.html.

ISO (International Organization for Standardization). 2005. New ISO committee will develop standards for nanotechnology. http://www.iso.org/iso/pressrelease.htm?refid=Ref978.

Kennedy, D. 2001. Breakthrough of the year. *Science* 294 (5521): 2429.

Kreyling, W.G., M. Semmler-Behnke, and W. Möller. 2006. Health implications of nanoparticles. *Journal of Nanoparticle Research* 8 (5): 543-52.

Kroto, W.W., J.R. Heath, S.C. O'Brien, R.F. Curl, and R.E. Smalley. 1985. C60: Buckminsterfullerene. *Nature* 318: 162-63.

Lux Research. 2006. Nanotechnology moves from discovery to commercialization: $50 billion in 2006 product sales, $12 billion in funding, November 20. http://www.luxresearchinc.com/press/2007-lux-research-nanotech-report-5.pdf.

Masciangioli T., and Zhang W-X. 2003. Environmental technologies at the nanoscale. *Environmental Science and Technology* (March 1) 37 (5): 102A-8A.

Maynard, A.D., R.J. Aitken, T. Butz, V. Colvin, K. Donaldson, G. Oberdorster, M.A. Philbert, et al. 2006. Safe handling of nanotechnology. *Nature* 444 (16): 267.

Maynard, A.D., and E.D. Kuempel. 2005. Airborne nanostructured particles and occupational health. *Journal of Nanoparticle Research* 7 (6): 587-614.

Mnyusiwalla, A., A.S. Daar, and P.A. Singer. 2003. Mind the gap: Science and ethics in nanotechnology. *Nanotechnology* 14 (3): R9-13.

NSET (Nanoscale Science, Engineering, and Technology Subcommittee) and the National Science and Technology Council. 2007. *The National Nanotechnology Initiative: Research and development leading to a revolution in technology and industry: Supplement to the president's FY 2008 budget.* National Nanotechology Initiative. http://www.nano.gov/NNI_08Budget.pdf.

Oberdörster, E. 2004. Manufactured nanomaterials (fullerenes, C60) induce oxidative stress in brain of juvenile largemouth bass. *Environmental Health Perspectives* 112 (10): S1058.

Oberdörster, G., A. Maynard, K. Donaldson, V. Castranova, J. Fitzpatrick, K. Ausman et al. 2005. Principles for characterizing the potential human health effects from exposure to nanomaterials: Elements of a screening strategy. *Particle and Fibre Toxicology* 2 (8): 1-35.

Oberdörster, G., E. Oberdörster, and J. Oberdörster. 2005. Nanotoxicology: An emerging discipline evolving from studies of ultra fine particles. *Environmental Health Perspectives* 113 (7): 823-39.

Office of the National Science Advisor. 2005, 2007. *Overview of nanotechnology research funding in Canada.* Ottawa: Government of Canada.

President's Council of Advisors on Science and Technology. 2005. *The National Nanotechnology Initiative at five years: Assessment and recommendations of the National Nanotechnology Advisory Panel.* http://www.ostp.gov/PCAST/PCASTreportFINALlores.pdf.

Priest, S. 2006. The North American opinion climate for nanotechnology and its products: Opportunities and challenges. *Journal of Nanoparticle Research* 8 (5): 563-68.

Roco, M.C. 2003. Broader societal issues of nanotechnology. *Journal of Nanoparticle Research* 5 (3): 181-89.

Royal Society and the Royal Academy of Engineering. 2004. *Nanoscience and nanotechnologies: Opportunities and uncertainties*. London: Royal Society.
Scheufele, D.A., and B.V. Lewenstein. 2005. The public and nanotechnology: How citizens make sense of emerging technologies. *Journal of Nanoparticle Research* 7 (6): 659-67.
Service, R.F. 2005. EPA ponders voluntary nanotechnology regulation. *Science* 309 (5731): 36.
Sheremeta, L. 2006. Nanotechnology: Legal and regulatory challenges. Paper presented at the Federal Workshop on the Health and Environmental Implications of Nanoproducts, Ottawa.
Sheremeta, L., and A.S. Daar. 2003. The case for publicly funded research on the ethical, environmental, economic, legal and social issues raised by nanoscience and nanotechnology (NE3LS). *Health Law Review* 12 (3): 74-77.
Shvedova A.A., E.R. Kisin, A.R. Murray, V.Z., Maynard A.D. Gandelsman, P.A. Baron, and V. Castranova. 2003. Exposure to carbon nanotube material: Assessment of the biological effects of nanotube materials using human keratinocyte cells. *Journal of Toxicology and Environmental Health* 66 (20): 1909-26.
Smalley, R.E. 1996. Discovering the fullerenes. Nobel lecture, December 7. Nobel Foundation. http://nobelprize.org/chemistry/laureates/1996/smalley-lecture.pdf.
–. 1999. Prepared written statement and supplemental material of R.E. Smalley, Rice University, Houston, May 12. http://www.sc.doe.gov//bes/senate/smalley.pdf.
Swiss Re. 2004. *Nanotechnology: Small matter, many unknowns*. Zurich: Swiss Re.
United Kingdom, Department for Environment, Food and Rural Affairs. 2006. *UK voluntary reporting scheme for engineered nanoscale materials*. http://www.defra.gov.uk/environment/nanotech/policy/pdf/vrs-nanoscale.pdf.
Waldron, A., D. Spencer, and C.A. Batt. 2006. The current state of public understanding of nanotechnology. *Journal of Nanoparticle Research* 8 (5): 569-75.
Wardak, A. 2003. Nanotechnology and regulation: A case study using the Toxic Substance Control Act. Woodrow Wilson International Center for Scholars, Foresight and Governance Project, publication 2003-6.
Warheit, D.B., B.R. Laurence, K.L. Reed, D.H. Roach, G.A.M. Reynolds, and T.R. Webb. 2004. Comparative pulmonary toxicity assessment of single-wall carbon nanotubes in rats. *Toxicological Sciences* 77: 117-25.
Zhang, W. 2003. Nanoscale iron particles for environmental remediation: An overview. *Journal of Nanoparticle Research* 5 (3 and 4): 323-32.

Part 3
Governance Challenges and Emerging Technologies

15
Technology, Democracy, and Ethics: Democratic Deficit and the Ethics of Public Engagement

Michael Burgess and James Tansey

Establishing a Democratic Deficit

Emerging technologies and their adoption demand an aggressive approach to public engagement.[1] Gaskell and colleagues (Gaskell et al. 2003, 19) call for enhanced public dialogue and debate as critical for managing the role of emerging new technologies in society: "Any platform of public debate between autonomous and responsible citizens is to be applauded. And if socially sustainable technological innovation is a societal goal, appropriate platforms for such debates will need to be established (on nanotechnology for example), if we are to avoid reliving the type of conflicts that raged over biotechnology in the mid to late 1990s."

Although conflicts over biotechnology are highly variable in their geographic distribution and may serve important social roles, it is likely that all parties to the debate agree on the need for more meaningful public engagement in the governance of emerging technologies. Citing Radder (1992, 1998) and Hamlett (2003), Keulartz and colleagues (Keulartz et al. 2004) argue that the avoidance of a normative stance comes at the cost of irrelevance to the social and ethical issues raised by technologies. Their "Ethics in Technological Culture: A Programmatic Proposal for a Pragmatist Approach" includes both the rational aspects of ethics and the process model related to developing public discourse that can explore a heterogeneity of perspectives and possible futures.

Public engagement is a pervasive feature of the political and research environment related to biotechnology and has become as difficult to oppose as truth telling or respecting people who are disadvantaged. Although public engagement is not necessarily intended to serve moral purposes (i.e., it could be intended to provide prudential advice to marketers or developers of products or policies), it is often considered to have ethical and political significance. Within representative democracies, it makes sense to assess the role that public engagement is intended to serve and what significance it should carry on the basis of its capacity to enhance representative democracy.

Representative democracy manifests several general problems in ethical discussions related to biotechnology and policy. First, it is not clear to what extent elected officials and public bureaucrats are able to be accountable to the public as citizens when organized interest groups and financial interests are so powerful. The ability to successfully stand for office or rise in government bureaucracy seems to be at least partly a function of being able to satisfy key interests. Second, it is no longer sufficient to ensure that policy is adequately informed through consultation with experts. The use of scientific information to inform policy choices requires prediction and a focus on known risks that necessarily introduces uncertainty and non-scientific judgments about what is important (McDaniels 1998; Jasanoff and Wynne 1998). Third, it has become increasingly apparent that policy choices in biotechnology are controversial primarily when strong interest groups are in conflict about what is in the best interest of the public. These conflicting accounts of the public interest are produced by vested interests, whose input is a necessary but not sufficient condition for comprehensive public input to policy formulation.

For example, consider a very simple overview of the debate about what should be regulated in the assessment of genetically modified (GM) foods or privacy and genetic database research. Safety and privacy standards existed before GM foods and the possibilities offered by computational approaches to genetic and genomic analysis in humans. Producers of foods and pharmaceuticals have, to a large extent, proliferated and built extensive organizations and profits within these rules. They have also exerted a powerful influence on these policies and the broader rules of the game. These industry stakeholders are now well situated to exert further influence on elected officials and government departments. These corporate entities are financially important to society's well-being, and, for some people, any interference with the activities of corporate interests must be justified by strong indications of harm to society. The recent World Trade Organization judgment against the de facto moratorium on GM products in Europe reflects this reasoning by insisting on strong scientific evidence of harms before restricting international trade. Buttressing these interests are claims that genetic modification of crops and foods will help society meet global food security and nutritional needs. Some health-related researchers claim that systematic collection, electronic storage, and computational analysis of large databases of health, epidemiological, and genomic data will enable major advances in human health for individuals and populations.

A key policy question for these areas is whether there is anything about the character of genetic sciences, genomics, or biotechnology that is beyond the capacity of existing rules in food safety and personal privacy. Some hold the view that GM foods and crops, as well as information about human genomes, pose unique risks that deserve stronger protection than is provided by the existing regulations. Advocates and scholars have articulated concerns about the complex interactions and systemic effects of GM crops or foods on the environment and in humans (Wynne 2001). Others are worried that genomic and genetic information about individuals or groups such as indigenous people may be used to make erroneous or unauthorized judgments about the allocation of social opportunities and resources (Glass and Kaufert 2001; Indigenous Peoples Council on Biocolonialism 2000); National Institutes of Health 2004; Weijer, Goldsand, and Emanuel 1999), with denial of–or higher premiums for–life, disability, and health insurance being the thin end of the wedge (Dahl et al. 2004; Drake, Reid, and Marteau 1996; Glover and Glover 1996; Kaplan 1993). These interests align with those of a wide range of other stakeholders. Concerns about the safety of GM crops align and overlap with the interests of groups that have concerns about maintaining less intensive forms of agriculture, prefer small and local suppliers, and advocate for environmental and species protection. Concerns about adequate protection from discrimination from genetic and genomic research align and overlap with broader concerns about civil liberties, disability advocacy, and contemporary activities to provide better and more nuanced protection from research-related harms for research participants and indigenous peoples. Although they may not have the financial power of some industrial stakeholders, these groups have strong influence on elected officials and policy makers.

In this context, the ability of representative electoral democracy to manage policy decisions encounters serious challenges. Stakeholder groups wield influence on the electoral process, the functioning of government departments, and on the conception of what kind of public interests should be served by policy. Similar scientific studies yield different policies in different jurisdictions, often influenced by the stakeholder groups that are most influential and by what Jasanoff (2005) has called "civic epistemologies," which reflect what a society or culture will support as a reasonable conclusion based on the evidence and accounts of the public interest (e.g., benefits of research or protection from poorly understood risks).

So, if there is a democratic deficit, it has to do with how the range of views of the public have less direct effects on policy than those of powerful stakeholders. Public engagement, to the extent it is a response to this deficit, must provide a means of enabling wider public involvement and must avoid capture by stakeholders with vested interests. From the perspective of theorists in justice and ethics, the democratic deficit is the product of unequal opportunities to shape how decisions are made about what constitutes important social goods and fair distribution in society (Buchanan et al. 2000, 263; Sherwin 2001).

A Practical Approach to Deliberative Democracy

One of the most practical and explicit (and ambitious) attempts to address this democratic deficit from the perspective of deliberative democracy is deliberative polling. Deliberative polling seeks to avoid the problems of capture by strong vested interests or superficial and phantom opinions that plague public consultation exercises by a design that balances inclusiveness with thoughtfulness (Fishkin 1997; Luskin, Fishkin, and Jowell 2002). Rather than review the literature on deliberative democracy, exploring the strengths and limitations of Fishkin's approach combines social science method and normative analysis to construct and execute an approach to addressing the democratic deficit. Deliberative polling recruits a large random sample of the population in order to be able to claim that the sample's deliberations and conclusions are representative of what the population would conclude if it were as well informed and deliberative as the participants in the deliberative polling event.[2] To demonstrate that the information and deliberation make a difference to participants' views, they complete a survey at the beginning and end of their involvement. The participants are provided with materials that a group of stakeholders related to the policy question agree represents the relevant information. They use the briefing materials to initiate a deliberation with each other in small, randomly assigned discussion groups. The groups are moderated to encourage listening and prevent domination of the discussion. Although most of the events have been face to face, there are also experiments with online discussions. The deliberation may last for a weekend or may be repeated over several weeks. Participants meet in large plenary sessions to ask additional questions of competing experts and politicians. These advisors do not give presentations. At the end of the event, the participants take the same survey as at the beginning of the event. The resulting judgments and edited discussions are broadcast, usually on television.

Deliberative polling and other methods of deliberative engagement are used when there is a policy decision to make. The policy decision implies a set of stakeholders, a body of relevant information, and a limited set of policy options. When there is a policy decision to be made and adequate time and resources to support deliberative polling, it may well be the best available option. The approach is not without criticism, but many of the critiques assume alternative reasons for engaging the public. For example, selecting the policy question necessarily rules out considerations that are beyond the scope of the policy or the responsibility of the policy makers. Selecting which stakeholders participate in framing the policy options and determining the relevant information further reinforces the scope and limits considerations beyond the policy and may fail to convince those who do not participate in the deliberations. Random recruitment excludes the option of selecting members of the public who may be uniquely affected by the policy but are not part of the stakeholder groups and may not appear in a scientifically designed statistical sample.

What Kinds of Deficits Are There?

Many characterizations of the purposes of public engagement or consultation have been proposed as a way of evaluating different methods (Arnstein 1969; Rowe and Frewer 2000; Levitt, Weiner, and Goodacre 2005), but the notion of democratic deficit provides another way to assess the relative merits and limitations of different approaches. If the problem is not merely that the wider public has expertise but that they need to be better able to provide oversight of the democratic structures that regulate and represent (Jasanoff 2003; Wynne 2003), then different kinds of representation might be appropriate for different aspects of the democratic deficit.

There are likely many ways to improve on the extent to which policies and structures of society more adequately represent the range of interests and perspectives reflected in the governed. For this discussion, we distinguish four kinds of concerns and related types of public engagement:

- Policy design: How best to incorporate informed, focused deliberation that adequately represents the population and avoids capture by strong interests?
- Policy support: How best to create opportunities for public assessment and challenges to policies and institutions?
- Public education: How best to access and use information so that public debates are based on relevant information?
- Public dialogue: How to maximize the inclusiveness of civil society's debate about the kind of society we want to be and the role of biotechnology in it?

These are fluid categories that run into each other and may be mutually supportive. Figure 15.1 suggests that engagement for the sake of policy design should still meet the standards for policy support, but the categories are not co-extensive. Public information provides a base for engagement related to policy design and support. Public dialogue emphasizes inclusiveness as important to determining relevant information and moves beyond informing to deliberative engagement on the broader issues than those likely to be practical in policy discussions. Distinguishing these roles of public engagement for addressing different aspects of democratic deficit supports recognition of where different approaches are more appropriate without claiming any approach is always the best.

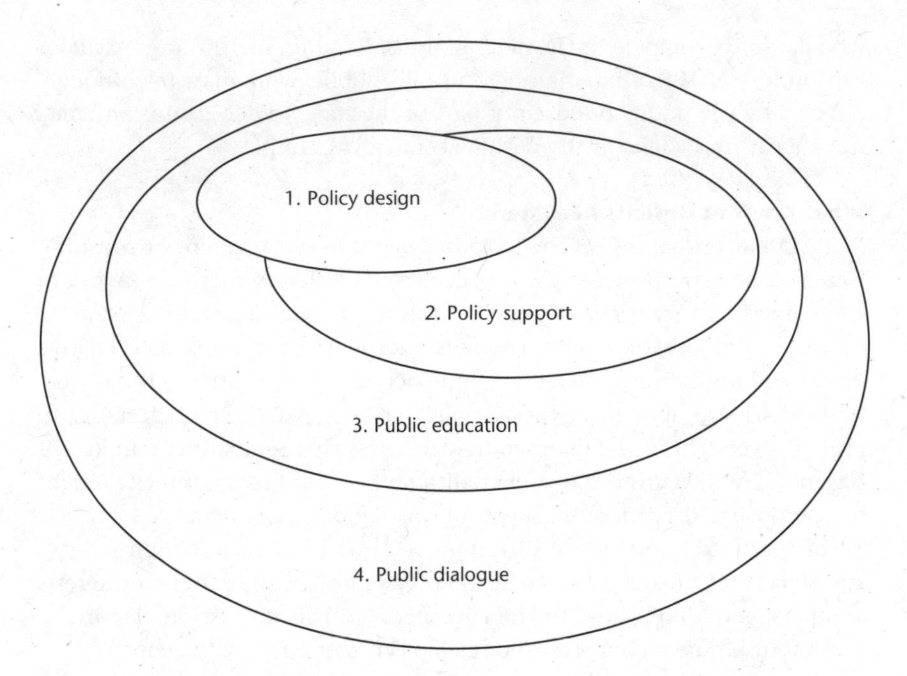

Figure 15.1 Purposes for engagement to address democratic deficit

Policy Design

Conflict and the obviousness of democratic deficit are most acute at the centre or policy design layer. The goal of designing policy requires that contributors accept the range of responsibility and scope of issues that can be reasonably captured by a policy proposal. Engagement of publics in this case will need to ensure that there is focused deliberation that is informed about the considerations that support different policy options. Direct or accountable input into the final policy is widely recognized as important for these kinds of consultations if they are to be respectful of, and motivating for, participants. Citizen juries, deliberative polling, and other deliberative forums are highly appropriate for deliberation on a focused topic and related information. The example of deliberative polling describes many of the important components for addressing democratic deficit related to policy design.

Policy Support

The decision not to consult may be based on beliefs that the norms related to a policy decision are well established, or because of difficulties in identifying or motivating the participation of people who may later be affected by a decision. But even in situations where policy is not anticipated to be controversial, the problem of the moral warrant of the policy when there is a democratic deficit is that there are people who have not had an opportunity to shape the values that inform the policy. Although any good policy process should identify and explain why alternative perspectives and policy options have been unpersuasive, the decisions are not always available for public scrutiny or appeal.

In the health ethics literature, Daniels and Sabin (1997, 1998) suggest that policies that will result in an allocation of benefits to some and not others must meet conditions for accountability for reasonableness. Concerned that there are differing views of what should constitute the substantive principles of justice for healthcare decision making in society, they claim that four conditions must be met for a process to be accountable for reasonableness:

1 Publicity: Rationales for coverage of new technologies must be transparent and publicly accessible.
2 Relevance: Rationales must be reasonable, that is, based on appeals to evidence or principles that fair-minded parties accept as relevant.
3 Appeals: There must be mechanisms for challenges, ongoing review, and revision of decisions as new information develops or the context changes.
4 Enforcement: Decision-making processes must be publicly regulated to ensure that the first three conditions are met. (Daniels and Sabin 1998, 57)

Daniels and Sabin provide strong support to the position that all policy decisions be required to provide their justification and appeal mechanisms so that any member of the public can be satisfied that the policy is reasonable, or know how to challenge it. Even appeals to the authority of elected officials and public offices must make available the justifications for policy decisions. The additional requirement of public engagement is sometimes unnecessary, impractical, and disproportionately costly. In determining whether to invest in public engagement, assessments of whether the policy decision requires additional democratic legitimacy to that provided by the electoral and policy institutions will often be considered. But the decision that policy has moral warrant without public engagement input can, of course, be incorrect because the range or nature of views understood by policy makers is insufficiently representative of the views actually held by the public. For this reason, the justification for the policy must be reasoned, and the reasons must be available for review, with the real opportunity to change the policy.

There are decisions for which accountability for reasonableness fails to provide sufficient opportunity for affected individuals and the public to adequately influence the decision, creating a democratic deficit related to legitimacy (Williams-Jones and Burgess 2004). Daniels and Sabin (1998, 61) accept that "consumer participation might improve deliberation about some matters" but doubt that participation would be sufficiently active.[3] Deliberative polling and other approaches mentioned above may provide representative participation but require a well-defined policy decision and considerable investment of time and resources. It is often only once policy is formulated that citizens can identify how their interests are affected and become motivated to deliberate. Accountability for reasonableness makes these policies available for input, but access to policy makers' reasons and justifications does not provide the public with access to relevant information and the views of other citizens about policies and the implications. This suggests that there is a role for public education.

Public Education

Public education is often maligned for having undisclosed vested interests, downplaying uncertainty, and implicitly promoting particular moral perspectives or political agendas. Education by scientists presumes that the development of knowledge is inherently valuable; ethics education presumes people want to be rational decision makers; nongovernmental organizations (NGOs) as information sources may increasingly be viewed as oriented to promote their agenda and persuade people to provide political and financial support. The mass media is touted as the primary source of public information about biotechnology (Roper Starch Worldwide 1997; Geller et al. 2005), but there are debates about the influence of powerful financial interests (Nelkin and Lindee 1995; Durant and Hanson 1995) and accuracy of reporting (Allan 2002; Leiss and Powell 2004; Kua, Reder, and Grossel 2004; Shuchman and Wilkes 1997; Bubela and Caulfield 2004). That said, there is a long-standing debate about the role of the media as a necessary component of an active democracy (Kovach and Rosenstiel 2001) and whether the role of news media is to provide information about what experts and publics say about events or to educate the public so they can better participate in democracy (Lippmann 1922; Ward 2004; Voakes 2004).[4]

Many approaches to public engagement have refined the presentation of a range of perspectives in a manner that permits participants to evaluate the claims and information for themselves. Materials produced for deliberative polls and citizen juries are sometimes made available as a form of public education. We have experience with deep online surveys (compare with Danielson et al. 2007; Ahmad et al. 2006) and online background documents (compare with Power 2003; Tansey 2003; Tansey and Burgess 2004) being useful to a broader audience as educational tools. Many NGOs and research agencies and institutions have explicit budgets and even branches dedicated to public education. It is reasonable to be concerned about the implicit perspectives of each source, the wide availability of these, and the considerable if uneven public awareness of the influence of vested interests on information sources. Public education addresses the components of democratic deficit related to identification of interests and information necessary to deliberate and participate in a democracy and in the other important ways in which civil society is active, including informed consumerism and collective action. Although public education as a goal avoids limiting information and perspectives to preconceived policy goals and responsibilities, it has serious limitations: public education will not provide much input for policy formation and direction, and the previous range of public engagements is still necessary for those goals. Accessibility of public education is an important issue that will influence how much public access to information will support previously unarticulated interests and perspectives being introduced into public discussion and policy. But there is one more broad approach to public engagement that involves enhancing the accessibility and deliberative functions necessary to address the democratic deficit.

Public Dialogue

The designers of the most comprehensive approach to policy consultation recognize the need to support broad public deliberative debate in addition to education. Deliberation Day is a detailed approach developed by Ackerman and Fishkin (2005), who propose to institute an American holiday dedicated to debating the issues related to presidential elections in presidential election years. Their proposal is that Americans meet in public spaces to engage in structured debates about issues that divide presidential candidates. The Deliberation Day proposal involves informing the public and deliberative engagement and helps citizens make better use of their opportunity to elect representatives. Selection of the candidates and issues, and to some extent the information and interests reflected by the selection of the issues, are less open for deliberative dialogue.

Probably the clearest call for a platform to support open-ended and ongoing public dialogue about the kind of society in which people want to live and the role of biotechnology in that society is the citation from Gaskell and colleagues (2003) in *Ambivalent GM Nation? Public Attitudes to Biotechnology in the UK, 1991-2002,* cited at the beginning of this chapter. Supporting this possibility is the accumulating evidence that laypersons are quite capable of learning enough technical information to meaningfully identify the issues and interests that are important to them, and often surprise themselves with the amount of knowledge and discussion they are able to generate in facilitated discussion groups (Kerr, Cunningham-Burley, and Amos 1998a, 1998b; Tansey and Burgess 2006; Longstaff, Burgess, and Lewis 2007).

When the policy consultation or support is sought without this broader discussion, the perception will be that the broader discussion is being precluded or closed. Broad social controversies or ethical issues inevitably persist beyond the conclusion of any policy process. When the policy process is the only venue for public engagement and deliberation, issues that are beyond the scope of the policy will motivate some people to prolong the policy debate. By acknowledging that there are persistent moral, social, and political issues, and by providing a venue for their continued debate, the scope of policies is more reasonably limited and concluded (Burgess 2004). Policies that raise contentious issues in the public dialogue can be proposed on the models of policy consultation or policy support as suggested above. Compromises and accumulating social effects of policies can be evaluated against the range of views and interests articulated in the public dialogue.

The literature on deliberative democracy is diverse, and this discussion cannot do justice to it. Among the very useful reflections on public dialogue is Dryzek's call (2005, 238) for a "discursive democracy in divided societies that emphasizes engagement in the public sphere only loosely connected to the state." He suggests that the development of such a distributed platform for public debate could develop from several influences:

- deliberative institutions at a distance from sovereign authority
- deliberative forums in the public sphere that focus on particular needs rather than general values
- issue-specific networks
- centripetal electoral systems
- a power-sharing state that does not reach too far into the public sphere
- the conditionality of sovereignty
- the transnationalization of political influence. (Dryzek 2005, 239)

Fung (2005) tries to derive guidance for current democratic struggles from the revolutionary notion that we need to change the structure of society to support deliberative democracy. He argues that ethical principles of action can be derived from the tension in deliberative democracy between insisting on deliberation and requiring confrontational political activism to develop a political system based on the principles of deliberative democracy. Drawing on Gutmann and Thompson's "middle ground of democracy," Fung (2005) suggests that supporting principles of deliberative dialogue might enable "incremental, nonrevolutionary steps" toward more deliberative civil society and government.

Conclusion

One component of the democratic deficit that requires public engagement is the identification and articulation of the diversity of views on the kind of society we want to inhabit and the role of genomics in that society. Public engagement exercises are sometimes designed to inform people about the science so their reflections are less fictional and more useful to policy formation. This requires careful assessment and presentation of scientific information. The emphasis on reliable expert information tends to structure an element of democratic deficit into the engagement by the necessary use of authoritative experts and setting a reasonable scope for policy. Such public engagement is not likely to expand the representativeness of the range of interests held by participants and citizens. But the views of various groups and the range of interests is also important information related to democratic legitimacy.

This is not to say that informed public discussion of policy options is unimportant, only that policy discussion can address all aspects of the deliberative deficit only by removing the practical emphasis on policy. These two goals for public engagement–seeking informed and inclusive participation in policy formation, and identifying the range of moral diversity in a society and its function in establishing broader agendas–are essential and complementary components of democratic participation. That said, they are likely mutually exclusive as goals for any one type of engagement. What is required is an approach that can develop a body of knowledge about the range of diverse moral views in a society and informed public engagement on public policy that captures that diversity.

Broader issues that ultimately direct how institutions of society make policies and govern are based on norms that cannot be easily taken up for evaluation within specific policies (e.g., what policies are necessary, what are important social goods, and what constitutes fair allocations). Engagement of the public for the purpose of education supports this element of democracy. The importance of a diversity of perspectives and the usefulness of specific contexts for understanding complex information argues for public education that is distributed, so that no one source is given inappropriate credibility or taken uncritically. Research needs to continue to identify what information people want and use and how it is best made accessible. Finally, there is an element of the democratic deficit that requires support for dialogue and the development of understanding of differences in civil society without pressure to prematurely conclude issues for the sake of a policy. Since policies have specific scopes and policy makers have particular responsibilities, policy advice and formation should be separated from, but supported by, the roles of public engagement for public education and dialogue. Here again there is an important role for research into ways to engage wider perspectives and encourage their articulation into ongoing open dialogue about the convergences and divergences in the values that shape society and the role of biotechnology. Critical evaluations of the role of the media, academics, funding agencies, industry, NGOs, and government in supporting public information and dialogue are important, particularly with recognition that no group can claim neutrality on ethical dialogue about the kind of society we wish to be.

Notes

1 The term *engagement* is intended to be inclusive and broader than *consultation*. Consultation has as a primary objective providing information useful for policy formation or, more cynically, convincing participants that a policy is appropriate or how to implement it. Public engagement involves participants in events or communications that can include objectives additional to deriving policy advice, such as understanding their views, how they use information, and supporting the development of associations and stimulating dialogue. See Castle and Culver (2006) for a useful discussion.

2 Fishkin is clear that this is a counterfactual situation that inevitably differs from the opinions held by the population that is not informed or deliberative.

3 Daniels and Sabin are writing about US health insurance companies, so the use of the term *consumers* may or may not imply a distinction from citizens or civil society. It would be more consistent with the other writings of Daniels (1985, 2001) to interpret their concerns as having to do with the entitlements of members of civil society.

4 The authors are grateful to Dave Secko and Stephen Ward for developing the relevance of these points for democracy.

References

Ackerman, B., and J.S. Fishkin. 2005. *Deliberation Day*. New Haven: Yale University Press.

Ahmad, R., J. Bailey, Z. Bornik, P. Danielson, H. Dowlatabadi, and E. Levy. 2006. A web-based instrument to model social norms: NERD design and results. *Integrated Assessment* 6 (2): 9-36.

Allan, S. 2002. *Media, risk and science*. Buckingham: Open University Press.

Arnstein, S.R. 1969. A ladder of citizen participation. *Journal of the American Planning Association* 35 (4): 216-24.

Bubela, T.M., and T.A. Caulfield. 2004. Do the print media "hype" genetic research? A comparison of newspaper stories and peer-reviewed research papers. *Canadian Medical Association Journal* 170 (9): 1399-407.

Buchanan, A., D.W. Brock, N. Daniels, and D. Wikler. 2000. *From chance to choice: Genetics and justice*. Cambridge, UK: Cambridge University Press.

Burgess, M.M. 2004. Public consultation in ethics: An experiment in representative ethics. *Journal of Bioethical Inquiry* 1 (1): 4-13.

Castle, D., and K. Culver. 2006. Public engagement, public consultation, innovation and the market. *Integrated Assessment* 6 (2): 137-52.

Dahl, E., K.D. Hinsch, B. Brosig, and M. Beutel. 2004. Attitudes towards preconception sex selection: A representative survey from Germany. *Reproductive Biomedicine Online* 9 (6): 600-3.

Daniels, N. 1985. Just health care. New York: Cambridge University Press.

–. 2001. Justice, health, and health care. *American Journal of Bioethics* 1 (2): 2-16.

Daniels, N., and J. Sabin. 1997. Limits to health care: Fair procedures, democratic deliberation and the legitimacy problem for insurers. *Philosophy and Public Affairs* 26 (4): 303-50.

–. 1998. The ethics of accountability in managed care reform. *Health Affairs* 17 (5): 50-64.

Danielson, P., R. Ahmad, Z. Bornik, H. Dowlatabadi, and E. Levy. 2007. Deep, cheap, and improvable: Dynamic democratic norms and the ethics of biotechnology. In *Ethics and the life sciences,* ed. F. Adams, 315-26. Charlottesville, VA: Philosophy and Documentation Center

Drake, H., M. Reid, and T. Marteau. 1996. Attitudes towards termination for foetal abnormality: Comparisons in three European countries. *Clinical Genetics* 49 (3): 134-40.

Dryzek, J.S. 2005. Deliberative democracy in divided societies: Alternatives to agonism and analgesia. *Political Theory* 33 (2): 218-42.

Durant, J., and A. Hanson. 1995. The role of the media. In *Parliaments and screening: Ethical issues arising from testing for HIV and genetic disease,* ed. W.K. Kennet, 89-121. Paris: John Libbey Eurotext.

Fishkin, J.S. 1997. *The voice of the people: Public opinion and democracy.* New Haven: Yale University Press.

Fung, A. 2005. Deliberation before the revolution: Toward an ethics of deliberative democracy in an unjust world. *Political Theory* 33 (2): 397-419.

Gaskell, G., N. Allum, M. Bauer, J. Jackson, S. Howard, and N. Lindsey. 2003. *Ambivalent GM nation? Public attitudes to biotechnology in the UK, 1991-2002.* London School of Economics and Political Science: Life Sciences in European Society report.

Geller, G., B.A. Bernhardt, M. Gardiner, J. Rodgers, and N.A. Holtzman. 2005. Scientists' and science writers' experiences reporting genetic discoveries: Toward an ethics of trust in science journalism. *Genetics in Medicine* 7 (3): 198-205.

Glass, K.C., and J.M. Kaufert, eds. 2001. *Research involving Aboriginal individuals and communities: Genetics as a focus.* Ottawa: National Council on Ethics in Human Research.

Glover, N.M., and S.J. Glover. 1996. Ethical and legal issues regarding selective abortion of foetuses with Down syndrome. *Mental Retardation* 34 (4): 207-14.

Hamlett, P.W. 2003. Technology theory and deliberative democracy. *Science, Technology, and Human Values* 28 (1): 112-40.

Indigenous Peoples Council on Biocolonialism. 2000. Indigenous Research Protection Act. http://www.ipcb.org/publications/policy/index.html.

Jasanoff, J. 2003. Breaking the waves in science studies: Comment on H.M. Collins and Robert Evans, "The third wave of social science studies." *Social Studies of Science* 33 (3): 389-400.

–. 2005. *Designs on nature: Science and democracy in Europe and the United States.* Princeton, NJ: Princeton University Press, 247-71.

Jasanoff, S., and B. Wynne. 1998. Science and decision-making. In *Human choice and climate change,* vol. 1., ed. S. Rayner and E. Malone, 1-87. Columbus, OH: Battelle Press.

Kaplan, D. 1993. Prenatal screening and its impact on persons with disabilities. *Foetal Diagnosis and Therapy* 8 (suppl. 1): 64-69.

Kerr, A., S. Cunningham-Burley, and A. Amos. 1998a. The new genetics and health: Mobilizing lay expertise. *Public Understanding of Science* 7 (2): 41-60.

–. 1998b. Drawing the line: An analysis of lay people's discussions about the new genetics. *Public Understanding of Science* 7 (2): 113-33.

Keulartz, J., M. Schermer, M. Korthals, and T. Swierstra. 2004. Ethics in technological culture: A programmatic proposal for a pragmatist approach. *Science, Technology, and Human Values* 29 (1): 3-29.

Kovach, B., and T. Rosenstiel. 2001. *The elements of journalism.* New York: Three Rivers Press.

Kua, E., M. Reder, and M. Grossel. 2004. Science in the news: A study of reporting genomics. *Public Understanding of Science* 13 (3): 309-22.

Leiss, W., and D. Powell. 2004. *Mad cows and mother's milk.* Montreal and Kingston: McGill-Queen's University Press.

Levitt, M., K. Weiner, and J. Goodacre. 2005. Gene Week: a novel way of consulting the public. *Public Understanding of Science* 14 (1): 67-79.

Lippmann, W. 1922. *Public Opinion.* New York: Macmillan.

Longstaff, H., M. Burgess, and P. Lewis. 2007. Comparing methods of ethical consultation for biotechnology related issues. *Health Law Review* 15 (1): 37-38

Luskin, R.C., J. Fishkin, and R. Jowell. 2002. Considered opinions: Deliberative polling in Britain. *British Journal of Political Science* 32 (2): 455-87.

McDaniels, T. 1998. Ten propositions for untangling descriptive and prescriptive lessons in risk perception findings. *Reliability Engineering and System Safety* 59 (1): 129-34.
National Institutes of Health. 2004. Points to consider when planning a genetic study that involves members of named populations. http://www.nih.gov/sigs/bioethics/named_populations.html.
Nelkin, D., and S. Lindee. 1995. *The DNA mystique: The gene as cultural icon.* New York, W.H. Freeman.
Power, M.D. 2003. Lots of fish in the sea: Salmon aquaculture, genomics and ethics. Electronic Working Papers Series, no. 004, W. Maurice Young Centre for Applied Ethics, University of British Columbia. http://www.ethics.ubc.ca/workingpapers/deg/deg004.pdf (accessed May 13, 2008).
Radder, H. 1992. Normative reflexions on constructivist approaches to science and technology. *Social Studies of Science* 22 (1): 141-73.
–. 1998. The politics of STS. *Social Studies of Science* 28 (2): 327.
Roper Starch Worldwide. 1997. Americans talk about science and medical news. The National Heath Council. http://www.nationalhealthcouncil.org/aboutus/about_index.htm.
Rowe, G., and L.J. Frewer. 2000. Public participation methods: A framework for evaluation. *Science, Technology and Human Values* 25 (1): 3-29.
Sherwin, S. 2001. *Toward setting an adequate ethical framework for evaluating biotechnology policy.* Ottawa: Canadian Biotechnology Advisory Committee.
Shuchman, M., and M.S. Wilkes. 1997. Medical scientists and health news reporting: A case of miscommunication. *Annals of Internal Medicine* 126 (12): 976-82.
Tansey, J. 2003. Prospects for the governance of biotechnology in Canada. Electronic Working Papers Series, no. 001, W. Maurice Young Centre for Applied Ethics, University of British Columbia. http://www.ethics.ubc.ca/workingpapers/deg/deg001.pdf.
Tansey, J., and M.M. Burgess. 2004. The foundations, applications and ethical dimensions of biobanks. Electronic Working Papers Series, no. 005, W. Maurice Young Centre for Applied Ethics, University of British Columbia. http://www.ethics.ubc.ca/workingpapers/deg/deg005.pdf.
–. 2006. Complexity of public interest in ethical analysis of genomics: Ethical reflections on salmon genomics/aquaculture. *Integrated Assessment* 6 (2): 37-57.
Voakes, P.S. 2004. A brief history of public journalism. *National Civic Review* 93 (3): 25-35.
Ward, S.J.A. 2004. *The invention of journalism ethics: The path to objectivity and beyond.* Montreal and Kingston: McGill-Queen's University Press, 175-82.
Weijer, C., G. Goldsand, and E.J. Emanuel. 1999. Protecting communities in research: Current guidelines and limits of extrapolation. *Nature Genetics* 23 (3): 275-79.
Williams-Jones, B., and M.M. Burgess. 2004. Social contract theory and just decision making: Lessons from genetic testing for the BRCA mutations. *Kennedy Institute of Ethics Journal* 14 (2): 115-42.
Wynne, B. 2001. Creating public alienation: Expert cultures of risk and ethics on GMOs. *Science as Culture* 10 (4): 445-81.
–. 2003. Seasick on the third wave? Subverting the hegemony of propositionalism: Comment on H.M. Collins and Robert Evans, "The third wave of social science studies." *Social Studies of Science* 33 (3): 401-17.

16
Impact Assessments and Emerging Technologies: From Precaution to "Smart Regulation"?

Jacopo Torriti

Despite their different scopes, stakeholders, from business corporations to nongovernmental organizations, unanimously demand more evidence-based decisions (Löfstedt 2005). In some circumstances they provide their own evidence to influence regulatory decisions (see the REACH case: European Commission 2003). When tackling the complexities related to new technologies, the regulator is compelled to fulfill the demand for an approach to decision making different from the precautionary principle (Majone 2004). Therefore, institutions are moving their focus from command and control policies to flexible, performance-oriented forms of regulation. They are entrenching the principle of "smart regulation" throughout the regulatory system (Hampton 2005). In this scenario, risk regulation and risk-analysis instruments (e.g., risk assessments, risk-benefit analysis, cost-benefit analysis, sensitivity analysis) are often seen as vanguard methods to decision making (Hahn 1996) or as supports to help make decisions under uncertainty (Hutter 2005). But do regulators base their decisions about new technologies on their potential risks and benefits? The European Union offers an appropriate setting to address this question. It has often been accused of using the precautionary principle as the main approach to regulating new technologies (Graham and Hsia 2002). More recently, it has instituted an impact assessment (IA) system where concepts of risk and safety, benefit and cost, hazard and safeguard levels all convene within the same reports. Deep-seated social and environmental values characterize the EU regulatory tradition (Thatcher 2001). As a consequence, new approaches to law making, different from the precautionary principle, are more challenging and significant to analyze. Observations and examples in this chapter are based on the analysis of approaches to emerging technologies in (1) the European Commission Communication on the precautionary principle; (2) the IA guidelines; (3) fifty-four preliminary IAs, or roadmaps, carried out in 2006; and (4) numerous specific recent (extended) IAs.

From Precaution to Smart Regulation?

Risk-based tools, such as IAs, are an example of the shift from informal qualitative-based standard setting toward a more calculative and formalized approach in decision making (Hood, Rothstein, and Baldwin 2001). Analysts interpret the introduction of (regulatory) IAs at the European level, for instance, as a move away of the regulatory philosophy from precaution (Löfstedt 2004). If one of the broad aims of the precautionary principle is to promote sustainable development, smart regulation's main purpose is to pursue competitiveness (Leiss 2003). Table 16.1 summarizes the resilient features that differentiate the precautionary approach from the "smart" philosophy of regulation. The term *smart regulation* has here a slightly different connotation from what Gunningam and Grabosky (1998) defined as the complementary combination of policy instruments, tailored to particular environmental goals and circumstances that will produce more effective and efficient policy outcomes. Indeed, smart regulation is often a retreat from command and control forms of policies. In this chapter, however, it is also intended to encompass a performance-oriented set of standards, driven by principles of proportionality, transparency, accountability, and consistency (Majone 1996; OECD 1997). In this sense, the nuance is rather close to the European concept of "better regulation" (Better Regulation Task Force 2005; European Commission 2005a).

In the past, European regulators seemed generally able to support their decisions in qualitative, even subjective, terms, in cases where no compelling numerical assessments of risks, costs, and benefits were provided (Jasanoff 2005). This situation differs significantly from the American regulators, who were not free to justify their actions by simply invoking delegated authority or superior expertise. This is not the place for comparative analysis. However, one has to bear in mind that the difference between the product-based US and the process-based European regulatory approaches is an important factor for understanding different decisions on new technology (Vig 2003). Americans, for example, believe that European policy took the wrong turn in targeting the technology of genetic manipulation rather than the actual risks, which should be the real focus of regulation (Jasanoff 2005). Europe's distinctive framing of the risks of biotechnology emerges as one step in the larger project of creating order in science and society. Recent studies point out that, until 2002, the precautionary principle was the main approach chosen in the European regulatory framework. After 2002, the use of this principle diminished significantly (Weinberg Group 2003). Arising technologies are often characterized by incomplete information about their scientific features and technical applications. Research may take too long to explore all the risks and side effects. This situation compels regulators to take decisions under uncertainty, without full knowledge of the technologies arising from the

Table 16.1

From precaution to smart regulation: Main features

	Precaution	Smart regulation
Far-reaching aim	Promoting sustainable development.	Pursuing competitiveness.
Principles	Taking a cautious approach to risk-management decisions when information is uncertain, unreliable, or inadequate.	Accountability, proportionality, transparency, and consistency are the main drivers.
Approach	Principle and process-based regulatory approach.	Flexible approach. Product-based process.
Instruments	ALARA (as low as reasonably achievable); BACT (best available control technology); containment in time and space; and constant monitoring of potential side effects.	Numerical thresholds, including quantitative safety goals, exposure limits, and standards.
Regulatory strategy	Command and control. Focus on provision. Strategies that do not necessarily encourage technological innovation and are not cost-effective (e.g., BREFs (best available techniques reference documents) prepared under the Integrated Pollution and Control Directive.	Flexible, performance-oriented standards. Focus on compliance. Strategies that indicate the required performance of the targeted population.
Risk management	Focus on hazards. The denotation of "precaution" implies prudent handling of uncertain or highly vulnerable situations.	Focus on risk-benefit analysis. The denotation of "risk" implies that management relies on the numerical assessment of probabilities and potential damages.

market. The choice is between a "better safe than sorry" tactic and reliance in risk-analysis instruments. There are reasons to believe that institutions are using analytical means emerging out of economics (cost-benefit approaches) and science (risk-assessment techniques) in order to prove a "smart" approach to regulation (Hutter 2005) and gain trust (Löfstedt and Vogel 2001). To understand how regulators deal with risk-related issues, one has to look at the principles and institutional methods that are in place to guide them through their decisions. For this reason, my analysis focuses first on the European regulator's leading principles (communication on precautionary principle) and guidelines (on IAs) for identifying, assessing, and thus regulating new technologies.

Emerging Technologies in the EU Communication on the Precautionary Principle

Few policies for risk management have created so much controversy as the application of the precautionary principle at the European level (Löfstedt, Fischhoff, and Fischhoff 2002). In its strongest formulations, the principle can be interpreted as calling for absolute proof of safety before allowing new technologies to be adopted (Sand 2000). Other formulations open the door to cost-benefit analysis and discretionary judgment (Wiener and Rogers 2002). The debate on the European approach on precaution has involved numerous scholars, Fischhoff, Graham, Klinke, Löfstedt, Renn, Rogers, and Wiener among them. As the application of the precautionary approach may have lasting implications and repercussions on regulatory decisions and international trade, the stakes in this debate are not merely theoretical or academic. Depending on the approach chosen, regulatory actions may vary considerably and shape economic competitiveness, public health levels, and environmental quality (Klinke and Renn 2002).

The European Commission standpoint on this principle is stated in an apposite communication (European Commission 2000), which confirms the EU position of applying the precautionary approach to hazard management. This overview is centred on how the communication directly addresses emerging technologies.

First, the starting assumption of the communication is that advances in technology have fostered a growing sensitivity to the emergence of new risks. Before scientific research is able to illuminate the problems, decision makers must take account of the fears generated by these perceptions and put preventive measures in place to eliminate the risk or at least reduce it to the minimum acceptable level. Therefore, the communication takes a cautious approach to risk-management decisions when information is uncertain, unreliable, or inadequate.

Second, the section of the communication on socio-ethical concerns envisages technology as a mere generator of risks. A radical interpretation of this section could be paralyzing; it may forbid regulation and lead to inaction (Sunstein 2003).

Third, the decision to take measures, without waiting until all the necessary scientific knowledge is available, is appropriate because "the dimension of the precautionary principle goes beyond the problems associated with a short or medium-term approach to risks. It also concerns the longer run and the well-being of future generations" (European Commission 2000, 8). In a sense, the precautionary principle is perceived as a long-term approach to risk management of new technologies. The broader aim is sustainable development and not short-term competition.

Fourth, reduction activities, such as ALARA (as low as reasonably achievable), BACT (best available control technology), containment in time and space, or constant monitoring of potential side effects, all derive from the precautionary approach. Numerical thresholds, including quantitative safety goals, exposure limits, and standards are considered as part of a more general risk-based approach (Stirling 1999).

Finally, some of the communication's key motivations are based on the Cartagena Protocol on Biosafety concerning the safe transfer, handling, and use of living modified organisms resulting from modern biotechnology. In accordance with this precautionary approach, the European Union has prohibited bioengineered crops and American beef treated with growth hormones (Vogel 2002) and is now crafting legislation that will require chemical companies to spend billions of dollars on safety-testing their products. On this matter, a World Trade Organization (WTO) dispute panel on February 7, 2006, issued a preliminary ruling suggesting that several aspects of the way that the European Union's process for assessing genetically modified organisms was not based on a risk assessment, in contravention of the WTO Agreement on the Application of Sanitary and Phytosanitary Measures (the SPS Agreement). The panel is largely favourable to the complaint brought in 2003 by the United States, Argentina, and Canada against what these countries alleged was an EU moratorium on the approval of new biotech products.

Pitfalls in the communication were already pointed out by some authors, along with the suggestions of introducing a systematic process of ranking hazards and targeting cost-effective protection opportunities (Graham and Hsia 2002). Other advocates of the risk-based approach argue that precautionary measures ignore scientific results and lead to arbitrary regulatory decisions (Cross 1996). The use of the precautionary principle, as it is intended in the communication, does not preclude science approaches at all (Klinke and Renn 2002). Advocates of the precautionary principle have argued that precaution does not automatically mean banning substances or activities but would imply a gradual, step-by-step diffusion of risky activities or technologies until more knowledge and experience is accumulated (Bennet 2000). Löfstedt (2004) highlights that although the precautionary principle has dominated the political and academic debate, it is likely that its application is seen as too costly, since EU regulation is driven not only by sustainable development and good governance principles but also by competitiveness. Therefore, the EU regulator in the future will prefer more and more use of IAs as a rational risk-management tool for devising risk regulation. He points out the divergence between Directorate-Generals (DGs) within the European Commission (namely, DG Environment, DG Enterprise and Industry) on chemical legislation. This concrete observation is extremely relevant, as it creates the grounds for understanding different uses of IAs and different approaches to new technologies within the European Commission.

Regulatory Impact Assessment and Risk-Based Regulation

Since its introduction in the United States in 1981, regulatory impact assessment is the analytical instrument used to appraise *ex ante* the effects of new legislation. Regulatory impact assessment was introduced to promote economic efficiency and therefore economic growth, to regulate only when the market fails, and to regulate by using cost-effective and market-based approaches. It was first used in the United States and later some pioneer OECD (Organisation for Economic Co-operation and Development) countries. It is an attempt by the regulator to understand what the consequences of such regulation will be in the proposal phase of new regulation. The scientific ambition of economics is heavily challenged: a crystal ball enabling a full vision of the future social, environmental, and economic impact of proposed regulations is nowhere to be found (Renda 2006). Regulatory impact assessment is also the main instrument in the regulator's hands to systematically identify the likely consequences of introducing emerging technologies in the market. In its US form, regulatory impact assessment is very much aligned with cost-benefit analysis (OMB 2000). The EU version is particularly interesting because it uses an integrated system that includes, simultaneously, environmental and social values in addition to economic analysis. However, the EU approach is far from unique. The incorporation of socio-ethical values in impact assessments is a common topic for all national regulators (Radaelli 2001). Another common dimension of numerous regulators beyond the European Union EU is the switch to risk analysis. Reviews of worldwide regulatory initiatives suggest that agencies that appear to have taken a much more total and systematic risk-based approach are visible in the United Kingdom, United States, and, increasingly, Australia and Canada (Hutter 2005). In these countries, the focus on risk-management strategies is a means of orienting regulatory activities and organizing governance attitudes and structures (Löfstedt 2004). In other circumstances, a limited buy-in to the risk philosophy is in evidence. On occasion, this is manifested simply by the use of the language of risk; in others, this is supported by the use of risk-based tools of assessment. Other Scandinavian and European regulators exemplify agencies adopting risk-based tools on an ad hoc basis by exhibiting a partial buy-in to broader risk philosophies (Swedish National Audit Office 1995). In Germany, for instance, there is evidence of a move in environmental and occupational health and safety regulation to develop more systematic quality targets and evaluation techniques. The European Union represents a significant unit of analysis because of the complex multiplicity of regulatory philosophies that can be observed in its legislative interventions. When the European Union decided to introduce the integrated system of IA in 2002 (European Commission 2002), the main focus was to produce higher quality regulation (Radaelli 2005). To date, research of the European IA system centred on the concept of regulatory quality (Radaelli and De Francesco 2004), the content (Vibert 2004) and procedural features (Lee and Kirkpatrick 2004) of the IA reports, the level of attention dedicated to sustainable development in IAs (Institute for European Environmental Policy 2004), and the role of economic analysis in IAs (Torriti 2007). However, no analysis has been carried out on emerging technologies in IAs.

Making Decisions on Emerging Technologies in the European Union: The Role of Impact Assessments

Since 2003, major draft laws issued by the European Commission have been accompanied by an impact assessment study. This means that new proposals need to be assessed against their costs and benefits for Europe's economy and be consistent with the commission's drive for improving business competitiveness. All IAs draw on the guidelines published by the Secretariat-General of the European Commission (European Commission 2005b). The guidelines define the appropriate level of action, in accordance with the principle of subsidiarity and proportionality. The option of taking no action at all at the EU level should always be considered to ensure consistency with other EU policies, along with the three pillars of sustainable development–economic, social, and environmental. Compared with previous versions (European Commission 2002), the 2005 guidelines stress the relative dominance of economic performance and competitiveness over social and environmental aspects. No other *ex ante* instrument is designed to support the decision making on technologies. The only exception is the Scientific and Technological Options Assessment (STOA) in the European Parliament. In a manner somewhat reminiscent of the Office of Technology Assessment (OTA) in the United States, the STOA is overseen by a panel made up of one member from each of the permanent committees of the European Parliament. The panel adopts an annual work plan of projects proposed by the committees, usually on the issues coming before the European Parliament that have a scientific or technological theme. These may be proposals directly relating to research or innovation policy, or measures concerning the many ways in which science and technology affect society, the economy, or the environment. STOA's aim is to assess emerging technologies, although it has no in-house research capacities and, therefore, usually contracts out all its projects to universities and research institutions. Its function is one of project definition and management, and it serves as a model for a bare bone, contract-out technology assessment service (Vig 2003). Hence, STOA is to be excluded from playing a significant role in decision making as much as the integrated system of IAs.

New Technologies in the Impact Assessment Guidelines

The European Commission guidelines on IAs (European Commission 2005a) include indications on how the European regulator should estimate the effects of emerging technologies.

The IA guidelines suggest that new products or services should be assessed depending on how they facilitate the introduction and dissemination of new production methods, technologies, and products. Innovative technologies are encouraged if they enhance value chains and network productions, leading to growth in economic performance. The guidelines also touch on aspects of business organization related to new technologies. In fact, proposals may directly or indirectly lead to impacts on the technological development and innovative activities of firms and the ways in which firms are organized. When impacts on technological development and innovation are likely, new legislation may affect the level and timing of research and development activities; for example, making it easier or harder to finance these activities. In these cases, the IA will establish whether the proposal stands in the way of–or promotes–firms' potential for innovation and so on (e.g., know-how and finance); the development and implementation of new technologies; the diffusion and take-up by users of new technologies; greater knowledge and know-how; and the exploitability of inventions and innovations.

Is there any evidence of the move from precaution to smart regulation? The guidelines do not welcome excessively precautionary environmental regulations that slow the diffusion of new technologies. Imposing tighter emission standards on or defining new sources of pollution for new installations may prolong the operating life of existing older, dirtier plants; raise the cost of entry to the industry; and thereby only reduce levels of pollution instead of leading to less pollution overall.

In the past, the European Commission used traditional "command and control" policies. These specify the use of certain technologies. They have always been considered advantageous for being relatively easy to monitor and enforce (Aalders and Wilthagen 1997). The guidelines on IAs emphasize that command-and-control policies do not encourage technological innovation and are not cost-effective. Examples can be found in the BREFs (best available techniques reference documents) prepared under the Integrated Pollution Prevention and Control Directive (1996/61/EC) (see also Royal Commission on Environmental Pollution 1998). According to this directive, to prevent disproportionate rigour, the European regulator introduced performance-oriented standards, which specify the required performance of the targeted population. Performance standards are expected to "increase flexibility to achieve the regulatory standard." If existing legislation restricts the introduction to the market of beneficial technologies, an alternative to tighter rules or regulations is a simplification of the rules with which compliance seems to be difficult.

Emerging Technologies in the 2006 Legislative Proposals (Roadmaps)

The guidelines provide a framework of various techniques and methodologies to move the focus of the regulator from precaution to smart regulation. To what extent are these suggestions taken into account in the actual IA reports that deal with new technologies? When asked about emerging technologies in IAs, European Commission officials at the Secretariat-General suggested looking at the 2006 roadmaps (European Commission 2006), the equivalent of preliminary IAs. The fifty-four roadmaps on the 2006 proposals refer to new technologies in very different fields of application, from digital rights management (2006/MARKT/008) to secure information (2006/INFSO/002), from internal market for postal services (2006/MARKT/006) to common market organization for wine (2006/AGRI/003), from public administration softwares (2006/INFSO/001 and 2006/INFSO/003) to satellite navigation (2006/TREN/025 and 2006/TREN/033). In some proposals, new technologies are considered a threat because of their potential use for copyright circumvention, terrorist activities, or terrorists' access to top-quality defence equipment (2006/ENTR/008). Numerous roadmaps reflect the excessively rigid regulatory regime of the past and therefore promote research (constitution of the European Institute of Technology, 2006/EAC/004) and endorsement (external aspects of competitiveness, 2006/TRADE/001) of new technologies. In other cases, new technologies have a positive valence. The proposal to improve postal services fosters technologies that contribute positively to European Community economic growth and competitiveness priorities. The wine proposal modifies substantially oenological practices allowing the use of new technologies.

However rooted in European citizens' everyday life wine and postal services may be, the technologies related to them are not destined to radically transform social behaviours. In other words, they are not on the same level as the emerging technologies discussed in this book. Moreover, the roadmaps do not entail any type of numerical assessments of risks, costs, or benefits within the reports. Their scientific and technical content is so low that they miss out on potentially interesting topics. For example, roadmaps on security and defence do not question how cost-effective biometric technologies might be in preventing unauthorized access to technology; roadmaps on satellite innovations do not include tests to demonstrate their functionality and effectiveness. These issues are expected to be analyzed in detail in extended IAs, which have not yet been published. Moreover, none of the 2006 roadmaps deals with pharmacogenomics, nanotechnologies, biotechnologies, transgenic animals, or stem cells, all topical subjects in this book. There may be two explanations for this. The first is that most of these topics are still being discussed in a pre-proposal phase. This is the case, for example, of the EU strategy on biofuels. The European Union is intending to set targets of 5.75 percent on biofuel content in fuels by 2010. The main ideas will be discussed by EU leaders before the commission drafts the proposal. In fact, the policy-making system in the European Union implies that the European Commission drafts proposals, whose principles and ideas actually generate from the European Council, namely, the board of the member states' prime ministers (Christiansen 2001; Draetta 1999). The second reason for this shortage of topical proposals on new technology in 2006 is that some of the issues may have already been considered in previous IAs. This brief overview of fifty-four roadmaps was nonetheless useful (1) to understand the current state of new technologies and (2) to identify the internal trends within a large regulator such as the European Commission. In 2006, proposals by Energy and Transport, as well as Information Society, Directorate-Generals present a favourable approach to emerging technologies.

Emerging Technologies in Individual Impact Assessment Reports

If the 2006 roadmaps do not contain a sufficient amount of information about the European approach on emerging technologies, it may be necessary to investigate other sources. Compared with roadmaps, IAs are supposed to encompass a higher use of risk analysis (European Commission 2005b). Do regulators base their decisions about new technologies on their potential risks and benefits? When asked which IAs best tackle emerging technologies, European researchers specializing in "better regulation" suggested a pool of three IAs: tissue engineering cell therapy and gene therapy (European Commission 2005b); visa information system (EPEC 2004); and INSPIRE (infrastructure for spatial information in Europe) (European Commission 2004).

Tissue Engineering Cell Therapy and Gene Therapy

The IA on tissue engineering cell therapy and gene therapy is remarkable because it takes into account ethical aspects. The proposed regulation (DG Enterprise and Industry) covers all advanced biotechnology-derived therapy products (gene therapy medicinal products, somatic cell therapy medicinal products, and tissue engineered products). The issue of embryonic stem cells was already at the centre of a debate during the adoption of the directive on the quality and safety of human tissues and cells (Directive 2004/23/EC). In this context, the legislator recognized that there is, to date, no consensus on the use or prohibition of embryonic stem cells. As a consequence, the European Union does not regulate or prohibit such use. This responsibility is remitted to member states. If, however, any member state authorizes the use of stem cells, the European Union should make "all provisions necessary to protect public health . . . throughout the Community" (European Commission 2005c, 35). The same approach is followed in the IA on tissue engineering cell therapy and gene therapy: the proposed regulation does not interfere with national legislation, prohibiting or restricting the use of specific types of human or animal cells; or with the sale, supply, or use of medicinal products based on such cells. The IA declares that the regulation respects fundamental human rights and "also takes into account, as appropriate, the Convention for the protection of human rights and dignity of the human being with regards to the application of biology and medicine" (European Commission 2005c, 31).

Visa Information System

The IA on the visa information system (Justice, Freedom and Security DG) has interesting implications for the application of biometrics. In essence, two of the policy options consider the benefits and disadvantages of introducing or not introducing biometric information (such as fingerprints and a digitalized image to facilitate facial recognition) of visa applicants. The IA–carried out by Justice, Freedom and Security DG–provides a rather detailed analysis of the costs and benefits related to these two options. Costs are divided into financial costs, opportunity costs, and retaliation costs. The estimate of the financial costs for member states for introducing biometrics is based on the average of the only two member states that provided cost estimates: France and Sweden. The starting assumption is that France is a very large issuer of visas, whereas Sweden is a small issuer of visas. This calculation of financial costs goes against the concept of differential costs and the estimate of expenses for national systems and is therefore statistically insignificant. Benefits are expressed in terms of increased efficiency in the implementation of a common visa policy, restrictions in fraud and visa shopping, a decrease in illegal migration, contributions to the internal security of member states, and spin-off benefits for the information technology industry in EU countries. The main risk contemplated in the IA is retaliation risk. Beyond the mishandling of cost-benefit analysis, the final decision of the European Commission will be based on the trade-off between avoiding the substantial financial costs of introducing biometrics and having a reliable visa system in Europe.

INSPIRE: Infrastructure for Spatial Information for Data in Public Sector

The INSPIRE initiative is a noteworthy example of how public administration appraises the state of the technologies it works with. The public sector is by far the largest producer of information in Europe. Environment DG estimates that between 15 percent and 25 percent of total data used in e-commerce trading is based on public sector information. The initiative is to be considered as part of the recent attempts to reduce paperwork requirements with technology-driven programs, such as the creation of the FirstGov website, electronic docketing, filing and reporting requirements, and the creation of other one-stop shops (Renda 2006). Although the internet is broadly considered an established technology, widespread access and use of spatial information is still considered problematic in Europe. The IA is characterized by a synthetic cost-benefit analysis. It is not prescriptive with regard to the pricing regimes adopted by public sector bodies, allowing national frameworks to remain in place and explicitly permitting charges based on costs of collection, production, reproduction, and dissemination, together with a reasonable return on investment. The IA does not include any assessment of the risks, excluding a brief legal mention of data protection and copyright. The focus of the IA is on the opportunities that will arise by producing a more streamlined and transparent set of conditions for the commercial exploitation and reuse of public-sector documents across Europe. The IA does not incorporate the concept of opportunity-loss risk: without an improvement in the conditions for the reuse of public sector information, there will be a considerable loss in the economic development of products and services based on this information. This may result in a loss both to the firms involved and to the end-users of the information and the value-added services that might have been built on it. On one side, if implementation is not effective, the benefits expected to flow from greater ease of reuse will not occur or will be reduced. On the other side, there is the risk that implementation will impose unnecessary costs through excessive regulatory burden on the public-sector bodies covered by the INSPIRE directive, and through the costs of excessive regulatory mechanisms. Table 16.2 summarizes the evidence gathered in this analysis about the shift of the European regulator from a precautionary approach to forms of smart regulation. In some cases, the regulator seems to maintain a neutral position.

Table 16.2

EU regulators' approach to new technologies

	Precautionary approach	Neutral approach	Competitive – "smart" approach	Use of risk-analysis instruments
Precautionary principle communication	Precautionary principle is applied to hazard management. Decision makers must take account of the fears generated by these perceptions and put in place preventive measures to eliminate the risk or at least reduce it to the minimum acceptable level.	(Precaution does not necessarily mean automatic ban of substances or activities. It would imply a gradual, step-by-step diffusion of risky activities or technologies until more knowledge and experience is accumulated.)	—	The precautionary principle is considered within a structured approach to the analysis of risk that comprises risk assessments.
Impact assessment guidelines	—	New products or services should be assessed depending on how they facilitate the introduction and dissemination of new production methods.	Innovative technologies are encouraged if they enhance competitiveness against precautionary measures that slow the diffusion of new technologies.	The in-house and external use of risk-assessment (and cost-assessment) techniques is encouraged.

▶

◀ *Table 16.2*

	Precautionary approach	Neutral approach	Competitive –"smart" approach	Use of risk-analysis instruments
Preliminary impact assessments (roadmaps): Cases	New technologies are considered a threat because of their potential use for copyright circumvention, terrorist activities, or terrorists' access to top-quality defence equipment.	Research (constitution of the European Institute of Technology) and endorsement (external aspects of competitiveness) of new technologies is promoted.	Technologies from niche markets are promoted for adding positive value to economic growth and competitiveness priorities (e.g., wine and postal service).	Roadmaps do not entail any type of numerical assessments of risks, costs, or benefits.
Extended impact assessments: Cases	—	Because of the lack of consensus on the use or prohibition of cell therapy and gene therapy, the regulator does not regulate or prohibit such use.	Biometrics may not be used because of their excessive costs. Use of spatial information is endorsed as "new technology."	Cost-benefit analysis on biometrics is based on questionable methodology on differential costs. Widespread spatial information is promoted without great detail of costs or risks.

Conclusion

Regulators use different instruments to assess the economic, social, and environmental impacts of emerging technologies. This chapter set out to explore the challenges that regulators face through regulatory proposals on emerging technologies.

The overview of the communication on the precautionary principle (European Commission 2000) confirms its extreme sensitivity to the risks associated with emerging technologies. The communication is permeated by socio-ethical concerns that may affect regulatory resolutions on new technologies. In 2003, the European Commission, namely the main European regulator, introduced an impact assessment system to fulfill business stakeholder demand for an approach to decision making different from the precautionary principle. Decisions are based on estimates of their potential risks and benefits, and technology is not merely targeted until evidence of its soundness is demonstrated. The overview concludes that the 2005 guidelines on IAs create the basis for a regulatory shift from precaution to smart regulation, as they

- promote the introduction and dissemination of new technologies
- are liable to appraise the impacts of emerging technologies also within business organizations
- dishearten the use of command-and-control policies that do not encourage the diffusion of new technologies
- do not welcome excessively precautionary environmental regulations when they slow the dissemination of new technologies.

The guidelines have the potential to wipe out the precautionary principle because of their focus on economic performance. Which approach to decision making prevails in practice? Has precaution been fully replaced by analytical instruments for smart regulation? The 2006 roadmaps show different approaches on new technologies from different regulators within the European Commission. Roadmaps comprise radically diverse cases where

- emerging technologies are considered mainly as a threat
- research and endorsement of new technologies is promoted to overcome the previous regulatory regime
- the impacts of new technologies are largely positive in some markets.

Consequently, the European Commission seems to entail different internal drivers when drafting proposals on new technologies. Several factors underpin the differences within the European Commission preliminary impact assessments (Torriti 2007). The Directorate-Generals function as different regulators and produce diverse regulatory outcomes.

The analysis of extended IAs confirmed the European regulator's heterogeneous attitude to new technologies. The line of intervention changes significantly from case to case, rendering it difficult to identify any shift from precaution to smart regulation. The analysis stressed three approaches in IAs that deal with new technologies:

- a neutral position (not regulating, nor prohibiting) on tissue engineering and gene therapy
- a cost-benefit analysis approach, although based on a rather dubious methodology, on the use of biometrics in the visa system
- a favourable standpoint on technologies for spatial information associated with low risks and costs.

The conclusions are that the European regulator has a system in place to assess the consequences of emerging technologies. Evidence suggests that the same is happening in a number of countries, including the United Kingdom, the United States, Australia, and Canada, where aggregate and methodical risk-based approaches have been established. In other words, regulators are equipping themselves with risk-analysis tools in order to be more responsive and competitive when regulating on emerging technologies. Moreover, this analysis emphasized cases where the EU regulator is not in complete control of risk-assessment and cost-assessment instruments. Further analysis is needed to understand whether these analytical instruments are being used in an appropriate way. The move away from precaution is not destined to happen promptly; for example, by introducing guidelines on IAs. The introduction of the IA system may foster the use of economic estimates (e.g., cost-benefit analysis) and scientific techniques (e.g., risk assessment) in the phase of proposals for new legislation. However, this does not directly guarantee a different regulatory outcome for new technologies.

References

Aalders, Marius, and Ton Wilthagen. 1997. Moving beyond command-and-control: Reflexivity in the regulation of occupational safety and health and the environment. *Law and Policy* 19 (4): 415-43.

Bennet, P.G. 2000. Applying the precautionary principle: A conceptual framework. In *Foresight and precaution,* vol. 1, ed. M.P. Cottam, D.W. Harvey, R.P. Paper, and J. Tait, 223-27. Rotterdam: Balkema.

Better Regulation Task Force. 2005. *Regulation—Less is more: Reducing burdens, improving outcomes.* http://www.brtf.gov.uk/docs/pdf/lessismore.pdf.

Christiansen, Thomas. 2001. The Council of Ministers: The politics of institutionalised intergovernmentalism. In *The European Union: Power and policy-making,* ed. J. Richardson, 95-115. London: Routledge.

Cross, F.B. 1996. Paradoxical perils of the precautionary principle. *Washington and Lee Law Review* 53 (3): 851-925.

Draetta, Ugo. 1999. *Elementi di diritto dell'Unione Europea*. Milan: Giuffre.

EPEC. 2004. *Study for the Extended Impact Assessment of the Visa Information System*. Brussels: European Policy Evaluation Consortium.

European Commission. 2000. *Communication on the precautionary principle*. Brussels: COM 2000.

–. 2002. *Communication on impact assessment*. Brussels: COM(2002) 276.

–. 2003. *Extended impact assessment of the economic, social and environmental impacts of the New Chemicals Policy proposals*. Brussels: SEC(2003) 1171/3.

–. 2004. *Directive on infrastructure for spatial information for data in public sector*. Brussels: SEC(2004) 980.

–. 2005a. *Guidelines on impact assessments*. Brussels: Secretariat-General of the European Commission.

–. 2005b. *Tissue engineering cell therapy and gene therapy*. Brussels: SEC(2005) 1444.

–. 2005c. *Proposal for a regulation on advanced therapy medicinal products–Impact assessment*. Brussels: COM(2005) 567.

–. 2006. *2006 Roadmaps*. Brussels: Secretariat-General of the European Commission.

Graham, John, and S. Hsia. 2002. Europe's precautionary principle: Promise and pitfalls. *Journal of Risk Research* 5 (4): 317-49.

Gunningham, Neil, and Peter Grabosky, eds. 1998. *Smart regulation: Designing environmental policy*. Oxford: Oxford University Press.

Hahn, Robert. 1996. *Risks, costs and lives saved: Getting better results from regulation*. New York: Oxford University Press.

Hampton, Philip. 2005. *Reducing administrative burdens: Effective inspections and enforcement*. London: HM Treasury.

Hood, Christopher, Henry Rothstein, and Robert Baldwin. 2001. *The government of risk: Understanding regulation regimes*. Oxford: Oxford University Press.

Hutter, Bridgitte. 2005. *The attractions of risk-based regulation: Accounting for the emergence of risk ideas in Regulation Discussion Paper 33*. London: Carr-LSE.

Institute for European Environmental Policy. 2004. *Sustainable development in the European Commission's integrated impact assessments for 2003*. London: IEEP.

Jasanoff, Sheila. 2005. *Designs on nature: Science and democracy in Europe and the United States*. Princeton: Princeton University Press.

Klinke, Andreas, and Ortwin Renn. 2002. A new approach to risk evaluation and management: Risk-based, precaution-based and discourse-based strategies. *Risk Analysis* 22 (6): 1071-94.

Lee, Norman, and Colin Kirkpatrick. 2004. A pilot study on the quality of European Commission extended impact assessments. IARC Working Paper 08/2004. University of Manchester.

Leiss, William. 2003. *Smart regulation and risk management*. Paper prepared for the Privy Council Office, External Advisory Committee on Smart Regulation. http://www.smartregulation.gc.ca.

Löfstedt, Ragnar E. 2004. The swing of the regulatory pendulum in Europe: From precautionary principle to (regulatory) impact analysis. *Journal for Risk and Uncertainty* 28 (3): 237-60.

–. 2005. *Risk management in post-trust societies*. Basingstoke, UK: Palgrave.

Löfstedt, Ragnar, Baruk Fischhoff, and Ilya Fischhoff. 2002. Precautionary principles: General definitions and specific applications to genetically modified organisms. *Journal of Policy Analysis and Management* 21 (3): 381-407.

Löfstedt, Ragnar, and David Vogel. 2001. The changing character of regulation: A comparison of Europe and the United States. *Risk Analysis* 23 (3): 411-21.

Majone, Giandomenico. 1996. *Regulating Europe*. London: Routledge.

–. 2004. *The principle of precaution in politico-institutional context*. Florence: European University Institute.

OECD. 1997. *Regulatory impact analysis: Best practice in OECD countries*. Paris: OECD Publications.

OMB (US Office of Management and Budget). 2000. *Report to Congress on the costs and*

benefits of federal regulations: Regulatory program of the United States Government. Washington, DC: US Office of Management and Budget.
Radaelli, Claudio M. 2001. *L'analisi di impatto della regolazione in prospettiva comparata*. Catanzaro: Rubbettino editore.
–. 2005. Diffusion without convergence: How political context shapes the adoption of regulatory impact assessment. *Journal of European Public Policy* 12 (5): 924-43. http://www.informaworld.com/smpp/title~content=t713685697~db=all~tab=issueslist~branches=12 - v12.
Radaelli, Claudio, and Fabrizio De Francesco. 2004. *Indicators of regulatory quality: Final report*. http://ec.europa.eu/enterprise/regulation/better_regulation/impact_assesment/docs_concluding_conference240105/radaelli_finalreport_executivesummary.pdf.
Renda, Andrea. 2006. *Impact assessment in the EU: The state of the art and the art of the state*. Brussels: CEPS.
Royal Commission on Environmental Pollution. 1998. *Setting environmental standards*. London: The Stationery Office.
Sand, P.H. 2000. The precautionary principle: A European perspective. *Human and Ecological Risk Assessment* 6 (3): 445-58.
Stirling, A. 1999. *On science and precaution in the management of technological risks: Final report of a project for the EC Forward Studies Unit under auspices of the ESTO Network*. Report EUR 19056 EN. Brussels: European Commission.
Sunstein, Cass. 2003. Beyond the precautionary principle. Working paper 149, University of Chicago.
Swedish National Audit Office. 1995. *Impact assessment*. Stockholm: Swedish National Audit Bureau.
Thatcher, Mark. 2001. European regulation. In *The European Union: Power and policy-making*, ed. J. Richardson, 303-21. London: Routledge.
Torriti, Jacopo. 2007. (Regulatory) Impact assessment in the EU: A tool for better regulation, less regulation or less bad regulation? *Journal of Risk Research* 10 (2): 239-76
Vibert, Frank. 2004. *The EU's new system of regulatory impact assessment: A scorecard*. London: European Policy Forum.
Vig, Norman. 2003. The European experience. In *Science and technology advice for Congress*, ed. M. Granger Morgan and Jon Peha, 90-98. Washington, DC: Resources for the Future Press.
Vogel, David. 2002. Risk regulation in Europe and United States. Berkeley: Haas Business School.
Weinberg Group. 2003. *A quantitative review of the trends in using the precautionary principle by non-governmental organizations in the context of European Regulatory Development*. Brussels: European Policy Centre.
Wiener, Jonathan B., and Michael Rogers. 2002. Comparing precaution in the United States and Europe. *Journal of Risk Research* 5 (4): 317-49.

17
Technology Ownership and Governance: An Alternative View of IPRs

Peter W.B. Phillips

There has been heightened interest and vociferous debate about how property systems should be structured to deal with intellectual advances–those ideas, recipes, formulae, and processes that have the greatest potential to generate increases in economic productivity and social well-being.

Most of the currently operating property rights systems simply define ownership and control. Intellectual property rights (IP rights or IPRs) are social constructs that confer exclusive rights to a specific person to exclude others from using specific products or processes that have been newly invented. However, because society usually benefits most from unrestricted access to any innovation that improves the state of the art, IP regimes for the most part offer only monopoly ownership in exchange for full disclosure of the invention and its recipe or formulae in order to facilitate follow-on research. Society has justified granting inventors (or their assignees) exclusive rights to their inventions as an incentive to private investment in research and in an effort to encourage inventors both to use and to disclose their new ideas. Nevertheless, these rights are seldom absolute. Most societies require that an owner's use of any product or process be constrained when it infringes on another person's rights or where there is a compelling public good (such as public health and safety or public morality). At the operational level, these exclusive use-rights are further limited by the definition of the scope and duration of the grant and the provision that the innovation must be "worked" for the rights to be maintained.

The balance between ownership and control of new ideas is socially embedded in a distributed governance system that involves a dispersion of power over a wide variety of actors and groups within the state, market, and collective action sectors. There are particular challenges related to the property regimes concerning convergent, transformative technologies–that is, those that draw on different epistemic bases of cutting-edge science; represent step-changes in the scale and direction of development of human capabilities; have consequences distributed widely over many areas of life; are of sufficiently high profile to attract the attention, interest, and risk perception of social movements, citizens, politicians, and regulators; precipitate public debate and draw the attention of the journalists and ethicists; and are within a few years or at most a decade of realization (6 2001). The ownership and control of these new technologies is complicated because there is seldom a single innovation process; rather, firms, regulators, and social-action groups are frequently involved in projects related to different waves of discovery and, within a single wave of discovery, in several projects that spread over different phases of research. Even relatively simple changes in these systems of ownership and control may produce complex, often non-linear, results. Such systems, while hard to model and forecast, exhibit high degrees of adaptability.

Management and, more importantly, reform, of the IP system as it relates to transformative technologies must be understood in the context of the complex system. What appear to be appropriate and actionable policies (e.g., changing the formal structures of IPRs or altering the regulatory structure) can lead to counterintuitive results. Two features of our current IP systems should be kept in mind when considering the structure, function, and reform of IP mechanisms. First, the nature of the compromise in our IP systems leads to a trade-off between the immediate costs of monopoly–which can include dead-weight losses from underutilization or increased transactional costs from the need to search, negotiate, and enforce use contracts and can lead to undesirable distributional effects–and the long-term cumulative gains from successive and compounded innovation. This fundamental trade-off is paramount; adding or subtracting from either the costs or benefits could have significant long-term impacts that may be hard to anticipate or accommodate in contemporary debate. Second, it is important to keep in mind that while IP rights systems simply draw boundaries of spheres of ownership and control, these are then frequently embedded and used in other governing systems. For example, the owner (or licensed agent) of a new technology is the only one with the responsibility or authority to develop and commercialize a new invention and is consequently accountable for any predictable or unanticipated adverse health, safety, or environmental effects resulting from production or use. As such, IPRs are fundamentally part of a larger distributed system of governing our economy and society, with all the attendant constraints and impacts of being part of a system.

State of the Debate

The IPR system for biotechnology inventions, especially the role of patents in protecting new ideas, has generated significant debate in recent years. Although the United States has allowed patents on asexually propagated plants (called plant patents) since 1930, and many countries have granted plant breeders' rights to new commercial plant varieties since the 1940s, the modern era can be dated to 1980, when the US Supreme Court overturned a decision of the Patent and Trademark Office and granted a utility patent for claims related to a genetically engineered microorganism that degraded multiple components of crude oil, based on defining the new organism as a new composition of matter.[1] This decision was shortly followed by grants of utility patents for genetically modified plants and for genetically modified animals (the United States Patent and Trademark Office patent on Harvard OncoMouse in 1988).[2] During that same period, the technology and science became more refined, enabling researchers to identify and isolate specific genes from various microbes, plants, animals, and humans. Many researchers have applied for and (mostly) been granted utility patents on those genes.

The end result is that there are now literally tens of thousands of patents for biotechnology tools (e.g., *Agrobacterium tumefaciens,* and polymerase chain reaction), for isolated genes, and for constructed single and multicellular organisms derived from the technologies. A recent study by Jensen and Murray (2005) concludes that 4,270 human genes (containing 4,382 claims), or nearly 20 percent of all known human genes, have been claimed in US patents. Nearly two-thirds (63 percent) of the human gene patents have been assigned to private firms.

Along the way, numerous controversies have emerged. The longest and most involved debate has been about the patents granted to Harvard's OncoMouse. These patents have raised issues of the morality and ethics of patenting higher life forms (leading to an extended review and debate in Europe), the appropriateness of patents on higher life forms (the subject of legal wrangling in Canada between the Harvard regents and the Canadian Commissioner of Patents), and the impact on freedom to operate and follow-on research. In an unrelated issue, Myriad Genetics, of Utah, holds patents on two genes implicated in breast cancer and have patented gene tests on those genes (known as BRCA1 and BRCA2). There is increased concern that these types of patenting constrain both research and public use. Cho et al. (2003) surveyed clinical laboratory directors in the United States and found that 53 percent decided not to develop new clinical genetic tests because of patent concerns and that clinical geneticists believed that their research was inhibited by existing human gene patents. Meanwhile, there has been significant concern in Canada and Europe about the commercial practices of patent holders (especially Myriad), with the result that some jurisdictions cease to use the contested technology, others conform to and duly license the technologies, while others decide to use the technology without concluding licences. Most recently, a consortium of European researchers successfully challenged those parts of the Myriad patent on

BRCA1 that restricted its use as a diagnostic test in research and health institutions, winning a favourable decision at the Opposition Division of the European Patent Office. The same consortium is challenging the claims related to BRCA2 as well.

Meanwhile, there have been various fights about IPRs and their use in the agri-food world. Although farmers' groups around the world have expressed concerns about plant patents, the debate has tended to focus on a few high-profile examples. Numerous groups have disputed–and at times formally challenged–utility patents issued in the United States and the European Union for apparently new cultivars that subsequently are determined to be traditional varieties from developing countries. High-profile cases include the Mexican yellow bean (patented as "Enola" by POD-NERS, in the United States and subsequently challenged in US courts by the International Center for Tropical Agriculture), Indian turmeric (a patent for its wound-healing properties was issued in the United States, challenged, and subsequently cancelled), the Indian neem tree (a series of patents was issued in the United States and the European Union and has been assigned to W.R. Grace for this traditional Indian tree with industrial and pharmacological properties; some of the patents have been challenged and struck down in Europe), and Basmati rice (patented in 1997 by RiceTec and challenged in the United States and subsequently amended). These and other proven or alleged acts of biopiracy have raised debate about the concept of invention, the definition of prior art, and the ethics of using traditional knowledge or genetic resources. The overall tenor of the debate is that the IP regime for plants is inimical to the interests of farmers. This issue was most recently engaged in a widely watched and cited case involving a Monsanto civil patent prosecution of Percy Schmeiser, a Saskatchewan farmer. In his defence, Schmeiser challenged the validity of the gene patents, challenged the right of private ownership of living organisms, and raised concerns about the economic, social, and ethical effect of private patents on self-propagating life forms. Although the Supreme Court of Canada sided with the plaintiff, Monsanto, the concerns about the case continue.

A significant number of NGOs, scholars, and practitioners have called for changes to the structure of patents and related IPRs to deal with both real and perceived flaws. Many have called for a review and change on the scope of what is patentable. There is virtually unanimous agreement that humans should not be patentable but, beyond that, there is a wide range of views about what other potential inventions should be excluded or have limited protection. Many environmental groups, development NGOs, church groups, and farmer advocates have called for patents to be removed from higher life forms, including animals and plants, arguing variously that it is immoral, inequitable, or inefficient for patents to be on inventions related to these organisms. Development NGOs and indigenous communities also have either called for much tighter control on patenting of indigenous knowledge and genetic resources or seek broad exemptions from patent enforcement for indigenous farmers and for humanitarian purposes. Although none of these recommendations has been formally taken up, there is some movement to consider changes to the system to address development and health considerations.

The World Trade Organization Agreement and its related Agreement on Trade-Related Aspects of Intellectual Property Rights (TRIPS Agreement), concluded in 1995, have provided one venue for debate. The TRIPS Agreement revolutionalized worldwide ownership of intangibles, especially in non-OECD (Organisation for Economic Co-operation and Development) countries. Because of the binding nature of the World Trade Organization, together with its winner-take-all structure, every country that signed onto the WTO now assumes the benefits and burdens arising from the TRIPS Agreement. The agreement provides that member states must provide any invention either patent protection or, in the case of living organisms, some form of sui generis protection such as plant breeders' rights. The TRIPS Agreement also permits an *ordre public* (public policy) provision to incorporate non-economic values into the patent system. Japan and some member states of the European Union have adopted related measures. On a case-by-case basis, patents can be refused should the commercial exploitation of the invention violate public order or morality. The European Directive 98/44 on the legal protection of biotechnological inventions, for one, explicitly states that processes to use human embryos for commercial purposes or processes to clone human beings violate public policy. In practice, the public policy provision is usually invoked by a third party in an opposition procedure after the patent has been granted. Some states provide a broader range of opportunities to challenge patents, including an opposition procedure that provides a forum for raising challenges, typically in terms of novelty and inventiveness, but also with respect to public policy. Australia, the European Patent Office, France, Germany, India, and Japan have opposition processes, and a recent US report has recommended establishing an opposition procedure in the US patent system. Even in absence of a more open opposition procedure, the United States has established the Court of Appeals for the Federal Circuit to increase uniformity in appeal decisions.

Over the past few years, many policy groups and government agencies have made formal recommendations about how the patent system should be reformed (e.g., Nuffield Council on Bioethics 2002; CBAC 2002; Australian Law Reform Commission 2004). Many of their recommendations are surprisingly similar. They all argue that there is a need to clarify research and use exemptions (particularly for medical applications), to consider some form of compulsory licensing in areas where patent rights are too restrictive or anti-competitive, and to tighten the utility requirements to avoid both biopiracy and the anti-commons.

Although none of those specific recommendations has seen action, there has been modest change in the domestic operation of patent systems in various OECD countries that have at least partly addressed some of the concerns raised. The United States, for example, revised its requirement of utility criterion in January 2001, now requiring patent claims to identify specific uses. The US *Manual of Patent Examining Procedure* (MPEP) provides detailed guidelines on the policies and procedures staff are to follow in examining patent applications. These guidelines are public and available for applicants and other stakeholders to better understand how patent criteria are applied. Meanwhile, numerous countries are addressing access questions more directly, through adoption of experimental use exemptions. The European Union, for instance, already has a liberal experimental use exemption and the OECD (2003) reports that other countries are contemplating adopting research or experimental use exemptions within their patent laws.

Finally, many governments and agencies are attempting to improve the practice of licensing. The US National Institutes of Health (2004) and the OECD (2006) have developed guidelines for the licensing of human genetic inventions. These guidelines aim at providing a non-binding but morally persuasive set of principles and best practices to assist industry and universities in negotiating licence arrangements that serve both the interests of industry and the public at large, including the healthcare sector. Likewise, the World Intellectual Property Organization and the International Trade Centre have a new, practical guide on negotiating technology-licensing agreements (WIPO/ITC 2005).

The difficulty with all of these issue-specific changes being offered is that none of them takes into account the reality that, first, patents are only one of many IPR mechanisms available and, second, patents are embedded into the broader governing system. In both instances, patent reform has the potential to either be counterproductive or produce counterintuitive effects.

Distributed Governing of Intellectual Property

The question of how complex changes are governed is extensively discussed in the literature. It is less clear how complex changes triggered by transformative technologies are governed. Because of their nature, IPR systems are vital parts in governing these technologies.

There has been a verbal deluge of catchphrases, terminology, and buzzwords used to describe the nature of technological change in recent years. Scholars and others have described and characterized technological change in two main ways. Some define the change in terms of the scale of the impact of change (revolutionary, radical/incremental, drastic/incremental, disruptive/sustaining, breakthrough, pervasive), while others focus on the scale and nature of the technology involved (hybrid, macro-invention/micro-invention, major, general purpose, enabling, and transformative). Governments and policy makers are most interested in those technologies that offer the greatest hope of gain–many call these technologies transformative (Phillips 2007).

Perri 6 (2001, 74) provides a comprehensive taxonomy of change: "Transformative technologies" that are "involved in large-scale, discontinuous change to societies." Woodley (2001) suggests that transformative technologies (such as printing, the steam engine, electricity, and atomic energy) are so fundamental because they have many varied purposes, have applications in a large part of society and the economy, exhibit significant scope for improvement initially, have strong complementarities with other technologies, and have a long-term effect on values, power structures, and ideas. Perry 6 (2001) explicitly distinguishes transformative changes from iterative ones, where there is merely incremental adjustment. He argues that transformative technologies draw on cutting-edge basic, as well as applied, science; represent step-changes in the scale and, in some cases, the direction of development of human capabilities (i.e., are not simply iterative, incremental enhancements of existing capabilities); have consequences distributed widely over many areas of life; are of sufficiently high profile to attract the attention, interest, and risk perception of social movements, citizens, politicians, and regulators; precipitate public debate and the attention of futurists, journalists, ethicists, and technologists; and are within a few years or at most a decade of realization.

Although there is quite a range of views about what transformative technology may be, looking at the definitions in their entirety offers a vibrant picture. A few elements jump out of the melee. Transformative technologies involve disjointed, step adjustments in our productive and institutional capacity, displacing, destabilizing, or overturning precursor systems. In technical terms, they often are driven by new epistemologies, they offer significant complementarities (within and across sectors), and they tend to involve convergent, recombined, or hybrid technologies (one might actually label them convergent technologies). Furthermore, although they often emerge wrapped in an optimistic fervour, they are hard to anticipate because they infiltrate and influence multiple sectors, markets, and domains over a long and variable timeline–in spite of the rhetoric, none of the identified technologies overwhelmed society quickly. This makes it difficult to manage or govern them through established markets or authorities. Finally, the nebulous nature of transformative technology creates uncertainty, which tends to generate debate and controversy.

IPRs are one of the first and most important institutions used to govern transformative change. Ultimately, governing involves several critical elements: a purpose, a set of actors (the governors and the governed), a domain to be governed, and a process of governing. A basic place to start, then, is with the concept of governing. The *Oxford English Dictionary* defines the verb *governing* as involving swaying, ruling, influencing, guiding, directing, or regulating the course of an event. This can be in human affairs or in nature. Governors thus may be mechanical devices (e.g., the governors on early steam engines), laws of nature (e.g., the role of genes in determining the heredity of traits), or any human-made institution (e.g., the criminal code and common law) that influences an outcome, while the governed may be nature, machines, people, organizations, and societies. The processes of governing thus involve a wide array of domains, actors, and processes. It is becoming increasingly clear that we cannot simply think of governing as what states or governments do; rather, governing involves the array of human-constructed processes that determine how different interactions will lead to specific outcomes. The concept of governing encompasses far more than the laws, regulations, and actions state governments engage in.

While governing involves all human constructs–both those that are formally and consciously constructed and adhered to and those that are informal and often part of our background consciousness–it is perhaps best to focus on those active forms of governing where there is a direct correlation between the governing system and the outcomes. The focus then is on those processes that involve controlling, directing, or regulating human affairs. Many would contest that approach, arguing instead that governing should also include the many intangibles that underpin society–the conventions and customs acquired over the millennia. While these factors are important, their import goes way beyond their role in governing systems. Hence, these factors are for the most part excluded from the analysis.

While Rosenau (1995) expresses reservations about the ability to create typologies and to identify commonalities among subgroups of governing systems, the literature offers places to start. The three convergent approaches–economic, political, and sociological–provide useful insights into the nature of the challenge and offer a few hints on how we can bring greater clarity to the options. First, there would appear to be three domains of governing (state, markets, and society), which operate on somewhat different aspects of life (political, economic, and social) and use different governing mechanisms (command and control, exchange, and voluntary association). That much would appear to be agreed on by many. Most would also agree that the resulting governing systems are distributed, multilayered, and interdependent.

Governing in such a world involves apparently self-organizing, inter-organizational networks that exhibit interdependence and sustained game-like exchanges where interactions are rooted in trust. In essence, single, central authority is replaced by a multiplicity of self-organizing networks. Amaral and Ottino (2004) argue that such a complex system will involve "a large number of elements, building blocks or agents, capable of interacting with each other and with their environment . . . The common characteristic of all complex systems is that they display organization without any external organizing principle being applied. The whole is much more that [sic] that sum of its parts" (148). Even simple systems may produce complex, often non-linear, outputs, which is the epitome of chaos. Network theory, in particular, offers insights into the various distributed authorities in a world of subsidiarity. Networks define the permutations of actors (or nodes) through relationships (connections or links), where networks of different density can have either central nodes or distributed relationships (the nature and stability, or volatility, of networks can often be modelled through small, discrete behavioural models, such as the "game of life" cellular automata model). Analyses of various small world networks–such as power grids, ecosystems, and epidemics–have revealed that most human-built systems "are not the result of a single design but an evolution and merging of designs, and that in many instances, self-organization can be used in profitable ways" (160). This gives rise to a distributed, centreless world. The state, market, and civil authorities each govern part of this distributed system.

Each of these institutions manages one or more social processes, which, when combined, generate complexity: competition and interdependence among actors create new roles for actors to fill (sometimes called co-evolutionary diversity), actors are at times encouraged to add new sub-systems (structural deepening), and large systems are made up of smaller, simpler sub-routines. These complex systems inevitably involve variably coupled (from loosely to tightly) interdependent, multiple entities within a dense web of causal connections that are open to external events. Although the synergies within these open systems may exhibit non-linear behaviours, they also generate synergies and feedback that may stimulate learning.

IPRs as Complex Systems

Although the contemporary focus is on the role of patents in our economy and society, they represent only the tip of the iceberg of the IPR system currently operating. A better way of looking at the current IPR system is as a complex system, with multiple levels of interdependent control, numerous subsystems, and many sub-routines (research, regulatory, and commercial) that will lead to small world effects. Furthermore, the current IP system is embedded in a range of other governing processes. Hence, small, discrete changes in any one part of the system may have little or no effect or, in some cases, may have profound, unanticipated consequences. Any changes in the patent and related IPR system must thus be considered carefully.

The Dense Network of IPRs in Biotechnology

From the beginning of recorded time, there are ample examples of inventors protecting their inventions by force or by guile. Over time, three generic mechanisms for IP protection have evolved. First, various legal mechanisms have been developed to protect IP, including patents, legally sanctioned trade secrets, plant breeders' rights, trademarks, and copyright. All of these mechanisms require the inventor to register or define his or her invention, are self-enforcing, and have limited durations. Second, commercial strategies have emerged that enable companies to backstop their rights, by linking their inventions to protected complementary technologies, engineering in environmental constraints, or using private contracts. Third, inventors can preclude others from claiming and exploiting their inventions by publishing the information widely, which for the most part precludes patents or other formal IP claims.

The oldest, yet least socially desirable, IP option is to keep new knowledge undisclosed in a trade secret. Trade secrets involve several features. First, an inventor must take steps to keep the secret, which usually involves forcing new employees or potential partners to sign contracts stating that they will not disclose the knowledge. Trade secrets are best used when they cannot be independently discovered or figured out by reverse engineering any products that use the secret. Trade secrets have two main benefits: If successful, they can protect an idea indefinitely in all markets simultaneously, which can at times lower the cost relative to patents. Furthermore, they can be used to protect non-patentable material. There are, however, risks involved with protecting IP as a trade secret: anyone who can find out how the invention works can use it; trade secrets are difficult to enforce because people have a great incentive to cheat and use the ideas; and, most importantly, trade secrets put property into the non-rival, excludable category permanently, with the result that social benefits are lower than if the knowledge was more widely disseminated.

The problem of non-disclosure of trade secrets led to the development of the modern patent system. Patents are essentially a bargain between society and inventors. Inventors get a specified period to exploit their inventions in exchange for full and complete disclosure of the recipe or formula. In this way, the knowledge at the end of the patent enters the public domain. For an invention to be patented, the technology, process, or product must meet three criteria. First, it must be novel (that is, never known before anywhere in the world). Second, it must be useful. It cannot be just a concept; it must be reduced to practice ("$E = MC^2$" is not patentable, but the formula for Roundup Ready canola is). Third, it must be nonobvious (others skilled in the art should not be able to conceive it without experimentation). Patents provide an incentive for invention by granting a twenty-year monopoly right to the inventor. In exchange, the inventor must disclose how the invention works. There are certain risks in patenting inventions. Disclosing the invention sometimes helps competitors leapfrog the inventor with a better product. Furthermore, patent rights are granted only in the country of application, and the rights must be enforced by the inventor through the national courts. Finally, although patents are extensively used in the biotechnology industry, some patent systems do not protect all relevant elements, including multicellular organisms such as whole plants and animals.

Plant breeders' rights (PBRs) are analogous to patents, except they concern crop varieties. They provide the developers of new, stable, and uniform plant varieties with rights to charge a royalty and to control the sale of propagating material. As with patents, these rules are nationally based–there are no automatic international rights. PBRs often are relatively cost-effective, but they also offer less protection than patents. By international agreement, they last for only eighteen years, and all PBR systems allow farmers to save seed for replanting and provide a research exemption to allow other breeders to gain access to elite germplasm in order to develop new varieties.

Trademarks are another way in which IP is protected. Trademarks are often used once a product enters a market as a way to identify and distinguish a product and to exclude others from that market segment. Trademarks offer protection over a word, symbol, or design that distinguishes a specific product or service. Popular examples are the brand name Coke, the slogan "Things go better with Coke," and the packaging of Coke–its recognizable bottle. Another example is the Nike swoosh logo. Trademarks offer national protection for a renewable fifteen-year term.

Copyright gives the creator the sole right to publish, produce, or reproduce the work; to perform the work in public; to communicate the work to the public by telecommunication; to translate the work; and to rent the work. The main benefit of copyright is that it is automatic–once the work has been produced, the creator has sole rights to it. Many creators deposit their written works with a national repository library to secure their claims to the work. The term for copyright is the life span of the creator plus fifty years (seventy-five years in some jurisdictions). One of the most relevant uses of copyright in the biotechnology industry is computer programs to manage the massive amounts of data involved in biotechnology research.

Keeping in mind that legally sanctioned IP rights are based on national rules, effective IP protection then varies depending on the country in which the property is protected. Only developed countries are uniformly involved in the International Union for the Protection of New Varieties of Plants (established by the International Convention for the Protection of New Varieties of Plants and in the World Intellectual Property Organization). Developing countries often lack any effective IP protection regimes. That may change with the recent negotiation of the World Trade Organization's TRIPs Agreement. Although this agreement requires that all developing countries establish effective IPR systems by 2006 and offers a binding dispute settlement system to enforce that requirement, now that the deadline has passed it is unclear how the intent will translate into practice.

There are many commercial strategies firms can use to protect IP. Some firms create technologies that are useful only if combined with another technology that is protected. For example, LibertyLink crops are useful only with gluphosinate produced under patent and protected by the trademark Liberty. Some breeders use biological control mechanisms, such as hybrids (e.g., corn), seedless varieties, transformation in targeted tissues, or genetic use-restriction technologies, sometimes called terminator genes. All these regulate the physiological functions of an organism so that the second generation is sterile or no longer economically valuable. Some target to do research and development only on products or technologies that the firm can exploit because of existing technologies and property rights. The environment in particular can restrict the use of specific technologies. Seeds that flourish only in specific conditions are fundamentally tied to the ecosystem in which they prosper. Finally, firms can always use private contracts (such as Monsanto's technology use agreements) to protect IP.

Ultimately, much of what is known is published, which, except within one year of release, virtually eliminates patent possibility. Publication allows dissemination but not exploitation, avails itself of copyright protection, and, depending on the circumstances, can either enhance or reduce use. Such knowledge, if published by a private company, is often placed in the public domain in support of regulatory approval and to help further define and delineate their IP from others.

No commercial firm or industry relies exclusively on one or even a narrow range of IPR mechanisms. Rather, all use a mix of formal IPR mechanisms (including patents, plant breeders' rights, or trademarks); trade secrets; open disclosure; private contracts; and industrial, technical, or organizational barriers to entry. This can perhaps be best illustrated through an example. Phillips and Khachatourians (2001) and Phillips (2002) examined the innovation system in the global canola industry in the period 1944-2000, noting that the rate of invention and innovation accelerated significantly over that time. Table 17.1 illustrates the key research, development, and commercialization elements of the global canola system and identifies the various IPR ownership structures and

Table 17.1

IPRs related to canola-breeding processes

	Key technologies	IPR regime
Basic science	17,995 source journal articles, from 1,294 journals, produced by approximately 28,800 authors in 3,816 organizations in 107 countries 4,908 canola-related articles in 650 journals produced by approximately 6,900 authors in approximately 1,500 organizations in 79 countries	Although articles are in the public domain, they are increasingly being accessed through the estimated sixty-nine research stars in thirteen countries and via proprietary, fee-for-service electronic databank servers, such as ISI, or servers run by the journal publishers
Germplasm	Eight to ten major public and private gene banks	All involve material transfer agreements, imposing restrictions on use, distribution, and commercialization
Genomic information	Arabidopsis genome, amplified fragment linkage polymorphing for gene mapping; and molecular markers	Although data are in the public domain, they are increasingly being mined through private, fee-for-service genome databanking services; AFLP technology and markers mostly patented
rDNA strands/ genes	More than fifty genes for herbicide tolerance, antifungal proteins, seed coat, fatty acids, and pharmaceutical proteins	100% private patents, often with marketing trademarks attached (e.g., Roundup Ready, InVigor)
Transformation technologies	At least four competing systems (*Agrobacterium,* whiskers, biolistics, and chemical mutagenesis)	100% private patents (often multiple patents per system) except mutagenesis; most also trademarked and subject to extensive private contractual obligations

►

◀ *Table 17.1*

	Key technologies	IPR regime
Selectable markers	A large number of markers for selecting specific transformants	100% private patents
Growth promoters	More than 100 constitutive and tissue-specific promoters	100% patented
Hybrid technologies	At least six competing hybrid systems	All but one patented; most subject to trademarks and extensive private contractual obligations or joint venture rules
Traditional breeding technologies	Double haploid process; backcrossing; gas liquid spectrometer analysis	All in public domain but require art available only via extensive learning-by-doing; economies of scale with some technologies
Seeds	More than two hundred commercially available varieties in 2005	93% were herbicide-tolerant varieties, with multiple layers of patents, about half with hybrids, virtually all with plant breeders' rights and names protected by the Seeds Act; most with trademarked/branded marketing programs; about half the market with some form of production and marketing contracts
Oil processing technologies	Numerous competing separating and purifying processes	100% patents or trade secrets; most also trademarked and often accessible only via joint venture

Source: Adapted from Phillips and Khachatourians (2001) and Phillips (2002)

strategies used by participants in the industry. No one firm or organization is completely autonomous in this respect; all require access to at least some property of others. The large multinationals all draw on others for primary and basic research and germplasm, and contract with each other to access specific technologies needed to advance their products. Smaller enterprises and the public sector–in competition–often have to expend a relatively greater effort assembling freedom to operate, as they often need to license a wider range of transformational technologies.

Although the above examination of canola provides some insight into the nature of the IPR system operating in modern agriculture, it is important not to take it as representative of all products. Each technology or product line is likely to have a different complex system of mechanisms. Each will have a different innovation process, with a different mix of interconnections between different phases of technological development and with a varying range of different technology tracks. Nesta and Mangematin (2004) argue that complexity evolves from the fact that firms are frequently involved in projects related to different waves of discovery, such as gene sequencing, genomics, or pharmacogenomics (each of which calls for a differentiated set of actors with a specific division of innovative labour and different IP structures), and that, within a single wave of discovery, firms frequently engage in several projects that spread over different phases of research. Put into the context of the more conventional lazy S-curve of the life cycle of a product, firms can be engaged at one or more points on any single curve, while at the same time involved in multiple different curves with different ranges and amplitudes. Nesta and Mangematin argue that this leads to a mix of actors and collaborative agreements and, by extension, will lead to a different mix of mechanisms and strategies in effect at any point.

The transaction costs of such an extensive web of IP rights are significant, yet it is not clear that any different system would be better. As with any complex system, discrete, linear changes in one part of the complex system may have little effect if there is some redundancy in the governing system. This is the case with patents for plants. While the *Monsanto v. Schmeiser* case did not invalidate the patents on the herbicide-tolerant genes in the disputed plant, the court reiterated its decision from *Harvard Regents v. Commissioner of Patents* that it did not find any basis in law for patents for higher life forms (such as plants or animals).[3] Nevertheless, the absence of patents on higher life forms would appear to be largely symbolic, as the patent rights to components in the organism (in that case the glyphosate-tolerant gene) provided adequate protection. Even if all patent rights were eliminated on those organisms, there is every chance that access would not increase. Rather, firms would be expected to rely more on PBRs, production and marketing contracts, and hybrids to control access to their proprietary technology.

Perhaps more troubling, small, partial changes to IP systems may very possibly make things worse. Currently there is almost full disclosure of new technologies (via the patent system and in the context of deposits of elite breeding materials for living organisms). If patents and other legally mandated property structures were withdrawn or limited, firms could either pursue the trade secret route, closing access to the technologies and concepts underlying their inventions, or they could impose much more expensive and time-consuming contractual arrangements on firms or farmers wishing to use their technology. The presence of legally enforceable rights to inventions makes it relatively straightforward for inventors to enforce their rights. In the absence of those rights, for-profit firms would likely reduce their investments, or, alternatively, would expend more on property protection mechanisms, which in the total scheme of things should simply be viewed as dead-weight losses to society. The end result would very likely be lower investment, lower usage, and lower value added from new inventions.

IPRs and Their Role in a Layered, Distributed Governing System

IPRs play a particularly important role as the key allocation of rights and obligations in broader society. They are one of the most important means of dividing responsibilities among actors in society. As a result, they are either formally or informally incorporated and embedded into the broader governing system.

Perhaps the most important role of IPRs is to provide a clearly defined asset that can be monetized in the market place. Clear ownership with well-defined boundaries and predictable (if costly) dispute settlement systems provides greater certainty for researchers and industry as they seek to collaborate in order to develop and commercialize new products. Although there is limited verification that IPRs have had any demonstrable effect on the rate of invention, there is significant circumstantial evidence that patents are a critical part of the commercialization process. Venture capital companies and lending organizations have traditionally had difficulty finding a way to monetize the knowledge in an enterprise. Patents may not in the end be the critical factor in a successful company, but they provide an important signal to markets about the nature of an enterprise and often are used by lenders and investors as a guide to investing. Because they can also be assigned as security, they furthermore provide one way to lock in key knowledge workers who may have a stake in the advancement of key technologies.

Beyond the commercialization process, patents and other IP mechanisms are important features in our systems designed to protect public health and safety. All of our key regulatory processes formally or informally rely on the ownership claims based on patents or other IP rights–the system is at root proponent-driven. The Seeds Act in Canada, for instance, protects the seed name, specifies quality standards, and sometimes establishes refugia conditions in order to provide a basis for fair competition (a benefit to those without market power, such as farmers and consumers) and to ensure that the quality of the product meets market needs. The company assigned the rights under the act is the only actor eligible to initiate regulatory review and, once a product is approved, has some obligations to protect, maintain, or manage each of its cultivars. Some cultivars, such as industrial oil rapeseed, are approved only under conditions of contract registration under the Seeds Act (in an effort to segregate inedible or dangerous products from the food system), which requires the owner to manage the isolation of the product in the environment and the food chain. The responsibility to manage the approved segregation system is delegated to the owner, and the owner would be liable if any of the conditions are not met. In addition, of course, the delegation of this responsibility to the owners of the seeds also helps strengthen their hold over their IP.

Meanwhile, environmental safety and public health regulations make significant use of the IP rights assigned to inventors. First, only the inventor or their assignee or agent has the authority to initiate the process of evaluation of new products. Although others may have concerns and interests, only the owner has the authority to initiate the process. In cases of disputed ownership, regulatory review is often held up or paused until clear ownership is determined. Furthermore, once a product is evaluated and a notice of acceptance (or no objection) is filed, the owners become liable for any expected or unanticipated adverse effects. Regulators look to owners to ensure the conformity of their product with the approvals granted. For example, in Canada, the chemical registration system operated by the Pest Management Regulatory Agency requires that all chemicals be certified by the licensed producer before sale, and that any stale dated product must be recertified by the manufacturer before it can be offered for sale. Similarly, pharmaceutical regulations in Canada (and most other countries) require that the owner of the products ensure authorized products in the market meet purity, health, and safety standards. Agents may be ultimately at fault in producing a flawed product, but the owner will ultimately be at least partly liable. This includes an obligation on the owner of any new technology to report any adverse effects observed in post-market use. Although it is often difficult to determine "who knew what when," the obligation remains and has at times been a critical factor in assigning blame for unintended adverse effects.

In practice, there is often a balance in any market between ownership and regulation. Governments in those markets that grant more clearly defined ownership rights often impose less rigorous regulatory oversight. Particularly where private tort law is strong (as in the United States and the United Kingdom), government can rely on industry to keep a close eye on any risks that could trigger adverse effects. In many cases, firms facing the threat of catastrophic commercial losses because of class-action lawsuits will be more conservative in their production and marketing choices than regulators might be. In countries where private ownership is less clear, or where tort laws are weak or deliver idiosyncratic results, states may be compelled to regulate more aggressively. In practical terms, greater private rights can replace government, or vice versa. So, in the United States, the presence of strong property rights and an effective tort system has allowed the US government to reduce its regulatory oversight over many new products. For instance, until very recently, all genetically modified (GM) foods were not compelled to undergo mandatory safety assessments. In practice, all new GM foods were submitted voluntarily by developers for Food and Drug Administration review and evaluation, though this was not required. In contrast, the relatively weak tort system in Canada (where class action lawsuits are difficult and costly to engage) has led the government to engage more extensively in regulatory oversight, including mandatory review and approval of all novel foods considered for release in Canada.

Changing the property rights systems, by, for example, allowing mandatory compulsory licensing for indigenous users, allowing generic production of a proprietary product, or simply by truncating or otherwise limiting the rights of an inventor or the assignee, could lead to ambiguous regulation of risks and benefits. Firms would probably use the dispersed ownership of such reforms in an effort to spread any resulting liabilities (regardless of whether any incremental adverse harms would result from the new structure). Furthermore, it would become a point of contention of what obligations, if any, would remain with the designated owner and what obligations would flow to others. There is a very real risk that some changes in the rights system could cause regulatory gaps or lapses, that liabilities might remain unmet, or that the state might become the agency of last resort in dealing with unanticipated adverse impacts of new technologies.

Conclusion

Thinking about IPRs as part of a complex, distributed, interconnected system of governing the creation, evaluation, and commercialization of new technologies suggests a new way of thinking about our current IPR system and any reforms to it. A better way of looking at the current IPR system is as a complex system–with multiple levels of interdependent control, numerous subsystems, and many subroutines involving a range of public, private, and collective actors–that exhibits small world effects. Small, discrete changes in any one part of the system may have little or no effect, or, in some cases, may have profound, unanticipated consequences.

Although policy makers and industry could use this in a simplistic way to argue either that changes will have no effect or that any changes would have counterproductive effects, this approach does not need to lead to inaction. Rather, it should lead to a different way of policy review and development.

Thinking of IPRs as part of complex systems would suggest that regulatory review should start first with an effort to determine the desired outcomes (e.g., support for innovation, regulatory efficiency, distributional effects, or ethical orientation) and then examine the entire complex system as it relates to the desired outcomes. This would in many cases bring in other regulatory and market measures as potential mechanisms to realize the policy outcome.

Any changes in the patent and related IPR system must thus be considered carefully. Simple solutions are unlikely to be effective–in this second-best world of nested governing routines, simple solutions can often lead to non-linear responses. Barring a rewrite of our entire socio-economic structure (which is unlikely in a non-revolutionary environment), solutions to the IPR challenges will need to be carefully constructed on a case-by-case basis.

Notes

1 *Diamond, Commissioner of Patents and Trademarks v. Chakrabarty*, 447 U.S. 303 (1980).
2 Ex parte Hibberd (1985), Lexsee 227 U.S.P.Q. (BNA) 443.
3 *President and Fellows of Harvard College v. Canada (Commissioner of Patents)* (C.A.) [2000] 4 F.C. 528, http://decisions.fca-caf.gc.ca/en/2000/a-334-98_6869/a-334-98.html (accessed May 23, 2008).

References

6, P. 2001. Governing by technique: Judgement and prospects for the governance of and with technology. In *Governance in the 21st century*, Organisation for Economic Co-operation and Development, 67-120. Paris: OECD. http://www.oecd.org/dataoecd/15/0/17394484.pdf.

Amaral, L., and J. Ottino. 2004. Complex networks: Augmenting the framework for the study of complex systems. *European Physical Journal* B 38 (2): 147-62.

Australian Law Reform Commission. 2004. *Genes and ingenuity: Gene patenting and human health*. Canberra: ALRC. http://www.austlii.edu.au/au/other/alrc/publications/reports/99/ (accessed May 23, 2008).

CBAC (Canadian Biotechnology Advisory Committee). 2002. *Improving the regulation of genetically modified foods and other novel foods in Canada: A report to the government of Canada biotechnology ministerial coordinating committee*. Final report. Ottawa: CBAC.

Cho, M., S. Illangasekare, M. Weaver, D. Leonard, and J. Merz. 2003. Effects of patents and licenses on the provision of clinical genetic testing services. *Journal of Molecular Diagnostics* 5 (1): 3-8.

Jensen, K., and F. Murray. 2005. Intellectual property landscape of the human genome. *Science* 310 (5746): 239-40.

National Institutes of Health. 2004. *Best practices for the licensing of genomic inventions*. Washington, DC: NIH.

Nesta, L., and V. Mangematin. 2004. The dynamics of innovation networks. SPRU Electronic Working Paper Series, no. 114. http://www.sussex.ac.uk/Units/spru/publications/imprint/sewps/sewp114/sewp114.pdf (accessed May 23, 2008).

Nuffield Council on Bioethics. 2002. The ethics of patenting DNA: A discussion paper. London: Nuffield Council on Bioethics.

OECD (Organisation for Economic Co-operation and Development). 2003. *Genetic inventions, IPRs and licensing practices: Evidence and policies*. Paris: OECD.

–. 2006. *Guidelines for the licensing of genetic invention*. Paris: OECD.

Phillips, P. 2002. Regional systems of innovation as a modern R&D entrepot: The case of the Saskatoon Biotechnology Cluster. In *Innovation, entrepreneurship, family business and economic development: A western Canadian perspective,* ed. J.J. Chrisman, J.A.D. Holbrook, and J.H. Chua, 31-58. Calgary: University of Calgary Press.

–. 2007. *Governing transformative technological innovation: Who's in charge?* Oxford: Edward Elgar.

Phillips, P., and G.G. Khachatourians. 2001. *The biotechnology revolution in global agriculture: Invention, innovation and investment in the canola sector.* Wallingford, UK: CABI.

Rosenau, J. 1995. Governance in the twenty-first century. *Global Governance* 1 (1): 13-43.

Stigliz, J. 2002. *Globalization and its discontents.* New York: W.W. Norton.

WIPO/ITC (World Intellectual Property Organization and International Trade Centre). 2005. *Exchanging value: Negotiating technology licensing agreements.* WIP/UPD/2005/2.

Woodley, B. 2001. *The impact of transformative technologies on governance: some lessons from history.* Ottawa: Institute on Governance, October. http://www.iog.ca/publications/transformative_tech.pdf (accessed August 11, 2005).

Conclusion: Reflections on Emerging Technologies

Edna F. Einsiedel

The futurist Arthur C. Clarke declared in his prognostications about future worlds that any sufficiently advanced technology is indistinguishable from magic. Magic conjures enchantment, bafflement, sleight of hand, unaccountable effects. It arouses an interest in explaining its phenomena, in trying to predict what will be produced–but always, we are astonished and surprised.

The emerging technologies we have described here and the various challenges and dilemmas they pose are magical in the different ways they enchant but at the same time puzzle, they challenge our ways of seeing, and they might offer surprises in the way futures unfold or are unveiled.

Proponents of technologies invoke their magical powers–promising silver bullets of cures for disease, hunger, and other maladies. Opponents inveigh against the threats, seeing plague instead of promise, perversion rather than possibility. Both accuse the other of superstition, the superstition of magical powers for proponents, of black magic for opponents.

Following the evolution of new technologies over the last several decades, the debates with each new set of transformative technologies raise a sense of déjà vu. As new sets of technologies emerge, the same passionate battles are fought over competing visions of what each might mean, each time pitting the magic of the silver bullet against the perils of the poison arrow. Nuclear power in the 1950s was set to result in "clean electricity too cheap to meter." Biotechnology was going to "end hunger in the third world." Pharmagenomics promises to bring drugs tailored to each individual, or to make medicine "personalized."

But the promised benefits have also come at social costs. Threats to biodiversity, social divides (e.g., health or digital divides), economic challenges, and ethical concerns illustrate the range of challenges different new technologies face.

By looking at a variety of new technologies in the related arenas of biotechnology, genomics, and nanotechnology, we begin to see a pattern:

each set of applications may develop its own unique set of benefits and challenges but, more importantly, the problems posed for societies have significant similarities. Below, I review the patterns briefly and consider the implications for the governance of technologies.

New Challenges for Knowledge Production Work

The scientific enterprise is work carried out not just by scientists. We find that science work is being impinged on by society. How is the research being carried out? To illustrate, we can consider the practice of informed consent obtained from research participants. Practices of informed consent for medical research are based on principles of individual autonomy. However, researchers have had to think about the sufficiency and appropriateness of such approaches when the bounds for genetic information go beyond the individual to familial kin or members of isolated or ethnic communities marked as carriers of disease.

In the case of stem cell research, work on human embryos has been at the epicentre of debates worldwide, and policy positions have been staked out on whether or not human embryos can be used for deriving stem cells and how they can be sourced. These policy and regulatory boundaries have demarcated the sorts of science particular societies or stakeholder groups are willing to tolerate. An underlying question here is what values permeate different perspectives? The discussions on human dignity presented by Bubela and Caulfield (Chapter 9) illustrate that taken-for-granted concepts such as rights, dignity, and democracy cannot be taken for granted. It is helpful to hold these up in the cold light of day to see what meanings are promoted and how these are contested.

Knowledge Production and Ownership

The principle of rewarding innovation by providing protections for the innovator to benefit remains a cornerstone of many industrial and post-industrial societies. Again, application of the principle is not quite as straightforward as we would like. What do we do when traditional knowledges that have always been shared, passed on, and based on the principle of collective benefits collide with intellectual property practices? Or when these practices contradict cultural norms? Different societies have arrived at different responses. Some welcome an exchange relationship where benefits are bestowed to a community or country that allows the exploitation of its resource through benefit-sharing arrangements. Others, such as India, have started to document indigenous knowledge and practices and have shared this information with patent offices, in an attempt to demonstrate that cultural knowledges have been openly shared for years, even centuries–and to stop in their tracks patent applications for their traditional knowledges. It succeeded at the European level when the

European Patent Office overturned a patent originally granted to the US Department of Agriculture and the W.R. Grace Company for an antifungal product derived from the neem tree and challenged by the government of India (Jayaraman 2000).

Even on this arcane subject of patents, publics and stakeholder groups have inserted their views on the issues. Although there may be little understanding or awareness of the technicalities of patenting, publics do understand the notion of ownership and the protection of the right to profit from one's inventive labour. At the same time, the balancing act remains the preferred standard, with this principle weighed against such values as access and distributive justice and even such inchoate and difficult-to-define discomforts around the "commodification of life." The European approach to public order and morality in patent considerations may be foreign to the North American patent psyche but nevertheless signals an attempt to engage in this balancing act.

In effect, these efforts to find solutions to the vexing problems of ownership of knowledge–however crude or clumsy–help address the complexities in this area.

Publics and Their Place

The public pulse on new technologies has been taken endless times. No topic has been the subject of so much polling as biotechnology. Beyond the polls, however, have been the increasing experiments with consultation mechanisms, and the marked interest in deliberations for lay citizens. The question has increasingly become not why publics should be engaged and consulted but who gets to participate and how different interests get represented. Having moved beyond the so-called deficits of publics on science and technology, the challenge today has become the democratic deficit that needs to be addressed on technology decision making.

At a broader level, the inclusion of publics in technological discussions is increasing recognition of the redefinition of expertise in technology. Addressing the democratic deficit *and* finding that lay perspectives can make important contributions could make for a more robust outcome.

At the same time, we recognize that we should not romanticize publics or their interests. Part of the spectrum of interests is the singular interests that may clash with a collective or public good. Sometimes, the interests are purely at a visceral level, exemplified, for instance, by Thompson's blind chicken in Chapter 4 or by the "yuck factor." Although banning smoking from public places may be good for public health, individual interests may oppose such moves, and availability of a legal product ensures individual preferences are served. BiDil, the first pharmacogenetic drug in the US market, remains a difficult issue, as discussed by Bates (Chapter 12). For African Americans and for other ethnic groups whose responses to drugs differentiate them from other groups, the double-edge of the sword remains.

Changing Contexts of Innovation: New Science-Society Relations

One of the innovation dogmas that has been increasingly criticized is its linear trajectory. The meandering and messy patterns that more frequently characterize innovations, particularly of controversial technologies, is more descriptive of the many actors and networks involved: the sometimes competing interests of policy and regulatory agencies; the turns taken by science at the lab, which in turn influence its performance; and the involvement of other forums (such as the media), which exert additional forces on innovation trajectories.

In this context, it is useful to think about technological innovation not as an outcome of a single idea but, rather, as forms emerging from bundles or ensembles: of ideas, information, technology, codified knowledge, and know-how. In turn, these may or may not be embodied by the new product or process. It is also often the case that new ideas are not born fully formed; they may emerge from articulations from several sources. What this describes is a process of innovation that generally arises from a portfolio or network of actors and relationships, of voluntary and involuntary linkages, of institutional structures that get shaped and reshaped as technical and social challenges arise.

This version of innovation is likewise reflective of the changing relations between science and society. Just as the scientific enterprise has been changing in terms of its competitiveness, collaboration, globalization, and modes of framing problems (Gibbons et al. 1994), so have its many and diverse publics (Nowotny, Scott, and Gibbons 2001). Science is increasingly being performed in the agora, and the values and standards to which science is held to account go increasingly beyond the techno-scientific domains. The patent system, operating within broader governance frameworks, is an illustration of the pushes and pulls exerted by other networks in society where other competing demands such as those exerted by the healthcare system run against society's interest in rewarding invention. The Myriad Genetics case discussed by Phillips (see Chapter 17), in which the company's patents on breast cancer genes have run up against competing interests to maintain the sustainability of healthcare system costs, illustrate this tension; at the level of the European Patent Office, restrictions on the patent's reach as a diagnostic tool have succeeded, illustrating the power of other groups in society to curb the increasing reach and grasp of the patent hand.

The pressures exerted by various social groups have also extended further upstream to where and how the science is being done. The debates over transgenic salmon continue to slow the commercialization path, and the traditional structures of regulation have not necessarily kept pace with that of knowledge production. A second view is that the regulatory hand has deliberately stayed cautiously slow in recognition of market unreceptiveness or anxieties.

As with many other transformative technologies, the emerging ones discussed in this book will result in products and processes that make risks and benefits the constant sides to the same coin; the desired consequences come with unintended and sometimes undesirable consequences. Genetic tests to determine one's pharmacological profile will provide information about relatives who may prefer the bliss of ignorance. The expected benefits of vaccines derived from plant factories must be weighed against the risks to food safety and the environment. The benefits of transgenic fish for producers may be viewed as high risk and anathema to consumers. Many of the new products of these emerging technologies–whether it is drugs from plants, genetically altered salmon, or stem cells–require new institutional structures and approaches to policy making, demand inclusion of a broader range of actors, and call for new ways of defining problems.

The Governance Challenges for Emerging Technologies

This book began by posing two questions: What have we learned from the experience of biotechnology, and what issues are being posed by emerging technologies?

Whether we are talking about pharmacogenomics, nanotechnology, molecular farming, or stem cell research, we see over and over a number of familiar issues.

We have already described the challenges to the traditional boundaries of scientific work. The questions of who governs and how are most pronounced with respect to the governance of emerging technologies. Several challenges underline this governance challenge. First, these new and revolutionary technologies are marked by increasing complexity, abstractness, uncertainty, or potentially significant economic value, weighted with environmental or social risks. Their potential geopolitical influence, stretching across political borders simultaneously, nudge these embryonic innovations beyond what many government institutions consider to be within their comfort spheres but not beyond their spheres of concern. Other groups have attempted to fill this gap between government capacity and concern and, consequently, play critical roles in guiding technological development.

Second, the science-policy link has been muddied by other considerations, not least of which is the increasing recognition of the uncertainties around science. The question of how to interpret the role of genes is underscored by discourses on race and personalized medicine. How the technicalities of scientific work–whether it is producing medical proteins in plants or deriving cell lines from cord blood or embryonic tissue–are deeply enmeshed in socio-cultural contexts and understandings is also an important dimension of these transformative technologies.

This is highlighted by the ongoing challenges to biotechnology posed by risk-focused approaches to policy and regulation. Regulations are made in order to achieve desirable outcomes, from ensuring human safety to achieving good environmental quality. Risk-based regulation assuredly has advantages, but it also has its limitations. There are epistemic, institutional, and normative constraints on risk-based regulation (Rothstein et al. 2006). Making decisions at the boundaries of scientific understanding may be required, raising the challenge of communicating uncertainty. Normative conflicts similarly put strains on institutional capacities and expertise and definitions of risk-management objectives. With the benefit of hindsight from the biotechnology regulatory experiences (which are still ongoing), the public spotlight has been trained on these challenges, even though they have not been solved. For the emerging technologies in biotechnology's wake, the centrality of risk as an organizing concept has at least opened up these problems earlier on rather than later.

A third strain in the governance fabric is the incessant demand for interdisciplinary thinking for the policy and regulatory communities. The challenges described for transgenic products similarly face policy makers who confront nanotechnology products and services. Keeping up with the scientific trains has always been challenging enough for these communities; new developments can be met only by incorporating social learning mechanisms into the governing institutions.

What to make of these conundrums? There is growing recognition that knowledge production is more socially distributed (Nowotny, Scott, and Gibbons 2001). This is reflected in the world of policy and regulation. Witness the consultations on plant molecular farming with various stakeholders. Although the case made in this book relates to consulting stakeholders as one way of understanding and forestalling liabilities, we would argue that such consultations are representative of two other elements: it is part of exercising foresight and experimentation in order to strengthen technological design, and it is part of social learning for better technology assessment.

Another important lesson is the recognition that science is an important arbiter for regulation–but it is not the final arbiter. The social dimensions can be ignored only at one's peril. This is the challenge of "building social credence into the scientific rationality approach" (Isaac 2002, 264). Regulations are not solely instruments for correcting market failures; consumers are not just economic agents (267). New criteria are needed. The Canadian government has engaged in an initiative to develop "smart" regulations, which argues for greater attention to publics and the diverse interests, represented by stakeholders. At the same time, some groups remain skeptical and maintain that the broadening of criteria could mean less attention to safety in favour of greater consideration for economic concerns.

As Sheila Jasanoff (2005) observed in her comparative analysis of biotechnology policy in the United States, the United Kingdom, and Germany, "making peace with biotechnology was not, in any of the three countries, simply a matter of applying old political routines to new agenda items; nor was it a case of creaky legal and political institutions playing catch-up with rapid developments in science and technology" (280). Through these countries' attempts to make biotechnology happen, their traditional understandings of technology and politics were challenged, and they had to reconstitute their notions of democratic politics through the frame of technology. Jasanoff concludes, "The politics of biotechnology proved to be constitutive of aspects of democratic politics writ large" (281).

This is a major legacy for the newer technologies that follow: In what ways can these technologies reconstitute society and, in turn, how can societies reshape technologies? The challenges we have documented in this book through our early experiences with the various emerging technologies are important steps in the learning society undertakes as it defines and redefines its changing relationships with transformative technologies

References

Gibbons, M., C. Limoges, H. Nowotny, S. Schwartzman, P. Scott, and M. Trow. 1994. *The new production of knowledge*. London: Sage Publications.

Isaac, G. 2002. *Agricultural biotechnology and transatlantic trade: Regulatory barriers to GM crops*. New York: CABI.

Jasanoff, S. 2005. *Designs on nature: Science and democracy in Europe and the United States*. Princeton, NJ: Princeton University Press.

Jayaraman, K.S. 2000. As India pushes ahead with plant database. *Nature* 405: 267.

Nowotny, H., P. Scott, and M. Gibbons. 2001. *Rethinking science: Knowledge and the public in an age of uncertainty*. Cambridge, UK: Polity Press.

Rothstein, R., P. Irving, T. Walden, and R. Yearsley. 2006. The risks of risk-based regulation: Insights from the environmental policy domain. *Environment International* 32: 1056-65.

Contributors

Benjamin R. Bates, Ph.D., is assistant professor in the School of Communication Studies at Ohio University. Dr. Bates' area of specialization is public understanding of health protection and risk reduction messages. His research domains have included health and risk related to genetics, second-hand smoke, and wildfire.

Tania Bubela, Ph.D., LL.B., is an assistant professor in the Department of Marketing, Economics, and Business at the University of Alberta. Her research has two principal foci: knowledge translation in biotechnology and new technologies in biomedicine, and impacts of commercialization/open science and intellectual property policies on scientific culture and knowledge and technology flows in biotechnology.

Michael M. Burgess, Ph.D., is a professor and chair in Biomedical Ethics at the W. Maurice Young Centre for Applied Ethics and in the Department of Medical Genetics at the University of British Columbia. His research combines qualitative and public engagement methods and social scientific literature review with ethical analysis.

Timothy Caulfield, LL.M., has been research director of the Health Law Institute at the University of Alberta since 1993. In 2001, he received a Canada Research Chair in Health Law and Policy. He is also a professor in the Faculty of Law and the School of Public Health. He is a Senior Health Scholar with the Alberta Heritage Foundation for Medical Research, the principal investigator for a Genome Canada project on the regulation of genomic technologies, the theme leader in the Stem Cell Network and the Advanced Foods and Materials Network (National Centres of Excellence) and has several projects funded by the Canadian Institutes of Health Research.

Edna F. Einsiedel, Ph.D., is University Professor and professor of Communication Studies at the University of Calgary. Her professional and academic work has focused primarily on the interactions between various publics and science and technology, and more specifically on strategic technologies of biotechnology and genomics. She is a principal investigator for a GE3LS Project

(Genomics, ethics, economic, environmental, legal and social studies) funded by Genome Canada. Her research has also been supported by the Social Sciences and Humanities Research Council, the Alberta Heritage Foundation for Medical Research, the Stem Cell Network, and the Canadian Biotechnology Secretariat.

Rose Geransar is a Ph.D. candidate in the Department of Community Health Sciences at the University of Calgary. Her thesis work focuses on parents' perspectives on the informed consent process in public umbilical cord blood banking. Rose has been working with GE3LS Genome Alberta since May 2003, and her research has spanned a broad range of topics, including genetically modified food controversies, stem cell research policy and public opinion, and direct-to-consumer advertising for genetic testing.

William K. Hallman, Ph.D., is a professor in the departments of Human Ecology and Psychology, and director of the Food Policy Institute at Rutgers, the State University of New Jersey. A psychologist, he has done research focused on public perceptions of risk, food, and health, and, with his research team, he has completed extensive studies of public understanding and perceptions of agricultural biotechnology, including genetically modified foods and animal cloning.

Ed Levy, Ph.D., is an adjunct professor at the W. Maurice Young Centre for Applied Ethics at the University of British Columbia, where he is one of the leaders of a research team funded by Genome Canada. The team is investigating alternative approaches to intellectual property in the field of genomics. He serves on the boards of Oncolytics Biotech, Inc., and of several not-for-profits: Pivot Foundation, Tides Canada, the Neil Squire Foundation, and the BC Civil Liberties Association.

Holly Longstaff is a Ph.D. candidate at the W. Maurice Young Centre for Applied Ethics at the University of British Columbia. Her work involves experimenting with and evaluating new methods of risk communication. The objective of her research is to help people make better health-related decisions concerning complex technologies. Holly is particularly proud of her work on innovative projects that explore new methods of public engagement. Although she remains grounded in conventional approaches to risk communication, she also explores the potential of theatre, genomic art, and other novel methods of discussing risk.

Emily Marden is a visiting scholar with the W. Maurice Young Centre for Applied Ethics at the University of British Columbia and is a practising attorney in the biotechnology/pharmaceutical arena. Emily's focus is on policy and regulatory strategy in genomics, and she has published widely in these areas.

Chika B. Onwuekwe, LL.B, LL.M, Ph.D., is a consultant and attorney-at-law (Canadian and Nigerian bar). Dr. Onwuekwe is an associate at MacPherson Leslie & Tyerman LLP, a foremost Canadian law firm. He practices, consults, and publishes in the areas of emerging and transformative technologies, intellectual property, business law, natural resources law, emerging democracies, governance, and policy issues around resource allocation and intellectual property.

Vural Ozdemir, M.D., Ph.D., DABCP, is a senior scientist and adjunct professor in the Bioethics Programs, Department of Social and Preventive Medicine, Faculty of Medicine, Université de Montréal. His current research concentrates on publication ethics, management of conflict of interests in personalized medicine (pharmacogenomics and nutrigenomics), and ethical analysis of the ways in which biomarker data are integrated in systems biology.

Peter W.B. Phillips, Ph.D., an international political economist, is a professor and head of Political Studies, acting director of the School of Public Policy at the University of Saskatchewan, Canada, and holds a concurrent faculty appointment as professor at large at the Institute for Advanced Studies, University of Western Australia, Perth. His research concentrates on issues related to governing transformative innovations.

Susanna Hornig Priest, Ph.D., is a professor in the Hank Greenspun School of Journalism and Media Studies at the University of Nevada, Las Vegas. She has studied the relationship between media coverage, media consumption, and public opinion for emerging technologies since the early 1990s. She currently serves as editor of the journal *Science Communication*.

Lorraine Sheremeta, LL.M., is a lawyer, a research officer at the National Research Council's National Institute for Nanotechnology, a research associate at the Health Law Institute at the Faculty of Law at the University of Alberta, and a special advisor to Alberta Ingenuity Fund's Strategic Programs. Lorraine's academic interests focus on the legal and regulatory issues implicated in new technologies, including genetics, genomics, regenerative medicine, and nanotechnology.

Stuart J. Smyth, Ph.D., is a research associate at the University of Saskatchewan, where he completed his Ph.D. in biotechnology. The focus of his research has been on how governments regulate innovation. This research has resulted in several publications on regulations for genetically modified crops and international liability issues resulting from the commercialization of these crops.

Patrick A. Stewart, Ph.D., is an assistant professor in the Department of Political Science and Public Administration at the University of Arkansas Fayetteville. His current research considers the influence of emotion on decision making in politics and policy making.

James Tansey, Ph.D., is an assistant professor and chair in Business Ethics at the W. Maurice Young Centre for Applied Ethics and Sauder School of Business, University of British Columbia. Dr. Tansey originally trained in environmental sciences and currently focuses on climate change and markets, new and emerging technology, and social enterprise.

Paul B. Thompson, Ph.D., holds the W.K. Kellogg Chair in Agricultural, Food, and Community Ethics at Michigan State University. He received his Ph.D. in philosophy from Stony Brook University in New York in 1980. He teaches in the Philosophy, Agricultural Economics, and Community Food Systems programs.

Jacopo Torriti, Ph.D., is a Jean Monnet Fellow at the European University Institute in Florence, Italy. His research focuses on economic and risk-analysis instruments for policy making, such as impact assessment, cost-benefit analysis, and the standard cost model.

Michele Veeman, Ph.D., is a professor emerita of agricultural and resource economics in the Department of Rural Economy at the University of Alberta, Canada. Her research has focused on the economics of food, agriculture, and rural resources. Her continuing collaborative research includes studies of consumers' responses and trade-offs relative to food biotechnology; how individuals' risk perceptions and decisions are modified by different types of information; and public assessments of research on different applications of plant molecular farming.

Bryn Williams-Jones, Ph.D., is an assistant professor in the Bioethics Programs of the Department of Social and Preventive Medicine, Faculty of Medicine, at the Université de Montréal. Bryn is an interdisciplinary scholar who employs analytic tools from applied ethics, health policy, and the social sciences to explore the socio-ethical implications of new technologies. His current research focuses on the commercialization of genetic tests and other biotechnologies, and the management of conflicts of interest that arise in the context of university-industry relations.

Index

Printed and bound in Canada by Friesens
Set in Stone by BookMatters
Copyedited by Judy Phillips
Proofread by Lesley Erickson

ENVIRONMENTAL BENEFITS STATEMENT

UBC Press saved the following resources by printing the pages of this book on chlorine free paper made with 100% post-consumer waste.

TREES	WATER	ENERGY	SOLID WASTE	GREENHOUSE GASES
11 FULLY GROWN	3,938 GALLONS	8 MILLION BTUs	506 POUNDS	949 POUNDS

Calculations based on research by Environmental Defense and the Paper Task Force. Manufactured at Friesens Corporation